Atlas of Fatigue Curves

Edited by

Howard E. Boyer

Senior Technical Editor
American Society for Metals

AMERICAN SOCIETY FOR METALS
Metals Park, Ohio 44073

Copyright © 1986
by the
AMERICAN SOCIETY FOR METALS
All rights reserved

First printing, January 1986
Second printing, May 1986

No part of this book may be reproduced, stored in a retrieval system, or transmitted, in any form or by any means, electronic, mechanical, photocopying, recording, or otherwise, without the prior written permission of the publisher.

Nothing contained in this book is to be construed as a grant of any right of manufacture, sale, or use in connection with any method, process, apparatus, product, or composition, whether or not covered by letters patent or registered trademark, nor as a defense against liability for the infringement of letters patent or registered trademark.

Editorial and production coordination by
Carnes Publication Services, Inc.
Project manager: Edward C. Huddleston

Library of Congress Catalog Card Number: 85-62430
ISBN: 0-87170-214-2
SAN: 204-7586

PRINTED IN THE UNITED STATES OF AMERICA

Preface

This Atlas was developed to serve engineers who are looking for fatigue data on a particular metal or alloy. In the past, the first step to locating this data was an expensive and time-consuming search through the technical literature. Now, many of the important and frequently referenced curves are presented together in this one volume. They are arranged by standard alloy designations and are accompanied by a textual explanation of fatigue testing and interpretation of test results. In each case, the individual curve is thoroughly referenced to the original source.

Having these important curves compiled in a single book will also facilitate the computerization of these data. Plans are currently under way also to make the data presented in this book available in ASCII files for analysis by computer programs.

The Atlas of Fatigue Curves is obviously not complete, in that many more curves could be included. Persons wishing to contribute curves to this compilation for inclusion in future revisions should contact the Editors, Technical Books, American Society for Metals, Metals Park, Ohio 44073.

Contents

Fatigue Testing 1

    Introduction   1
    Fatigue Crack Initiation   4
    Fatigue Crack Propagation   12

SECTION 1: *S-N* Curves That Typify Effects of Major Variables   27

    1-1. *S-N* Curves Typical for Steel   27
    1-2. *S-N* Curves Typical for Medium-Strength Steels   28
    1-3. *S-N* Diagrams Comparing Endurance Limit for Seven Alloys   30
    1-4. Steel: Effect of Microstructure   31
    1-5. Steel: Influence of Derating Factors on Fatigue Characteristics   32
    1-6. Steel: Correction Factors for Various Surface Conditions   33
    1-7. Fatigue Behavior: Ferrous vs Nonferrous Metals   34
    1-8. Comparison of Fatigue Characteristics: Mild Steel vs Aluminum Alloy   35
    1-9. Carbon Steel: Effect of Lead as an Additive   36
    1-10. Corrosion Fatigue: General Effect on Behavior   37
    1-11. Effect of Corrosion on Fatigue Characteristics of Several Steels   38
    1-12. Steel: Effect of Hydrogen on Fatigue Crack Propagation   39
    1-13. Relationship of Stress Amplitude and Cycles to Failure   40
    1-14. Strain-Life and Stress-Life Curves   41
    1-15. Fatigue Plot for Steel: Ultrasonic Attenuation vs Number of Cycles   42

SECTION 2: Low-Carbon Steels: Flat-Rolled, Weldments and Tubes   43

    2-1. Typical *S-N* Curve for Low-Carbon Steel Under Axial Tension   43
    2-2. AISI 1006: Effects of Biaxial Stretching and Cold Rolling   44
    2-3. AISI 1006: Weldment; FCAW, TIG Dressed   45
    2-4. AISI 1006: Weldment; Shear Joints   46
    2-5. AISI 1006: Weldment; Lap-Shear Joints   47
    2-6. AISI 1015: Effect of Cold Working   48
    2-7. A533 Steel Plate: Fatigue Crack Growth Rate   49
    2-8. A514F Steel Plate: Fatigue Crack Growth Rates   50
    2-9. A514F and A633C: Variation in Fatigue Crack Growth Rate With Orientation   51
    2-10. A514F: Scatterbands of Fatigue Crack Growth Rate   52
    2-11. A633C Steel Plate: Scatterbands of Fatigue Crack Growth Rates   53
    2-12. Low-Carbon Steel Weldment: Effects of Various Weld Defects   54
    2-13. Low-Carbon Steel Weldment: Effect of Weld Reinforcement and Lack of Inclusions   55
    2-14. Low-Carbon Steel Weldment: Effect of Weld Reinforcement and Lack of Penetration   56
    2-15. Low-Carbon Steel Weldment: Computed Fatigue Strength; Weldment Contained Lack of Fusion   57
    2-16. Low-Carbon Steel Weldment: Effect of Reinforcement and Undercutting   58
    2-17. Low-Carbon Steel: Transverse Butt Welds; Effect of Reinforcement   59
    2-18. A36/E60S-3 Steel Plate: Butt Welds   60
    2-19. A514F/E110 Steel: Bead on Plate Weldment   61
    2-20. A36 and A514 Steel Plates: Butt Welded   62
    2-21. A36 Plate Steel: Butt Welded   63
    2-22. Low-Carbon Steel Tubes: Effect of Welding Technique   64
    2-23. Low-Carbon Steel: Effect of Applied Anodic Currents in 3% NaCl   65
    2-24. Low-Carbon Steel: Effect of pH in NaCl and NaOH   66
    2-25. Low-Carbon Steel: Effect of Carburization and Decarburization   67

Contents

2-26. A514B Steel: Effect of Various Gaseous Environments on Fatigue Crack Propagation 68
2-27. Cast 1522 and 1541 Steels: Effect of Various Surface Conditions 69
2-28. Cast A216 (Grade WCC) Steel: Fatigue Crack Growth Rate 70

SECTION 3: Medium-Carbon Steels, Wrought and Cast 71

3-1. AISI 1030 (Cast) Compared With AISI 1020 (Wrought) 71
3-2. AISI 1035: Effect of Gas and Salt Bath Nitriding 72
3-3. AISI 1040: Cast vs Wrought 73
3-4. AISI 1045: Relationship of Hardness and Strain-Life Behavior 74
3-5. AISI 1141: Effect of Gas Nitriding 75
3-6. Medium-Carbon Steels: Interrelationship of Hardness, Strain Life and Fatigue Life 76
3-7. Medium-Carbon Steel: Effect of Fillet Radii 77
3-8. Medium-Carbon Steel: Effect of Keyway Design 78
3-9. Medium-Carbon Steel: Effect of Residual Stresses 79
3-10. Medium-Carbon Cast Steel: Effect of Changes in Residual Stress 80
3-11. Medium-Carbon Cast Steel: S-N Projection (Effect of Applied Stress) 81
3-12. Medium-Carbon Cast Steel: Effect of Applied Stress (Shot Blasting) 82

SECTION 4: Alloy Steels: Low- to High-Carbon, Inclusive 83

4-1. Medium-Carbon Alloy Steels, Five Grades: Effect of Martensite Content 83
4-2. Medium-Carbon Alloy Steels, Six Grades: Hardness vs Endurance Limit 84
4-3. Medium-Carbon Alloy Steels: Effect of Specimen Orientation 85
4-4. 4027 Steel: Carburized vs Uncarburized 86
4-5. 4120 Steel: Effect of Surface Treatment in Hydrogen Environment 87
4-6. 4120 Steel: Effect of Surface Treatment in Hydrogen Environment 88
4-7. 4120 Steel: Effect of Various Surface Treatments on Fatigue Characteristics in Air vs Hydrogen 89
4-8. 4130 Steel: Fatigue Crack Growth Rate vs Temperature in Hydrogen 90
4-9. 4135 and 4140 Steels: Cast vs Wrought 91
4-10. 4135 and 4140 Steels: Cast vs Wrought 92
4-11. 4140, 4053 and 4063 Steels: Effect of Carbon Content and Hardness 93
4-12. 4140 Steel: Effect of Direction on Fatigue Crack Propagation 94
4-13. 4140 Steel: Effect of Cathodic Polarization 95
4-14. Cast 4330 Steel: Effects of Various Surface Conditions 96
4-15. 4340 Steel: Scatter of Fatigue Limit Data 97
4-16. 4340 Steel: Strength vs Fatigue Life 98
4-17. 4340 Steel: Total Strain vs Fatigue Life 99
4-18. 4340 Steel: Stress Amplitude vs Number of Reversals 100
4-19. 4340 Steel: Effect of Periodic Overstrain 101
4-20. 4340 Steel: Estimation of Constant Life 102
4-21. 4340 Steel: Effect of Strength Level on Constant-Life Behavior 103
4-22. 4340 Steel: Notched vs Unnotched Specimens 104
4-23. 4340 Steel: Effect of Decarburization 105
4-24. 4340H Steel: Effect of Inclusion Size 106
4-25. 4340 Steel: Influence of Inclusion Size 107
4-26. 4340 Steel: Effect of Hydrogenation; Static Fatigue 108
4-27. 4340 Steel: Effect of Hydrogen 109
4-28. 4340 Steel: Effect of Nitriding 110
4-29. 4340 Steel: Effect of Nitriding and Shot Peening 111
4-30. 4340 Steel: Effect of Induction Hardening and Nitriding 112
4-31. 4340 Steel: Effect of Surface Coatings 113
4-32. 4340 Steel: Effect of Temperature on Constant-Lifetime Behavior 114
4-33. 4520H Steel: Effect of Type of Quench 115
4-34. 4520H Steel: Effect of Shot Peening 116
4-35. 4620 Steel: Effect of Nitriding 117
4-36. 4620 Steel: P/M-Forged 118
4-37. 4620 Steel: P/M-Forged at Different Levels 119

Contents

4-38. 4625 Steel: P/M vs Ingot Forms   120
4-39. 4640 Steel: P/M-Forged   121
4-40. High-Carbon Steel (Eutectoid Carbon): Pearlite vs Spheroidite   122
4-41. 52100 EF Steel: Surface Fatigue; Effect of Finish and Additives   123
4-42. 52100 EF Steel: Surface Fatigue; Effect of Surface Finish and Speed   124
4-43. 52100 EF Steel: Surface Fatigue; Effect of Lubricant Additives   125
4-44. 52100 EF Steel: Surface Fatigue; Effect of Lubricant Viscosity, Slip Ratio and Speed   126
4-45. 52100 EF Steel: Rolling Ball Fatigue; Effect of Oil Additives   127
4-46. 52100 Steel: Carburized vs Uncarburized   128
4-47. 8620H Steel: Carburized; Results From Case and Core   129
4-48. 8620H Steel: Effect of Variation in Carburizing Treatments   130
4-49. 8620 Steel: Effect of Nitriding   131
4-50. 8622 Steel: Effect of Grinding   132
4-51. Cast 8630 Steel: Goodman Diagram for Bending Fatigue   133
4-52. Cast 8630 Steel: Effect of Shrinkage   134
4-53. Cast 8630 Steel: Effect of Shrinkage on Torsion Fatigue   135
4-54. Cast 8630 Steel: Effect of Shrinkage on Torsion Fatigue   136
4-55. Cast 8630 Steel: Effect of Shrinkage on Plate Bending   137
4-56. Cast 8630 vs Wrought 8640   138
4-57. 8630 and 8640 Steels: Effect of Notches on Cast and Wrought Specimens   139
4-58. Nitralloy 135 Steel: Effect of Nitriding   140
4-59. AMS 6475: Effects of Welding   141
4-60. Medium-Carbon, 1Cr-Mo-V Steel Forging: Effect of Cycling Frequency   142
4-61. EM12 Steel: Effect of Temperature on Low-Cycle Fatigue   143
4-62. Cast 0.5Cr-Mo-V Steel: Effects of Dwell Time in Elevated-Temperature Testing   144
4-63. Cast 0.5Cr-Mo-V Steel: Effect of Environment at 550 °C (1022 °F)   145
4-64. Cast C-0.5Mo Steel: Effect of Temperature and Dwell Period on Cyclic Endurance at Various Strain Amplitudes   146

SECTION 5: HSLA Steels   147

5-1. HI-FORM 50 Steel vs 1006   147
5-2. HI-FORM 50 Steel vs 1006: Stress Response   148
5-3. HI-FORM 50 Steel Compared With 1006, DP1 and DP2   149
5-4. HSLA vs Mild Steel: Torsional Fatigue   150
5-5. Proprietary HSLA Steel vs ASTM A440   151
5-6. Comparison of HSLA Steel Grades BE, JF and KF for Plastic Strain Amplitude vs Reversals to Failure   152
5-7. Comparison of HSLA Steel Grades BE, JF and KF for Total Strain Amplitude vs Reversals to Failure   153
5-8. Comparison of a Dual-Phase HSLA Steel Grade With HI-FORM 50: Total Strain Amplitude vs Reversals to Failure   154
5-9. AISI 50 XF Steel: Effects of Cold Deformation   155
5-10. AISI 80 DF Steel: Effects of Cold Deformation   156
5-11. Comparison of Three HSLA Steel Grades, Cb, Cb-V and Cb-V-Si: Strain Life From Constant Amplitude   157
5-12. Comparison of Stress Responses: DP1 vs DP2 Dual-Phase HSLA Steels   158
5-13. Dual-Phase HSLA Steel Grade: Stress Response for As-Received vs Water-Quenched   159
5-14. Dual-Phase HSLA Steel Grade: Stress Response for As-Received vs Gas-Jet-Cooled   160
5-15. S-N Comparison of Dual-Phase HSLA Steel Grades DP1 and DP2 With 1006   161
5-16. Comparison of Dual-Phase HSLA Steel DP2 With HI-FORM 50   162
5-17. Comparison of Cyclic Strain Response Curves for Cb, Cb-V and Cb-V-Si Grades of HSLA Steel   163
5-18. Fatigue Crack Propagation Rate: Effect of Temperature for Two HSLA Steel Grades   164

5-19. Effect of *R*-Ratio and Test Temperature on Crack Propagation of HSLA Steel Grade 1   165
5-20. Effect of Test Temperature on Fatigue Crack Propagation Behavior for Two HSLA Steel Grades   166
5-21. Stress-Cycle Curves for Weldments of Different HSLA Steel Grades   167
5-22. Weldments (FCAW): SAE 980 X Steel vs 1006   168
5-23. Weldments (TIG): DOMEX 640 XP Steel Welded Joints vs Parent Metal   169
5-24. Weldments (FCAW Dressed by TIG): Fatigue Life Estimates Compared With Experimental Data for SAE 980 X Steel   170
5-25. SAE 980 X Steel Weldment (FCAW): Smooth Specimen vs TIG-Dressed vs As-Welded   171
5-26. SAE 980 X Steel Weldment (FCAW): Lap-Shear Joints   172
5-27. Microalloyed HSLA Steels: Properties of Fusion Welds   173
5-28. Microalloyed HSLA Steels: Properties of Spot Welds   174

SECTION 6: High-Strength Alloy Steels   176

6-1. HY-130 Steel: Effect of Notch Radii   176
6-2. 300 M Steel: Effect of Notch Severity on Constant-Lifetime Behavior   177
6-3. TRIP Steels Compared With Other High-Strength Grades   178
6-4. Corrosion Fatigue: Special High-Strength Sucker-Rod Material   179
6-5. Corrosion Fatigue Cracking of Sucker-Rod Material   180
6-6. Hydrogenated Steel: Effect of Baking Time on Hydrogen Concentration   181
6-7. Hydrogenated Steel: Effect of Notch Sharpness   182

SECTION 7: Heat-Resisting Steels   183

7-1. 0.5%Mo Steel: Effect of Hold Time in Air and Vacuum at Different Temperatures   183
7-2. DIN 14 Steel (1.5 Cr, 0.90 Mo, 0.25 V): Effect of Liquid Nitriding   184
7-3. 2.25Cr-1.0Mo Steel: Influence of Cyclic Strain Range on Endurance Limit in Various Environments   185
7-4. 2.25Cr-1.0Mo Steel: Effect of Elevated Temperature   186
7-5. 2.25Cr-1.0Mo Steel: Effect of Elevated Temperature and Strain Rate   187
7-6. 2.25Cr-1.0Mo Steel: Effect of Temperature on Fatigue Crack Growth Rate   188
7-7. 2.25Cr-1.0Mo Steel: Effect of Cyclic Frequency on Fatigue Crack Growth Rate   189
7-8. 2.25Cr-1.0Mo Steel: Fatigue Crack Growth Rates in Air and Hydrogen   190
7-9. 2.25Cr-1.0Mo Steel: Effect of Holding Time   191
7-10. Cast 2.25Cr-1.0Mo Steel, Centrifugally Cast: Fatigue Properties at 540 °C (1000 °F)   192
7-11. H11 Steel: Crack Growth Rate in Water and in Water Vapor   193
7-12. 9.0Cr-1.0Mo Steel: Creep-Fatigue Characteristics   194
7-13. 9.0Cr-1.0Mo Modified Steel: Stress Amplitudes Developed in Cycling   195
7-14. 9.0Cr-1.0Mo Modified Steel: Effect of Deformation   196

SECTION 8: Stainless Steels   197

8-1. Type 301 Stainless Steel: Scatter Band for Fatigue Crack Growth Rates   197
8-2. Type 301 Stainless Steel: Effects of Temperature and Environment on Fatigue Crack Growth Rate   198
8-3. Type 304 Stainless Steel: Effect of Temperature on Frequency-Modified Strains   199
8-4. Type 304 Stainless Steel: Fatigue Crack Growth Rate—Annealed and Cold Worked   200
8-5. Type 304 Stainless Steel: Effect of Humidity on Fatigue Crack Growth Rate   201
8-6. Type 304 Stainless Steel: Effect of Aging on Fatigue Crack Growth Rate   202
8-7. Type 304 Stainless Steel: Effect of Temperature on Fatigue Crack Growth Rate   203
8-8. Type 304 Stainless Steel: Damage Relation at 650 °C (1200 °F)   204

Contents

8-9. Type 304 Stainless Steel: Fatigue Crack Growth Rate at Room and Subzero Temperatures 205
8-10. Types 304 and 304L Stainless Steel: Effect of Cryogenic Temperatures on Fatigue Crack Growth Rate 206
8-11. Type 304 Stainless Steel: Fatigue Crack Growth Rate in Air With Variation in Waveforms 207
8-12. Type 304 Stainless Steel: Effect of Hold Time on Cycles to Failure 208
8-13. Type 304 Stainless Steel: Effect of Hold Time and Continuous Cycling on Fatigue Crack Growth Rates 209
8-14. Type 304 Stainless Steel: Effect of Cyclic Frequency on Fatigue Crack Growth Rate 210
8-15. Type 304 Stainless Steel: Effect of Frequency on Fatigue Crack Growth Behavior 211
8-16. Type 304 Stainless Steel Welded With Type 308: Fatigue Crack Growth Rates 212
8-17. Types 304 and 310 Stainless Steel: Effect of Direction on S-N 213
8-18. Types 304, 316, 321, and 348 Stainless Steel: Effects of Temperature on Fatigue Crack Growth Rates 214
8-19. Type 309S Stainless Steel: Effect of Grain Size on Fatigue Crack Growth Rate 215
8-20. Type 310S Stainless Steel: Effect of Temperature on Fatigue Crack Growth Rate 216
8-21. Type 316 Stainless Steel: Growth Rate of Fatigue Cracks in Weldments 217
8-22. Type 316 Stainless Steel: Fatigue Crack Growth Rates—Aged vs Unaged 218
8-23. Type 316 Stainless Steel: Fatigue Crack Growth Rates—Effect of Aging 219
8-24. Type 316 Stainless Steel: Effect of Temperature on Fatigue Crack Growth Rate 220
8-25. Type 316 Stainless Steel: Effect of Cyclic Frequency on Fatigue Crack Growth Rate 221
8-26. Type 316 Stainless Steel: Fatigue Crack Growth Rate in the Annealed Condition 222
8-27. Type 316 Stainless Steel: Effect of Environment (Sodium, Helium, and Air) on Cycles to Failure 223
8-28. Types 316 and 321 Stainless Steel: Effects of Gaseous Environments on Fatigue Crack Growth Rates 224
8-29. Type 321 Stainless Steel: Effect of Hold Time on Fatigue Crack Growth Rates 225
8-30. Type 403 Stainless Steel: Effect of Environment on Fatigue Crack Growth Rate 226
8-31. Type 403 Modified Stainless Steel: Scatter of Fatigue Crack Growth Rates 227
8-32. Type 422 Stainless Steel: Fatigue Crack Growth Rates in Precracked Specimens 228
8-33. Type 422 Stainless Steel: Fatigue Strength—Longitudinal vs Transverse 229
8-34. Type 422 Stainless Steel: Effect of Temperature on Fatigue Strength 230
8-35. Type 422 Stainless Steel: Effects of Delta Ferrite on Fatigue Strength 231
8-36. 17-4 PH Stainless Steel: Fatigue Crack Growth Rates in Air vs Salt Solution 232
8-37. 15-5 PH Stainless Steel: Fatigue Crack Growth Rates in Air vs Salt Solution 233
8-38. PH 13-8 Mo Stainless Steel: Fatigue Crack Growth Rates at Room Temperature 234
8-39. PH 13-8 Mo Stainless Steel: Fatigue Crack Growth Rates in Air and Sump Tank Water 235
8-40. PH 13-8 Mo Stainless Steel: Fatigue Crack Growth Rates at Subzero Temperatures 236
8-41. PH 13-8 Mo Stainless Steel: Constant-Life Fatigue Diagram 237
8-42. Types 600 and 329 Stainless Steel: S-N Curves for Two Processing Methods 238
8-43. Grade 21-6-9 Stainless Steel: Effect of Temperature on Fatigue Crack Growth Rates 239
8-44. Kromarc 58 Stainless Steel: Effect of Cryogenic Temperatures on Weldments 240
8-45. Pyromet 538 Stainless Steel: Effects of Welding Methods on Fatigue Crack Growth Rates 241
8-46. Duplex Stainless Steel KCR 171: Corrosion Fatigue 242

SECTION 9: Maraging Steels   243

9-1. Grades 200, 250, and 300 Maraging Steel: S-N Curves for Smooth and Notched Specimens   243
9-2. Grade 300 Maraging Steel: Fatigue Life in Terms of Total Strain   244

SECTION 10: Cast Irons   245

10-1. Fatigue of Cast Irons as a Function of Structure-Sensitive Parameters   245
10-2. Gray Iron: Fatigue Life, and Fatigue Limit as a Function of Temperature   246
10-3. Gray Iron: S-N Curves for Unalloyed vs Alloyed   247
10-4. Gray Iron: Effect of Environment   248
10-5. Class 30 Gray Iron: Modified Goodman Diagram for Class 30   249
10-6. Class 30 Gray Iron: Fatigue Crack Growth Rates for Class 30   250
10-7. Gray Irons: Torsional Fatigue for Various Tensile Strength Values   251
10-8. Gray Irons: Torsional Fatigue Data for Five Different Compositions   252
10-9. Gray Irons: Thermal Fatigue—Effect of Aluminum Additions   253
10-10. Gray Irons: Thermal Fatigue—Effect of Chromium and Molybdenum Additions   254
10-11. Gray Irons: Thermal Fatigue—Room Temperature and 540 °C (1000 °F)   255
10-12. Gray Irons: Thermal Fatigue Properties—Comparisons With Ductile Cast Iron and Carbon Steel   256
10-13. Cast Irons: Thermal Fatigue Properties for Six Grades   257
10-14. Ductile Iron: Effect of Microstructure on Endurance Ratio–Tensile Strength Relationship   258
10-15. Ductile Iron: Effect of Microstructure on Endurance Ratio–Tensile Strength Relationship   259
10-16. Ductile Iron: S-N Curves for Ferritic and Pearlitic Grades, Using V-Notched Specimens   260
10-17. Ductile Iron: S-N Curves for Ferritic and Pearlitic Grades, Using Unnotched Specimens   261
10-18. Ductile Iron: Fatigue Diagrams for Bending Stresses and Tension-Compression Stresses   262
10-19. Ductile Iron: Effect of Surface Conditions—As-Cast vs Polished Surface   263
10-20. Ductile Iron: Fatigue Limit in Rotary Bending as Related to Hardness   264
10-21. Ductile Iron: Effect of Rolling on Fatigue Characteristics   265
10-22. Ductile Iron: Effect of Notches on a 65,800-psi-Tensile-Strength Grade   266
10-23. Ductile Iron: Fatigue Crack Growth Rate Compared With That of Steel   267
10-24. Malleable Iron: S-N Curve Comparisons of Four Grades   268
10-25. Pearlitic Malleable Iron: Effect of Surface Conditions on S-N Curves   269
10-26. Pearlitic Malleable Iron: Effect of Nitriding   270
10-27. Ferritic Malleable Iron: Effect of Notch Radius and Depth   271

SECTION 11: Heat-Resisting Alloys   272

11-1. A286: Effect of Environment   272
11-2. A286: Effect of Frequency on Life at 593 °C (1095 °F)   273
11-3. A286: Fatigue Crack Growth Rates at Room and Elevated Temperatures   274
11-4. Astroloy: S-N Curves for Powder vs Conventional Forgings   275
11-5. Astroloy: Powder vs Conventional Forgings Tested at 705 °C (1300 °F)   276
11-6. FSX-430: Effect of Grain Size on Cycles to Cracking   277
11-7. FSX-430: Effect of Grain Size on Fatigue Crack Propagation Rate   278
11-8. HS-31: Effect of Testing Temperature   279
11-9. IN 738 LC Casting Alloy: Standard vs HIP'd Material   280
11-10. IN 738 LC: Effect of Grain Size on Cycles to Failure   281
11-11. IN 738 LC: Effect of Grain Size on Cycles to Cracking   282
11-12. IN 738 LC: Effect of Grain Size on Fatigue Crack Propagation Rate   283
11-13. IN 738 LC: Fatigue Crack Growth Rate at 850 °C (1560 °F)   284
11-14. Inconel 550: Axial Tensile Fatigue Properties in Air and Vacuum at 1090 K   285

Contents

11-15. Inconel 625: Effect of Temperature on Cycles to Failure   286
11-16. Inconel 706: Effect of Temperature on Fatigue Crack Growth Rate   287
11-17. Inconel "713C": Effect of Elevated Temperatures on Fatigue Characteristics   288
11-18. Inconel "713C" and As-Cast HS-31: Comparison of Two Alloys for Number of Cycles in Thermal Fatigue to Initiate Cracks   289
11-19. Inconel 718: Effect of Frequency on Fatigue Crack Propagation Rate   290
11-20. Inconel 718: Relationship of Fatigue Crack Propagation Rate With Stress Intensity   291
11-21. Inconel 718: Relationship of Fatigue Crack Growth Rate With Load/Time Waveforms   292
11-22. Inconel 718: Fatigue Crack Growth Rate in Air vs Helium   293
11-23. Inconel 718: Effect of Environment on Fatigue Crack Growth Rate   294
11-24. Inconel 718: Fatigue Crack Growth Rate in Air Plus 5% Sulfur Dioxide   295
11-25. Inconel 718: Fatigue Crack Growth Rate in Air at Room Temperature   296
11-26. Inconel 718: Fatigue Crack Growth Rate in Air at 316 °C (600 °F)   297
11-27. Inconel 718: Fatigue Crack Growth Rate in Air at 427 °C (800 °F)   298
11-28. Inconel 718: Fatigue Crack Growth Rate in Air at 538 °C (1000 °F)   299
11-29. Inconel 718: Fatigue Crack Growth Rate in Air at 649 °C (1200 °F)   300
11-30. Inconel 718: Fatigue Crack Growth Rates at Cryogenic Temperatures   301
11-31. Inconel 718 and X-750: Fatigue Crack Growth Rates at Cryogenic Temperatures   302
11-32. Inconel X-750: Effect of Temperature on Fatigue Crack Growth Rates   303
11-33. Jethete M152: Interrelationship of Tempering Treatment, Alloy Class, and Testing Temperature With Fatigue Characteristics   304
11-34. Lapelloy: Interrelationship of Hardness and Strength With Fatigue Characteristics   305
11-35. MAR-M200: Effect of Atmosphere on Cycles to Failure   306
11-36. MAR-M509: Correlation of Initial Crack Propagation and Dendrite Arm Spacing   307
11-37. MAR-M509: Correlation Between Number of Cycles Required to Initiate a Crack and Dendrite Arm Spacing   308
11-38. MERL 76, P/M: Axial Low-Cycle Fatigue Life of As-HIP'd Alloy at 540 °C (1000 °F)   309
11-39. Nickel-Base Alloys: Effect of Solidification Conditions on Cycles to Onset of Cracking   310
11-40. René 95 (As-HIP): Cyclic Crack Growth Behavior Under Continuous and Hold-Time Conditions   311
11-41. René 95: Effect of Temperature on Fatigue Crack Growth Rate   312
11-42. S-816: Effect of Notches on Cycles to Failure at 900 °C (1650 °F)   313
11-43. Udimet 700: Fatigue Crack Growth Rates at 850 °C (1560 °F)   314
11-44. U-700 and MAR-M200: Comparison of Fatigue Properties   315
11-45. Waspaloy: Stress-Response Curves   316
11-46. X-40: Effect of Grain Size and Temperature on Fatigue Characteristics   317
11-47. Cast Heat-Resisting Alloys: Ranking for Resistance to Thermal Fatigue   318

SECTION 12: Aluminum Alloys   319

12-1. Corrosion-Fatigue Properties of Aluminum Alloys Compared With Those of Other Alloys   319
12-2. Comparisons of Aluminum Alloys With Magnesium and Steel: Tensile Strength vs Endurance Limit   320
12-3. Aluminum Alloys (General): Yield Strength vs Fatigue Strength   321
12-4. Comparison of Aluminum Alloy Grades for Crack Propagation Rate   322
12-5. Alloy 1100: Relationship of Fatigue Cycles and Hardness for H0 and H14 Tempers   323
12-6. Alloy 1100: Interrelationship of Fatigue Cycles, Acoustic Harmonic Generation and Hardness   324
12-7. Alloy 2014-T6: Notched vs Unnotched Specimens; Effect on Cycles to Failure   325
12-8. Alloy 2024-T3: Effect of Air vs Vacuum Environments on Cycles to Failure   326
12-9. Alloy 2024-T4 Alclad Sheet: Effect of Bending on Cycles to Failure   327

12-10. Alloy 2024-T4: High-Cycle vs Low-Cycle Fatigue   328
12-11. Alloy 2024-T4: Relationship of Stress and Fatigue Cycles   329
12-12. Alloy 2024-T4: Dependence of the Average Rocking Curve Halfwidth $\bar{\beta}$ on Distance From the Surface   330
12-13. Alloys 2024 and X2024: Effect of Alloy Purity on Cycles to Failure   331
12-14. Alloys 2024 and 2124: Relationship of Particle Size and Fatigue Characteristics   332
12-15. Alloys 2024-T4 and 2124-T4: Comparison of Resistance to Fatigue Crack Initiation   333
12-16. Alloys 2024-T3 and 7075-T6: Summary of Fatigue Crack Growth Rates   334
12-17. Alloys 2024-T4 and 7075-T6: Effect of Product Form and Notches   335
12-18. Alloys 2024-T351 and 7075-T73XXX: Comparison of P/M Extrusions and Rod   336
12-19. Alloy 2048-T851: Longitudinal vs Transverse for Axial Fatigue   337
12-20. Alloy 2048-T851: Notched vs Unnotched Specimens at Room and Elevated Temperatures   338
12-21. Alloy 2048-T851: Fatigue Crack Propagation Rates in LT and TL Orientations   339
12-22. Alloy 2048-T851: Modified Goodman Diagram for Axial Fatigue   340
12-23. Alloy 2219-T851: Dependence of Relaxation Behavior on the Cyclic Hardening Parameter   341
12-24. Alloy 2219-T851: Effect of Strain Amplitude on the Relaxation of Residual Surface Stress With Fatigue   342
12-25. Alloy 2219-T851: Relationship of Fatigue Cycles to Different Depth Distributions of Surface Stress   343
12-26. Alloy 2219-T851: Probability of Fatigue Failure   344
12-27. Alloys 3003-O, 5154-H34 and 6061-T6: Effect of Alloy on Fatigue Characteristics of Weldments   345
12-28. Alloy 5083-O Plate: Effect of Orientation on Fatigue Crack Growth Rates   346
12-29. Alloy 5083-O Plate: Effect of Temperature and Humidity on Fatigue Crack Growth Rates   347
12-30. Alloys 5086-H34, 5086-H36, 6061-T6, 7075-T73 and 2024-T3: Comparative Resistance to Axial-Stress Fatigue   348
12-31. Alloys 5083-O/5183: Fatigue Life Predictions and Experimental Data Results for Double V-Butt Welds   349
12-32. Alloys 5083-O/5183: Predicted Effect of Stress Relief and Stress Ratio on Fatigue Life of Butt Welds   350
12-33. 7XXX Alloys: Cyclic Strain vs Crack Initiation Life   351
12-34. Alloy 7050: Influence of Alloy Composition and Dispersoid Effect on Mean Calculated Fatigue Life   352
12-35. Alloy 7050: Effect of Grain Shape on Cycles to Failure   353
12-36. Alloy 7075 (TMP, T6 and T651): Effect of Thermomechanical Processing on Cycles to Failure   354
12-37. Alloys 7075 and 7475: Effect of Inclusion Density on Cycles to Failure   355
12-38. Alloy 7075: Effect of TMT on Cycles to Failure   356
12-39. Alloys 7075 and 7050: Relative Ranking for Constant Amplitude and Periodic Overload   357
12-40. Alloy 7075: Effect of Environment and Mode of Loading   358
12-41. Alloy 7075-T6: Effects of Corrosion and Pre-Corrosion   359
12-42. Alloy 7075-T73: Effect of a 3.5% NaCl Environment on Cycles to Failure   360
12-43. Alloy 7075: Effect of Cathodic Polarization on Fatigue Behavior   361
12-44. Alloy 7075-T6: Effect of Surface Treatments and Notch Designs on Number of Cycles to Failure   362
12-45. Alloy 7075-T6: Effect of R-Ratio on Fatigue Crack Propagation   364
12-46. Alloy 7075: Effect of Predeformation on Fatigue Crack Propagation Rates   365
12-47. Alloys 7075 and 2024-T3: Comparative Fatigue Crack Growth Rates for Two Alloys in Varying Humidity   366
12-48. Alloy 7075-T651: Fatigue Life as Related to Harmonic Generation   367
12-49. Alloys 7075-T6 and 7475-T73: Effect of Laser-Shock Treatment on Fatigue Properties   368
12-50. Alloy 7075-T6: Effect of Laser-Shock Treatment on Hi-Lok Joints   369

Contents

12-51. Alloy 7075 (High Purity): Effect of Iron and Silicon on Cycles to Failure   370
12-52. Alloy X-7075: Effect of Grain Size on Cycles to Failure   371
12-53. Alloy X-7075: Effect of Grain Size on Stress-Life Behavior   372
12-54. Alloy X-7075: Effect of Environment; Air vs Vacuum   373
12-55. Alloy X-7075: Effect of Environment on Two Different Grain Sizes   374
12-56. Alloy X-7075: Effect of Grain-Boundary Ledges on Cycles to Failure   375
12-57. Alloys X-7075 and 7075: Effects of Chromium Inclusions on Fatigue Crack Propagation   376
12-58. Alloy 7475-T6: S-N Diagram for a Superplastic Fine-Grain Alloy   377
12-59. Alloy 7475: Effect of Alignment of Grain Boundaries on Cycles to Failure   378
12-60. Alloy 7475-T6: Superplastic vs Nonsuperplastic, as Related to Fatigue Crack Growth   379
12-61. Alloys X-7075 and 7075: Effect of Chromium-Containing Inclusions on Cycles to Failure   380
12-62. Aluminum Forging Alloys: Stress Amplitude vs Reversals to Failure   381
12-63. Al-5Mg-0.5Ag: Effect of Condition on Fatigue Characteristics   382
12-64. Al-Zn-Mg and Al-Zn-Mg-Zr: Effect of Grain Size on Strain-Life Behavior   383
12-65. Al-Zn-Mg: Strain-Life Curves of a Large-Grained Alloy   384
12-66. Aluminum With a Copper Overlay: Stress Amplitude vs Cycles to Failure   385
12-67. P/M Alloys 7090 and 7091 vs Extruded 2024   386
12-68. P/M Alloys 7090 and 7091 vs I/M 7050 and 7075 Products   387
12-69. P/M Aluminum Alloys: Typical Fatigue Behavior   388
12-70. P/M Aluminum Alloys: Comparison With Specimens Made by Ingot Metallurgy   389
12-71. P/M Aluminum Alloys: Comparison With Forged 7175 for Cycles to Failure   390
12-72. Various Aluminum Alloys: Comparison of Grades for Corrosion-Fatigue Crack Growth Rates; Air vs Salt Water   391
12-73. Various Aluminum Alloys: Comparison of Grades for Corrosion-Fatigue Crack Growth Rates in Salt Water   392
12-74. Various Aluminum Alloys: Wrought vs Cast, and Influence of Casting Method on Fatigue Life   393
12-75. Aluminum Casting Alloy AL-195: Interrelationship of Fatigue Properties With Degree of Porosity   394
12-76. Aluminum Casting Alloy LM25-T6: Squeeze Formed vs Chill Cast; Effect on Reversals to Failure   395

SECTION 13: Copper Alloys   396

13-1. Copper: Effect of Air and Water Vapor on Cycles to Failure   396
13-2. Copper: Applied Plastic-Strain Amplitude vs Fatigue Life   397
13-3. Copper Alloy C11000 (ETP Wire): Effect of Temperature on Fatigue Strength   398
13-4. Copper Alloy C26000 (Cartridge Brass): Influence of Grain Size and Cold Work on Cycles to Failure   399
13-5. Copper Alloy C83600 (Leaded Red Brass): S-N Curves; Scatter Band   400
13-6. Copper Alloy C86500 (Manganese Bronze): S-N Curves; Scatter Band   401
13-7. Copper Alloys C87500 and C87800 (Silicon Brasses): S-N Curves; Scatter Band   402
13-8. Copper Alloy C92200 (Navy "M" Bronze): S-N Curves; Scatter Band   403
13-9. Copper Alloy C93700 (High-Leaded Tin Bronze): S-N Curves; Scatter Band   404
13-10. Copper Alloy No. 192: Effect of Salt Spray on Tubes   405
13-11. Copper Alloy 955: Goodman-Type Diagram   406

SECTION 14: Magnesium Alloys   407

14-1. Magnesium Casting Alloy QE22A-T6: Effects of Notches and Testing Temperature   407
14-2. Magnesium Casting Alloy QH21A-T6: S-N Curves; Effects of Notches and Testing Temperature   408
14-3. Mg-Al-Zn Casting Alloys: Effects of Surface Conditions on Fatigue Properties   409

SECTION 15: Molybdenum   410

15-1. Molybdenum: Fatigue Limit Ratio vs Temperature   410

SECTION 16: Tin Alloys   411

16-1. Tin-Lead Soldering Alloy: S-N Data for Soldered Joints   411
16-2. Babbitt: Variation of Bearing Life With Babbitt Thickness   412
16-3. SAE12 Bearing Alloy: Effect of Temperature on Fatigue Life   413

SECTION 17: Titanium and Titanium Alloys   414

17-1. Unalloyed Titanium, Grade 3: S-N Curves for Annealed vs Cold Rolled   414
17-2. Unalloyed Titanium, Grade 4: S-N Curves for Three Testing Temperatures   415
17-3. Ti-24V and Ti-32V: Stress Amplitude vs Cycles to Failure   416
17-4. Ti-5Al-2.5Sn: Effects of Notches and Types of Surface Finish   417
17-5. Ti-5Al-2.5Sn and Ti-6Al-4V: Fatigue Crack Growth Rates   418
17-6. Ti-6Al-6V-2Sn: Effects of Machining and Grinding   419
17-7. Ti-6Al-6V-2Sn (HIP): S-N Curves for Titanium Alloy Powder Consolidated by HIP   420
17-8. Ti-6Al-6V-2Sn (HIP): S-N Curves for Annealed Plate vs HIP   421
17-9. Ti-6Al-2Sn-4Zr-2Mo: Bar Chart Presentation on Effects of Machining and Grinding   422
17-10. Ti-6Al-2Sn-4Zr-2Mo: Constant-Life Fatigue Diagram   423
17-11. Ti-6Al-2Sn-4Zr-6Mo: Low-Cycle Axial Fatigue Curves   424
17-12. Ti-8Mo-2Fe-3Al: S-N Curves; Solution Treated and Aged Condition   425
17-13. Ti-10V-2Fe-3Al: S-N Curves; Notched vs Unnotched Specimens in Axial Fatigue   426
17-14. Ti-10V-2Fe-3Al and Ti-6Al-4V: Comparison of Fatigue Crack Growth Rates   427
17-15. Ti-10V-2Fe-3Al: S-N Curve; Notched Bar Fatigue Life for a Series of Forgings Compared With Ti-6Al-4V Plate   428
17-16. Ti-13V-11Cr-3Al: Constant-Life Fatigue Diagrams   429
17-17. Ti-6Al-4V: Effect of Condition and Notches on Fatigue Characteristics   430
17-18. Ti-6Al-4V: Effect of Direction on Endurance   431
17-19. Ti-6Al-4V: Effect of Isothermally Rolled vs Extruded Material on Cycles to Failure   432
17-20. Ti-6Al-4V: Comparison of Wrought vs Isostatically Pressed Material for Cycles to Failure   433
17-21. Ti-6Al-4V: Effect of Fretting and Temperature on Cycles to Failure   434
17-22. Ti-6Al-4V (Beta Rolled): Effect of Finishing Operations on Cycles to Failure   435
17-23. Ti-6Al-4V: Effect of Yield Strength on Stress-Life Behavior   436
17-24. Ti-6Al-4V: Effect of Stress Relief on Cycles to Failure   437
17-25. Ti-6Al-4V: Interrelationship of Machining Practice and Cutting Fluids on Cycles to Failure   438
17-26. Ti-6Al-4V: Relative Effects of Machining and Grinding Operations on Endurance Limit   439
17-27. Ti-6Al-4V: Effects of Various Metal Removal Operations on Endurance Limit   440
17-28. Ti-6Al-4V: Effect of Texture on Fatigue Strength   441
17-29. Ti-6Al-4V: Effect of Complex Texture on Cycles to Failure   422
17-30. Ti-6Al-4V: Effect of Texture and Environment on Cycles to Failure   443
17-31. Ti-6Al-4V: Fatigue Crack Growth Rates   444
17-32. Ti-6Al-4V: Fatigue Crack Growth Rates for ISR Tee, and Extrusions   445
17-33. Ti-6Al-4V: Fatigue Crack Growth Rates   446
17-34. Ti-6Al-4V: Effect of Final Cooling on Fatigue Crack Growth Rates   447
17-35. Ti-6Al-4V: Effect of Dwell Time on Fatigue Crack Growth Rates   448
17-36. Ti-6Al-4V: Fatigue Crack Growth Data   449
17-37. Ti-6Al-4V P/M: Comparison of HIP'd Material With Alpha-Beta Forgings for Cycles to Failure   450

Contents

17-38. Ti-6Al-4V P/M: Comparisons of HIP'd Material With Annealed Plate for Cycles to Failure   451
17-39. Ti-6Al-4V P/M: Effect of Powder Mesh Size on Fatigue Properties   452
17-40. Ti-6Al-4V P/M: Comparison of Blended Elemental, Prealloyed and Wrought Material for Effect on Cycles to Failure   453
17-41. Ti-6Al-4V: P/M Compacts vs I/M Specimens: Cycles to Failure   454
17-42. Ti-6Al-4V: Comparison of Specimens Processed by Various Fabrication Processes for Cycles to Failure   455
17-43. Ti-6Al-4V: Comparison of Fatigue Crack Growth Rate, P/M vs I/M   456
17-44. Ti-6Al-4V: Base Metal vs SSEB-Welded Material for Cycles to Failure   457
17-45. Ti-6Al-4V: Base Metal vs SSEB-Welded Material for Cycles to Failure   458
17-46. Ti-6Al-4V EB Weldments: Base Metal Compared With Flawless Weldments   459
17-47. Ti-6Al-4V EB Weldments: Effects of Porosity on Cycles to Failure   460
17-48. Ti-6Al-4V Gas Metal-Arc Weldments: Effects of Porosity on Cycles to Failure   461
17-49. Ti-6Al-4V: Unwelded vs Electron Beam Welded Material for Cycles to Failure   462
17-50. Ti-6Al-4V: S-N Diagram for Laser-Welded Sheet   463
17-51. Ti-6Al-4V (Cast): S-N Diagram for Notched Specimens   464

SECTION 18: Zirconium   465

18-1. Zirconium 702: Effects of Notches and Testing Temperature on Cycles to Failure   465

SECTION 19: Steel Castings   466

(For other data on steel castings see Sections 3, 4 and 5, on carbon and alloy steels.)

19-1. Steel Castings (General): Effect of Design and Welding Practice on Fatigue Characteristics   466
19-2. Steel Castings (General): Effects of Discontinuities on Fatigue Characteristics   467

SECTION 20: Closed-Die Forgings   468

(See also under specific grades of alloys.)

20-1. Closed-Die Steel Forgings: Effect of Surface Condition on Fatigue Limit   468

SECTION 21: Powder Metallurgy Parts   469

(See also under specific alloys.)

21-1. P/M: Relation of Density to Fatigue Limit and Fatigue Ratio   469
21-2. P/M: Relation of Fatigue Limit to Tensile Strength for Sintered Steels   470
21-3. P/M (Nickel Steels): As-Sintered vs Quenched and Tempered for Cycles to Failure   471
21-4. P/M (Nickel Steels): Relation Between Fatigue Limit and Tensile Strength for Sintered Steels   472
21-5. P/M (Nickel Steels): Effect of Notches on Cycles to Failure for the As-Sintered Condition   473
21-6. P/M (Nickel Steels): Effect of Notches on Cycles to Failure for the Quenched and Tempered Condition   474
21-7. P/M (Low-Carbon, 1-5%Cu): Effects of Notches and Nitriding on Cycles to Failure   475
21-8. P/M (Sintered Iron, Low-Carbon, No Copper): Effect of Density and Nitriding on Cycles to Failure   476
21-9. P/M: Effect of Nitriding on Ductile Iron and Sintered Iron (3%Cu) for Cycles to Failure   477

SECTION 22: Composites   478

22-1. Brass/Mild Steel Composite: Comparison of Brass-Clad Mild Steel With Brass and Mild Steel for Cycles to Failure   478
22-2. Stainless Steel/Mild Steel Composite: Comparison of Stainless-Clad Mild Steel With Stainless Steel and Mild Steel for Cycles to Failure   479

SECTION 23: Effects of Surface Treatments 480

23-1. Carbon and Alloy Steels (Seven Grades): Effects of Nitrocarburizing on Fatigue Strength 480
23-2. Carbon and Alloy Steels (Seven Grades): Effects of Tufftriding on Fatigue Characteristics 481
23-3. Carbon and Alloy Steels (Six Grades): Effects of Nitriding on Fatigue Strength 482
23-4. Carbon-Manganese Steel: Effects of Nickel Coating on Fatigue Strength 483

SECTION 24: Test Results for Component Parts 484

24-1. Coil Springs, Music Wire (Six Sizes): Data Presented by Means of a Goodman Diagram 484
24-2. Coil Springs: S-N Data for Oil-Tempered and Music Wire Grades 485
24-3. Coil Springs: Effects of Shot Peening on Cycles to Failure 486
24-4. Coil Springs, 8650 and 8660 Steels: Relation of Design Stresses and Probability of Failure 487
24-5. Coil Springs, HSLA Steels: Effects of Corrosion on Cycles to Failure 488
24-6. Leaf Springs, 5160 Steel: Maximum Applied Stress vs Cycles to Failure 489
24-7. Front Suspension Torsion Bar Springs, 5160H Steel: Distribution of Fatigue Results for Simulated Service Testing 490
24-8. Gears, Carburized Low-Carbon Steel: Relation of Life Factor to Required Life 491
24-9. Gears, Carburized Low-Carbon Steel: Bending Stress vs Cycles to Failure 492
24-10. Gears, Carburized Low-Carbon Steel: Effect of Shot Peening on Cycles to Failure 493
24-11. Gears, Carburized Low-Carbon Steel: Probability-Stress-Life Design Curves 494
24-12. Gears, 8620H Carburized: Bending or Contact Stress vs Cycles to Fracture or Pitting 495
24-13. Gears, 8620H Carburized: A Weibull Analysis of Bending Fatigue Data 496
24-14. Gears, 8620H Carburized: T-N Curves for Six-Pinion, Four-Square Tests 497
24-15. Hypoid Gears, 8620H Carburized: Minimum Confidence Level; Stress vs Cycles to Rupture 498
24-16. Hypoid, Zerol and Spiral Bevel Gears, 8620H Carburized: S-N Scatter Band and Minimum Confidence Level 499
24-17. Spiral Bevel and Zerol Bevel Gears, 8620H Carburized: S-N Scatter Band and Minimum Confidence Level 500
24-18. Gears, 8620H Case Hardened: Relation of Life Factor to Cycles to Rupture 501
24-19. Bevel Gears, Low-Carbon Steel Case Hardened: Relation of Life Factor to Cycles to Rupture for Various Confidence Levels 502
24-20. Gears, AMS 6265: S-N Data for Cut vs Forged 503
24-21. Spur Gears, 8620H: S-N Data for Cut vs Forged 504
24-22. Gears and Pinions: P/M 4600V vs 4615; Weibull Distributions 505
24-23. Gears and Pinions: P/M Grades 4600V and 2000 vs 4615; Percent Failure vs Time 506
24-24. Gear Steel AMS 6265: Parent Metal vs Electron Beam Welded 507
24-25. Gears, 42 CrMo4 (German Specification): S-N Curves for Various Profiles 508
24-26. Gears, 42 CrMo4 (German Specification): Endurance Test Results in the Weibull Distribution Diagram 509
24-27. Bolts, 1040 and 4037 Steels: Maximum Bending Stress vs Number of Stress Cycles 510
24-28. Bolts: S-N Data for Roll Threading Before and After Heat Treatment 511
24-29. Power Shafts, AMS 6382 and AMS 6260: Electron Beam Welded vs Silver Brazed Joints 512
24-30. Axle Shafts, 1046, 1541 and 50B54 Steels: S-N Data for Induction Hardening vs Through Hardening 513
24-31. Steel Rollers, 8620H Carburized: Effects of Carburizing Temperature and Quenching Practice on Surface Fatigue 514

24-32. Steel Rollers, 8620H Carburized: Effects of Carburizing Temperature and Quenching Practice on Surface Fatigue   515
24-33. Linkage Arm, Cast Low-Carbon Steel: Starting Crack Size vs Cycles to Failure   516
24-34. Notched Links, Hot Rolled Low-Carbon Steel: S-N Data for Component Test Model   517
24-35. Fuselage Brace, Ti-6Al-6V-2Sn: Fatigue Endurance of HIP-Consolidated Powder   518

Fatigue Testing

Introduction

Fatigue is the progressive, localized, permanent structural change that occurs in materials subjected to fluctuating stresses and strains that may result in cracks or fracture after a sufficient number of fluctuations. Fatigue fractures are caused by the simultaneous action of cyclic stress, tensile stress and plastic strain. If any one of these three is not present, fatigue cracking will not initiate and propagate. The cyclic stress starts the crack; the tensile stress produces crack growth (propagation). Although compressive stress will not cause fatigue, compression load may do so.

The process of fatigue consists of three stages:

- Initial fatigue damage leading to crack nucleation and crack initiation
- Progressive cyclic growth of a crack (crack propagation) until the remaining uncracked cross section of a part becomes too weak to sustain the loads imposed
- Final, sudden fracture of the remaining cross section

Fatigue cracking normally results from cyclic stresses that are well below the static yield strength of the material. (In low-cycle fatigue, however, or if the material has an appreciable work-hardening rate, the stresses also may be above the static yield strength.)

Fatigue cracks initiate and propagate in regions where the strain is most severe. Because most engineering materials contain defects and thus regions of stress concentration that intensify strain, most fatigue cracks initiate and grow from structural defects. Under the action of cyclic loading, a plastic zone (or region of deformation) develops at the defect tip. This zone of high deformation becomes an initiation site for a fatigue crack. The crack propagates under the applied stress through the material until complete fracture results. On the microscopic scale, the most important feature of the fatigue process is nucleation of one or more cracks under the influence of reversed stresses that exceed the flow stress, followed by development of cracks at persistent slip bands or at grain boundaries.

Prediction of Fatigue Life

The fatigue life of any specimen or structure is the number of stress (strain) cycles required to cause failure. This number is a function of many variables, including stress level, stress state, cyclic wave form, fatigue environment, and the metallurgical condition of the material. Small changes in the specimen or test conditions can significantly affect fatigue behavior, making analytical prediction of fatigue life difficult. Therefore, the designer may rely on experience with similar components in service rather than on laboratory evaluation of mechanical test specimens. Laboratory tests, however, are essential in understanding fatigue behavior, and current studies with fracture mechanics test specimens are beginning to provide satisfactory design criteria.

Laboratory fatigue tests can be classified as crack initiation or crack propagation. In crack initiation testing, specimens or parts are subjected to the number of stress cycles required for a fatigue crack to initiate and to subsequently grow large enough to produce failure.

In crack propagation testing, fracture mechanics methods are used to determine the crack growth rates of preexisting cracks under cyclic loading. Fatigue crack propagation may be caused by cyclic stresses in a benign environment, or by the combined effects of cyclic stresses and an aggressive environment (corrosion fatigue).

Fatigue Crack Initiation

Most laboratory fatigue testing is done either with axial loading, or in bending, thus producing only tensile and compressive stresses. The stress usually is cycled either between a maximum and a minimum tensile stress, or between a maximum tensile stress and a maximum compressive stress.

The latter is considered a negative tensile stress, is given an algebraic minus sign, and therefore is known as the minimum stress.

The stress ratio is the algebraic ratio of two specified stress values in a stress cycle. Two commonly used stress ratios are the ratio, A, of the alternating stress amplitude to the mean stress ($A = S_a/S_m$) and the ratio, R, of the minimum stress to the maximum stress ($R = S_{min}/S_{max}$).

If the stresses are fully reversed, the stress ratio R becomes -1; if the stresses are partially reversed, R becomes a negative number less than 1. If the stress is cycled between a maximum stress and no load, the stress ratio R becomes zero. If the stress is cycled between two tensile stresses, the stress ratio R becomes a positive number less than 1. A stress ratio R of 1 indicates no variation in stress, making the test a sustained-load creep test rather than a fatigue test.

Applied stresses are described by three parameters. The mean stress, S_m, is the algebraic average of the maximum and minimum stresses in one cycle, $S_m = (S_{max} + S_{min})/2$. In the completely reversed cycle test, the mean stress is zero. The range of stress, S_r, is the algebraic difference between the maximum and minimum stresses in one cycle, $S_r = S_{max} - S_{min}$. The stress amplitude, S_a, is one half the range of stress, $S_a = S_r/2 = (S_{max} - S_{min})/2$.

During a fatigue test, the stress cycle usually is maintained constant so that the applied stress conditions can be written $S_m \pm S_a$, where S_m is the static or mean stress, and S_a is the alternating stress, which is equal to half the stress range. Nomenclature to describe test parameters involved in cyclic stress testing are shown in Fig. 1.

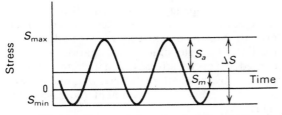

Fig. 1 Nomenclature to describe test parameters involved in cyclic stress testing

S-N Curves. The results of fatigue crack initiation tests usually are plotted as maximum stress, minimum stress, or stress amplitude to number of cycles, N, to failure using a logarithmic scale for the number of cycles. Stress is plotted on either a linear or a logarithmic scale. The result-

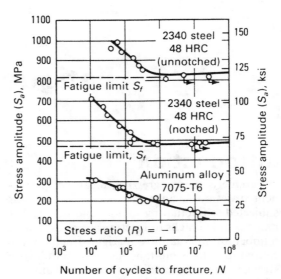

Fig. 2 Typical S-N curves for constant amplitude and sinusoidal loading

ing plot of the data is an S-N curve. Three typical S-N curves are shown in Fig. 2.

The number of cycles of stress that a metal can endure before failure increases with decreasing stress. For some engineering materials such as steel (see Fig. 2) and titanium, the S-N curve becomes horizontal at a certain limiting stress. Below this limiting stress, known as the fatigue limit or endurance limit, the material can endure an infinite number of cycles without failure.

Fatigue Limit and Fatigue Strength. The horizontal portion of an S-N curve represents the maximum stress that the metal can withstand for an infinitely large number of cycles with 50% probability of failure and is known as the fatigue (endurance) limit, S_f. Most nonferrous metals do not exhibit a fatigue limit. Instead, their S-N curves continue to drop at a slow rate at high numbers of cycles, as shown by the curve for aluminum alloy 7075-T6 in Fig. 2.

For these types of metals, fatigue strength rather than fatigue limit is reported, which is the stress to which the metal can be subjected for a specified number of cycles. Because there is no standard number of cycles, each table of fatigue strengths must specify the number of cycles for which the strengths are reported. The fatigue strength of nonferrous metals at 100 million (10^8) or 500 million (5×10^8) cycles is erroneously called the fatigue limit.

Low-Cycle Fatigue. For the low-cycle fatigue region ($N < 10^4$ cycles) tests are conducted with controlled cycles of elastic plus plastic strain,

Introduction

rather than with controlled load or stress cycles. Under controlled strain testing, fatigue life behavior is represented by a log-log plot of the total strain range, $\Delta \epsilon_t$, versus the number of cycles to failure (Fig. 3).

The total strain range is separated into elastic and plastic components. For many metals and alloys, the elastic strain range, $\Delta \epsilon_e$, is equal to the stress range divided by the modulus of elasticity. The plastic strain range, $\Delta \epsilon_p$, is the difference between the total strain range and the elastic strain range.

Stress-Concentration Factor. Stress is concentrated in a metal by structural discontinuities, such as notches, holes, or scratches, which act as stress raisers. The stress-concentration factor, K_t, is the ratio of the area test stress in the region of the notch (or other stress concentrators) to the corresponding nominal stress. For determination of K_t, the greatest stress in the region of the notch is calculated from the theory of elasticity, or equivalent values are derived experimentally.

The fatigue notch factor, K_f, is the ratio of the fatigue strength of a smooth (unnotched) specimen to the fatigue strength of a notched specimen at the same number of cycles.

Fatigue notch sensitivity, q, for a material is determined by comparing the fatigue notch factor, K_f, and the stress-concentration factor, K_t, for a specimen of a given size containing a stress concentrator of a given shape and size. A common definition of fatigue notch sensitivity is $q = (K_f - 1)/(K_t - 1)$, in which q may vary between zero (where $K_f = 1$) and 1 (where $K_f = K_t$). This value may be stated as percentage.

Fatigue Crack Propagation

In large structural components, the existence of a crack does not necessarily imply imminent failure of the part. Significant structural life may remain in the cyclic growth of the crack to a size at which a critical failure occurs. The objective of fatigue crack propagation testing is to determine the rates at which subcritical cracks grow under cyclic loadings prior to reaching a size critical for fracture.

The growth or extension of a fatigue crack under cyclic loading is principally controlled by maximum load and stress ratio. However, as in crack initiation, there are a number of additional factors that may exert a strong influence, including environment, frequency, temperature, and grain direction. Fatigue crack propagation testing usually involves constant-load-amplitude cycling of notched specimens that have been precracked in fatigue. Crack length is measured as a function of elapsed cycles, and these data are subjected to numerical analysis to establish the rate of crack growth, da/dN.

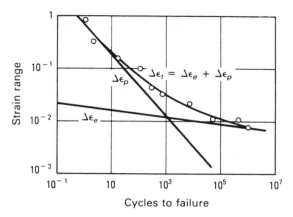

Fig. 3 Typical plot of strain range versus cycles-to-failure for low-cycle fatigue

Crack growth rates are expressed as a function of the crack tip stress-intensity factor range, ΔK. The stress-intensity factor is calculated from expressions based on linear elastic stress analysis and is a function of crack size, load range, and cracked specimen geometry. Fatigue crack growth data are typically presented in a log-log plot of da/dN versus ΔK (Fig. 4).

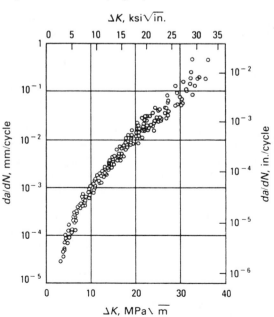

Fig. 4 Fatigue crack propagation rate data in 7075-T6 aluminum alloy ($R < 0$)

Fatigue Crack Initiation

Crack initiation tests are procedures in which a specimen or part is subjected to cyclic loading to failure. A large portion of the total number of cycles in these tests is spent initiating the crack. Although crack initiation tests conducted on small specimens do not precisely establish the fatigue life of a large part, such tests do provide data on the intrinsic fatigue crack initiation behavior of a metal or alloy. As a result, such data can be utilized to develop criteria to prevent fatigue failures in engineering design. Examples of the use of small-specimen fatigue test data can be found in the basis of the fatigue design codes for boilers and pressure vessels, complex welded, riveted, or bolted structures, and automotive and aerospace components.

Fatigue Cracking

Fatigue cracks normally result from cyclic stresses that are below the yield strength of the metal. In low-cycle fatigue, however, the cyclic stress may be above the static yield strength, especially in a material with an appreciable work-hardening rate. Generally, a fatigue crack is initiated at a highly stressed region of a component subjected to cyclic loading of sufficient magnitude. The crack then propagates in progressive cyclic growth through the cross section of the part until the maximum load cannot be carried, and complete fracture results.

Crack Nucleation. A variety of crystallographic features have been observed to nucleate fatigue cracks. In pure metals, tubular holes that develop in persistent slip bands, slip band extrusion-intrusion pairs at free surfaces, and twin boundaries are common nucleation sites. Grain boundaries in polycrystalline metals, even in the absence of inherent grain boundary weakness, are crack nucleation sites. At high strain rates, this appears to be the preferred site. Nucleation at grain boundaries appears to be a geometrical effect, whereas nucleation at twin boundaries is associated with active slip on crystallographic planes immediately adjacent and parallel to the twin boundary.

The foregoing processes also occur in alloys and heterogeneous materials. However, alloying and commercial production practices introduce segregation, inclusions, second-phase particles, and other features that disturb the structure. All of these phenomena have a significant influence on the crack nucleation process. In general, alloying that (1) enhances cross slip, (2) enhances twinning, or (3) increases the rate of work hardening will stimulate crack nucleation. On the other hand, alloying usually raises the flow stress of a metal, thus offsetting its potentially detrimental effect on fatigue crack nucleation.

Crack Initiation. Fatigue cracks initiate at points of maximum local stress and minimum local strength. The local stress pattern is determined by the shape of the part and by the type and magnitude of the loading. In addition to the geometric features of a part, features such as surface and metallurgical imperfections can act to concentrate stress locally. Surface imperfections such as scratches, dents, burrs, cuts, and other manufacturing flaws are the most obvious sites at which fatigue cracks initiate. Except for instances where internal defects or special surface-hardening treatments are involved, fatigue cracks initiate at the surface.

Relation to Environment. Corrosion fatigue describes the degradation of the fatigue strength of a metal by the initiation and growth of cracks under the combined action of cyclic loading and a corrosive environment. Because it is a synergistic effect of fatigue and corrosion, corrosion fatigue can produce a far greater degradation in strength than either effect acting alone or by superposition of the singular effects. An unlimited number of gaseous and liquid mediums may affect fatigue crack initiation in a given material. Fretting corrosion, which occurs from relative motion between joints, may also accelerate fatigue crack initiation.

Fatigue Testing Regimes

The magnitude of the nominal stress on a cyclically loaded component frequently is measured by the amount of overstress—that is, the amount by which the nominal stress exceeds the fatigue limit or the long-life fatigue strength of the material used in the component. The number of load cycles that a component under low overstress can endure is high; thus, the term high-cycle fatigue is often applied.

As the magnitude of the nominal stress increases, initiation of multiple cracks is more likely. Also, spacing between fatigue striations, which indicate the progressive growth of the crack front, is increased, and the region of final fast fracture is increased in size.

Fatigue Crack Initiation

Low-cycle fatigue is the regime characterized by high overstress. The arbitrary, but commonly accepted, dividing line between high-cycle and low-cycle fatigue is considered to be about 10^4 to 10^5 cycles. In practice, this distinction is made by determining whether the dominant component of the strain imposed during cyclic loading is elastic (high cycle) or plastic (low cycle), which in turn depends on the properties of the metal as well as the magnitude of the nominal stress.

Presentation of Fatigue Data. High-cycle fatigue data are presented graphically as stress (S) versus cycles-to-failure (N) in S-N diagrams or S-N curves. These are described in the Introduction to this Section along with the symbols and nomenclature commonly applied in fatigue testing. Because the stress in high-cycle fatigue tests is usually within the elastic range, the calculation of stress amplitude, stress range, or maximum stress on the S-axis is made using simple equations from mechanics of materials; i.e., stress calculated using the specimen dimensions and the controlled load or deflection applied axially, in flexure, or in torsion.

Figure 5 illustrates a stress-strain loop under controlled constant-strain cycling in a low-cycle fatigue test. During initial loading, the stress-strain curve is O-A-B. Upon unloading, yielding begins in compression at a lower stress C due to the Bauschinger effect. In reloading in tension, a hysteresis loop develops. The dimensions of this loop are described by its width $\Delta\epsilon$ (the total strain range) and its height $\Delta\sigma$ (the stress range). The total strain range $\Delta\epsilon$ consists of an elastic strain component $\Delta\epsilon_e = \Delta\sigma/E$ and a plastic strain component $\Delta\epsilon_p$.

The width of the hysteresis loop depends on the level of cyclic strain. When the level of cyclic strain is small, the hysteresis loop becomes very narrow. For tests conducted under constant $\Delta\epsilon$, the stress range $\Delta\sigma$ usually changes with an increasing number of cycles.

The common method of presenting low-cycle fatigue data is to plot either the plastic strain range, $\Delta\epsilon_p$, or the total strain range, $\Delta\epsilon$, versus N. When plotted using log-log coordinates, a straight line can be fit to the $\Delta\epsilon_p$-N plot. The slope of this line in the region where plastic strain dominates has shown little variation for the large number of metals and alloys tested in low-cycle fatigue, the average value being ½. This power-law relationship between $\Delta\epsilon_p$ and N is known as the Coffin-Manson relationship. Figure 6 is an example of the typical presentation of low-cycle fatigue test results.

Classification of Fatigue Testing Machines

Fatigue test specimens are primarily described by the mode of loading:

- Direct (axial) stress
- Plane bending
- Rotating beam
- Alternating torsion
- Combined stress

Testing machines, however, may be universal-type machines that are capable of conducting all of the above modes of loading, depending on the fixturing used.

Fatigue Testing Machine Components

Whether simple or complex, all fatigue testing machines consist of the same basic components: a load train, controllers, and monitors. The load train consists of the load frame, gripping devices, test specimen, and drive (loading) system. Typical load train components in an electrohydraulic axial fatigue machine are shown in Fig. 7.

The load frame is the structure of the machine that reacts to the forces applied to the specimen by the drive system.

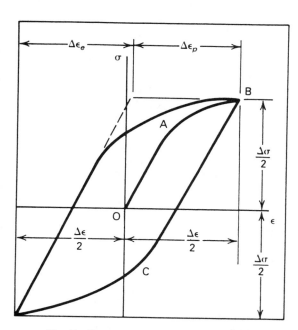

Fig. 5 Stress-strain loop for constant-strain cycling

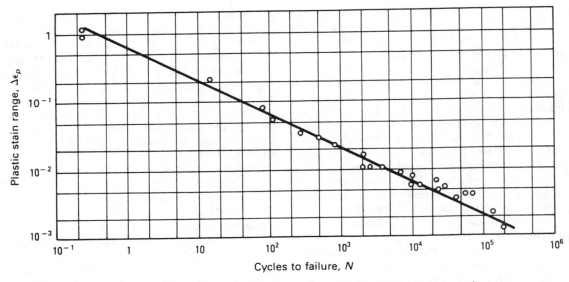

Fig. 6 Low-cycle fatigue curve ($\Delta\epsilon_p$ versus N) for type 347 stainless steel

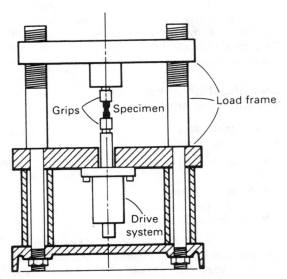

Fig. 7 Schematic of the load train in an electrohydraulic axial fatigue machine

The drive system is the most significant feature of a fatigue testing system and usually is electrically powered. The simplest systems use electric motors to act on test specimens via cams, levers, or rotating grips. In electrohydraulic machines, the motors drive hydraulic pumps to provide service pressure for control of the motion and force of a hydraulic piston actuator. Electromagnetic excitation can be used to excite a mass or inertia system to load a specimen.

Control Systems. The controls and controllers manually or automatically initiate power and test, adjust, and maintain the controlled test parameter(s). Controllers also terminate the test at a predefined status (failure, load drop, extension, or deflection limit). The control of time-varying deflection or displacement can be obtained in mechanical systems by cam-operated deflection levels, a rotating eccentric mass, or hydraulically through a piston limited by stops.

Control in most simple machines and drive systems is obtained via the open-loop mode. In such systems, the magnitude of force and displacement initially set by the control system remains constant throughout the test.

Sensors are required to measure the load, strain, displacement, deflection, and cycle count. Some devices provide an output signal to the controller, or to a readout device in the case of uncontrolled parameters. Common sensors are load cells (resistance strain gage bridges calibrated to load) inserted in the load train. Pressure transducers are used in hydraulic or pneumatic actuator devices.

Loading fixtures to alter the mode of loading provide versatility. Fixtures can be designed to convert the axial force provided by a hydraulic actuator to perform four-point bending or torsion testing. Similarly, fixtures attached to an oscillating platen of a rotating-eccentric-mass-type machine can facilitate axial, bending, and torsion fatigue testing of specimens.

Fatigue Crack Initiation

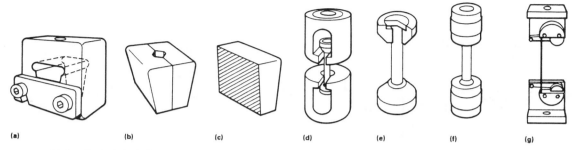

(a) Standard grip body for wedge-type grips. (b) V-grips for rounds for use in standard grip body. (c) Flat grips for specimens for use in standard grip body. (d) Universal open-front holders. (e) Adapters for special samples (screws, bolts, studs, etc.) for use with universal open-front holders. (f) Holders for threaded samples. (g) Snubber-type wire grips for flexible wire or cable.

Fig. 8 Grip designs used for axial fatigue testing

Grips. Proper gripping is not simply the attachment of the test specimen in the load train. Grip failure sometimes occurs prior to specimen failure. Frequently, satisfactory gripping evolves after specimen design development. Care must be taken in grip design and specimen installation in the grips to prevent misalignment. The grips shown in Fig. 8 are typical of those used for axial fatigue tests.

Axial (Direct-Stress) Fatigue Testing Machines

The direct-stress fatigue testing machine subjects a test specimen to a uniform stress or strain through its cross section. For the same cross section, an axial fatigue testing machine must be able to apply a greater force than a static bending machine to achieve the same stress.

Electromechanical systems have been developed for axial fatigue studies. Generally, these are open-loop systems, but often have partial closed-loop features to continuously correct mean load.

In crank and lever machines, a cyclic load is applied to one end of the test specimen through a deflection-calibrated lever that is driven by a variable-throw crank. The load is transmitted to the specimen through a flexure system, which provides straight-line motion to the specimen. The other end of the specimen is connected to a hydraulic piston that is part of an electrohydraulically controlled load-maintaining system that senses specimen yielding. This system automatically and steplessly restores the preset load through the hydraulic piston.

Servohydraulic closed-loop systems offer optimum control, monitoring, and versatility in fatigue testing systems. These can be obtained as component systems and can be upgraded as required. A hydraulic actuator typically is used to apply the load in axial fatigue testing.

Electromagnetic or magnetostrictive excitation is used for axial fatigue testing machine drive systems, particularly when low-load amplitudes and high-cycle fatigue lives are desired in short test durations. The high cyclic frequency of operation of these types of machines enables testing to long fatigue lives ($>10^8$ cycles) within weeks.

Bending Fatigue Machines

The most common types of fatigue machines are small bending fatigue machines. In general, these simple, inexpensive systems allow laboratories to conduct extensive test programs with a low equipment investment.

Cantilever beam machines, in which the test specimen has a tapered width, thickness, or diameter, result in a portion of the test area having uniform stress with smaller load requirements than required for uniform bending or axial fatigue of the same section size.

Rotating Beam Machines. Typical rotating beam machine types are shown in Fig. 9. The R.R. Moore-type machines (Fig. 9a) can operate up to 10 000 rpm. In all bending-type tests, only the material near the surface is subjected to the maximum stress; therefore, in a small-diameter specimen, only a very small volume of material is under test.

Torsional Fatigue Testing Machines

Torsional fatigue tests can be performed on axial-type machines using the proper fixtures if the maximum twist required is small. Specially

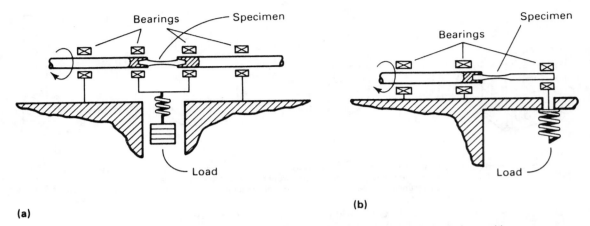

(a) Four-point loading R.R. Moore testing machine. (b) Single-end rotating cantilever testing machine.

Fig. 9 Schematic of rotating beam fatigue testing machines

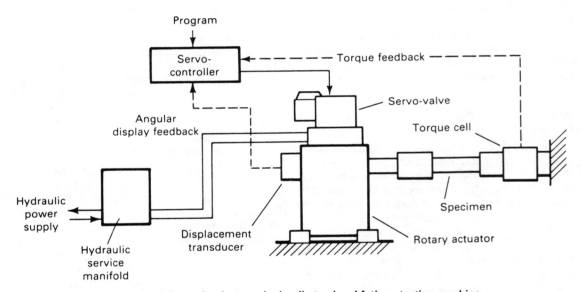

Fig. 10 Schematic of a servohydraulic torsional fatigue testing machine

designed torsional fatigue testing machines consist of electromechanical machines, in which linear motion is changed to rotational motion by the use of cranks, and servohydraulic machines, in which rotary actuators are incorporated in a closed-loop testing system (Fig. 10).

Special-Purpose Fatigue Testing Machines

To perform fatigue testing of components that are prone to fatigue failure (gears, bearings, wire, etc.), special devices have been used, sometimes as modifications to an existing fatigue machine. Wire testers are a modification of rotating beam machines, in which a length of the test wire is used as the beam and is deflected (buckled) a known amount and rotated.

Rolling contact fatigue testers usually are constant-load machines in which a Hertzian contact stress between two rotating bearings is applied until occurrence of fatigue failure by pitting or spalling is indicated by a vibration or noise level in the system. Rolling contact fatigue of ball and roller bearings under controlled lubrication conditions is a specialized field of fatigue testing.

Multiaxial Fatigue Testing Machines

Many special fatigue testing machines have been designed to apply two or more modes of loading, in or out of phase, to specimens to de-

Fatigue Crack Initiation

termine the properties of metals under biaxial or triaxial stresses.

Fatigue Test Specimens

A typical fatigue test specimen has three areas: the test section and the two grip ends. The grip ends are designed to transfer load from the test machine grips to the test section and may be identical, particularly for axial fatigue tests. The transition from the grip ends to the test area is designed with large, smoothly blended radii to eliminate any stress concentrations in the transition.

The design and type of specimen used depend on the fatigue testing machine used and the objective of the fatigue study. The test section in the specimen is reduced in cross section to prevent failure in the grip ends and should be proportioned to use the upper ranges of the load capacity of the fatigue machine; i.e., avoiding very low load amplitudes where sensitivity and response of the system are decreased. Several types of fatigue test specimens are illustrated in Fig. 11.

Effect of Stress Concentration

Fatigue strength is reduced significantly by the introduction of a stress raiser such as a notch or hole. Because actual machine elements invariably contain stress raisers such as fillets, keyways, screw threads, press fits, and holes, fatigue cracks in structural parts usually initiate at such geometrical irregularities.

An optimum way of minimizing fatigue failure is the reduction of avoidable stress raisers through careful design and the prevention of accidental stress raisers by careful machining and fabrication. Stress concentration can also arise from surface roughness and metallurgical stress raisers such as porosity, inclusions, local overheating in grinding, and decarburization.

The effect of stress raisers on fatigue is generally studied by testing specimens containing a notch, usually a V-notch or a U-notch. The presence of a notch in a specimen under uniaxial load introduces three effects: (1) there is an increase or concentration of stress at the root of the notch, (2) a stress gradient is set up from the root of the notch toward the center of the specimen, and (3) a triaxial state of stress is produced at the notch root.

The ratio of the maximum stress in the region of the notch (or other stress concentration) to the corresponding nominal stress is the stress-con-

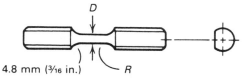

4.8 mm (³⁄₁₆ in.) R
D, selected on basis of ultimate strength of material R, 12.7 mm (0.50 in.)

(a)

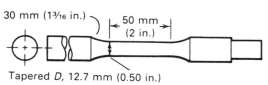

30 mm (1³⁄₁₆ in.)   50 mm (2 in.)
Tapered D, 12.7 mm (0.50 in.)

(b)

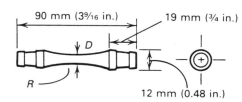

90 mm (3⁹⁄₁₆ in.)   19 mm (¾ in.)
D
R   12 mm (0.48 in.)
D, 5 to 10 mm (0.20 to 0.40 in.) selected on basis of ultimate strength of material R, 90 to 250 mm (3.5 to 10 in.)

(c)

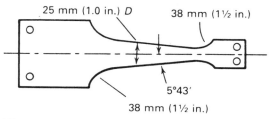

25 mm (1.0 in.) D   38 mm (1½ in.)
5°43'
38 mm (1½ in.)

(d)

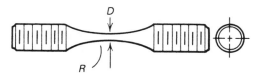

D
R
D, selected on basis of ultimate strength of material R, 75 to 250 mm (3 to 10 in.)

(e)

(a) Torsional specimen. (b) Rotating cantilever beam specimen. (c) Rotating beam specimen. (d) Plate specimen for cantilever reverse bending. (e) Axial loading specimen.

Fig. 11 Typical fatigue test specimens

centration factor, K_t (see the Introduction to this Section). In some situations, values of K_t can be calculated using the theory of elasticity, or can be measured using photoelastic plastic models.

The effect of notches on fatigue strength is determined by comparing the S-N curves of notched and unnotched specimens. The data for notched specimens usually are plotted in terms of nominal stress based on the net cross section of the specimen. The effectiveness of the notch in decreasing the fatigue limit is expressed by the fatigue-notch factor, K_f. This factor is the ratio of the fatigue limit of unnotched specimens to the fatigue limit of notched specimens.

For materials that do not exhibit a fatigue limit, the fatigue-notch factor is based on the fatigue strength at a specified number of cycles. Values of K_f have been found to vary with (1) severity of the notch, (2) type of notch, (3) material, (4) type of loading, and (5) stress level.

Effect of Test Specimen Size

It is not possible to predict directly the fatigue performance of large machine members from the results of laboratory tests on small specimens. In most cases, a size effect exists; i.e., the fatigue strength of large members is lower than that of small specimens. Precise determination of this phenomenon is difficult. It is extremely difficult to prepare geometrically similar specimens of increasing diameter that have the same metallurgical structure and residual stress distribution throughout the cross section. The problems in fatigue testing of large specimens are considerable, and few fatigue machines can accommodate specimens with a wide range of cross sections.

Changing the size of a fatigue specimen usually results in variations of two factors. First, increasing the diameter increases the volume or surface area of the specimen. The change in amount of surface is significant, because fatigue failures usually initiate at the surface. Secondly, for plain or notched specimens loaded in bending or torsion, an increase in diameter usually decreases the stress gradient across the diameter and increases the volume of material that is highly stressed.

Experimental data on the size effect in fatigue typically show that the fatigue limit decreases with increasing specimen diameter. Horger's data for steel shafts tested in reversed bending (Table 1) show that the fatigue limit can be appreciably reduced in large section sizes.

Table 1 Effect of specimen size on the fatigue limit of normalized plain carbon steel in reversed bending

Specimen diameter		Fatigue limit	
mm	in.	MPa	ksi
7.6	0.30	248	36
38	1.50	200	29
152	6.00	144	21

Surface Effects and Fatigue

Generally, fatigue properties are very sensitive to surface conditions. Except in special cases where internal defects or case hardening is involved, all fatigue cracks initiate at the surface. Factors that affect the surface of a fatigue specimen can be divided into three categories: (1) surface roughness or stress raisers at the surface, (2) changes in the properties of the surface metal, and (3) changes in the residual stress condition of the surface. Additionally, the surface may be subjected to oxidation and corrosion.

Surface Roughness. In general, fatigue life increases as the magnitude of surface roughness decreases. Decreasing surface roughness minimizes local stress raisers. Therefore, special attention must be given to the surface preparation of fatigue test specimens. Typically, a metallographic finish, free of machining grooves and grinding scratches, is necessary. Figure 12 illustrates the effects that various surface conditions have on the fatigue properties of steel.

Effect of Mean Stress

A series of fatigue tests can be conducted at various mean stresses, and the results can be plotted as a series of S-N curves. A description of applied stresses and S-N curves can be found in the Introduction to this Section. For design purposes, it is more useful to know how the mean stress affects the permissible alternating stress amplitude for a given life (number of cycles). This usually is accomplished by plotting the allowable stress amplitude for a specific number of cycles as a function of the associated mean stress.

At zero mean stress, the allowable stress amplitude is the effective fatigue limit for a specified number of cycles. As the mean stress increases, the permissible amplitudes steadily decrease. At a mean stress equal to the ultimate tensile strength of the material, the permissible amplitude is zero.

The two straight lines and the curve shown in

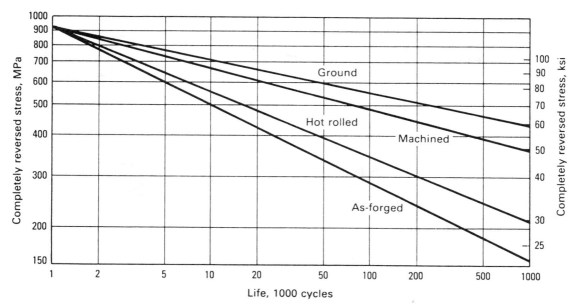

Fig. 12 Effect of surface conditions on the fatigue properties of steel (302 to 321 HB)

Fig. 13 represent the three most widely used empirical relationships for describing the effect of mean stress on fatigue strength. The straight line joining the alternating fatigue strength to the tensile strength is the modified Goodman law. Goodman's original law included the assumption that the fatigue limit was equal to one third of the tensile strength; this has since been generalized to the relation shown in Fig. 13, using the fatigue strength as determined experimentally.

Stress Amplitude. Because stress amplitude varies widely under actual loading conditions, it is necessary to predict fatigue life under various stress amplitudes. The most widely used method of estimating fatigue under complex loading is provided by the linear damage law. This is a hypothesis first suggested by Palmgren and restated by Miner, and is sometimes known as Miner's rule.

The assumption is made that the application of n_i cycles at a stress amplitude S_i, for which the average number of cycles to failure is N_i, causes an amount of fatigue damage that is measured by the cumulative cycles ratio n_i/N_i, and that failure will occur when $\Sigma(n_i/N_i) = 1$.

This method is not applicable in all cases, and numerous alternative theories of cumulative linear damage have been suggested. Some considerations of redistribution of stresses have been clarified, but there is as yet no satisfactory approach for all situations.

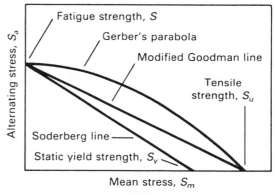

As shown by the modified Goodman line, Gerber's parabola, and Soderberg line. See text for discussion.

Fig. 13 Effect of mean stress on the alternating stress amplitude

The effect of varying the stress amplitude (linear damage) can be evaluated experimentally by means of a test in which a given number of stress cycles are applied to a test piece at one stress amplitude. The test is then continued to fracture at a different amplitude. Alternatively, the stress can be changed from one stress amplitude to another at regular intervals; such tests are known as block, or interval, tests. These tests do not simulate service conditions, but may serve a useful purpose for assessing the linear damage law and indicating its limitations.

Corrosion Fatigue

Corrosion fatigue is the combined action of repeated or fluctuating stress and a corrosive environment to produce progressive cracking. Usually, environmental effects are deleterious to fatigue life, producing cracks in fewer cycles than would be required in a more inert environment. Once fatigue cracks have formed, the corrosive aspect also may accelerate the rate of crack growth.

In corrosion fatigue, the magnitude of cyclic stress and the number of times it is applied are not the only critical loading parameters. Time-dependent environmental effects also are of prime importance. When failure occurs by corrosion fatigue, stress-cycle frequency, stress-wave shape, and stress ratio all affect the cracking processes.

Fatigue Crack Propagation

Fatigue failure of structural and equipment components due to cyclic loading has long been a major design problem and the subject of numerous investigations. Although considerable fatigue data are available, the majority has been concerned with the nominal stress required to cause failure in a given number of cycles—namely, S-N curves. Usually, such data are obtained by testing smooth or notched specimens. With this type of testing, however, it is difficult to distinguish between fatigue crack initiation life and fatigue crack propagation life.

Preexisting flaws or crack-like defects within a material reduce or may eliminate the crack initiation portion of the fatigue life of the component. Fracture mechanics methodology enhances the understanding of the initiation and propagation of fatigue cracks and assists in solving the problem of designing to prevent fatigue failures.

Fatigue Crack Propagation Test Methods

The general nature of fatigue crack propagation using fracture mechanics techniques is summarized in Fig. 14. A logarithmic plot of the crack growth per cycle, da/dN, versus the stress-intensity factor range, ΔK, corresponding to the load cycle applied to a specimen is illustrated. The da/dN versus ΔK plot was constructed of

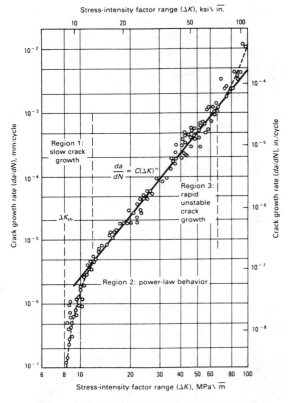

Yield strength of 470 MPa (70 ksi). Test conditions: $R = 0.10$; ambient room air, 24 °C (75 °F).

Fig. 14 Fatigue crack growth behavior of ASTM A533 B1 steel

data on five specimens of ASTM A533 B1 steel tested at 24 °C (75 °F). A plot of similar shape is anticipated with most structural alloys; the absolute values of da/dN and ΔK, however, are dependent on the material.

Results of fatigue crack growth rate tests for nearly all metallic structural materials have shown that the da/dN versus ΔK curves have three distinct regions. The behavior in Region I (Fig. 14) exhibits a fatigue crack growth threshold, ΔK_{th}, which corresponds to the stress-intensity factor range below which cracks do not propagate.

At intermediate values of ΔK (Region II in Fig. 14), a straight line usually is obtained on a log-log plot of ΔK versus da/dN. This is described by the power-law relationship:

$$\frac{da}{dN} = C(\Delta K)^n$$

where C and n are constants for a given material and stress ratio.

Fatigue Crack Propagation

Fatigue crack growth rate data for some steels show that the primary parameter affecting growth rate in Region II is the stress-intensity factor range and that the mechanical and metallurgical properties of these steels have negligible effects on the fatigue crack growth rate in a room-temperature air environment. Data for four martensitic steels fall within a single band, as shown in Fig. 15. The upper bound of scatter can be obtained from:

$$\frac{da}{dN} = 0.66 \times 10^{-8}(\Delta K)^{2.25}$$

where a is given in inches, and ΔK is given in ksi$\sqrt{\text{in}}$.

For some steels, the stress ratio and mean stress have negligible effects on the rate of crack growth in Region II. Also, the frequency of cyclic loading and the waveform (sinusoidal, triangular, square, trapezoidal) do not affect the rate of crack propagation per cycle of load for some steels in benign environments.

At high ΔK values (Region III in Fig. 14), unstable behavior occurs, resulting in a rapid increase in the crack growth rate just prior to complete failure of the specimens. There are two possible causes of this behavior. First, the increasing crack length during constant load testing causes the peak stress intensity to reach the fracture toughness, K_{Ic}, of the material, and the unstable behavior is related to the early stages of brittle fracture. Second, the growing crack reduces the uncracked area of the specimen sufficiently for the peak load to cause fully plastic limit load behavior. The first possibility is operative for high-strength, low-toughness metals, in which specimen sizes normally used for fatigue crack growth rate testing behave in a linear elastic manner at K levels equal to K_{Ic}. The second possibility, plastic limit load behavior, is common for ductile metals, particularly if K_{Ic} is high.

When plastic limit load behavior causes unstable crack growth, ΔK values have no meaning, because the limitations of linear elastic fracture mechanics have been exceeded. Here, the use of the J-integral concept, crack-opening displacement, or some other elastic-plastic fracture mechanics approach is more appropriate than ΔK for correlating the data.

Standardized testing procedures for measuring fatigue crack growth rates are described in ASTM Standard E 647. This method applies to medium to high crack growth rates—that is, above 10^{-8} m/cycle (3.9×10^{-7} in./cycle). Procedures for growth rates below 10^{-8} m/cycle are

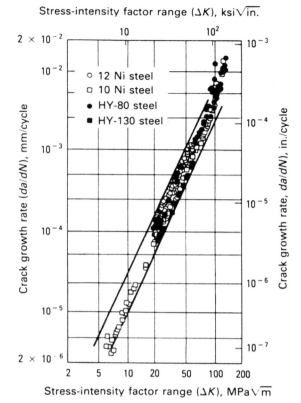

Fig. 15 Summary of fatigue crack growth data for martensitic steels

under consideration by ASTM. For applications involving fatigue lives of up to about 10^6 load cycles, the procedures recommended in ASTM E 647 can be used. Fatigue lives greater than about 10^6 cycles correspond to growth rates below 10^{-8} m/cycle, and these require special testing procedures, which are related to the threshold of fatigue crack growth illustrated in Fig. 14.

ASTM E 647 describes the use of center-cracked specimens and compact specimens (Fig. 16 and 17). The specimen thickness-to-width ratio, B/W, is smaller than the 0.5 value for K_{Ic} tests; the maximum B/W values for center-cracked and compact specimens are 0.125 and 0.25, respectively. With the thinner specimens, crack length measurements on the sides of the specimens can be used as representations of through-thickness crack growth behavior.

For tension-tension fatigue loading, the K_{Ic} loading fixtures frequently can be used. For this type of loading, both the maximum and minimum loads are tensile, and the load ratio, $R = P_{min}/P_{max}$, is in the range $0 < R < 1$. A ratio of

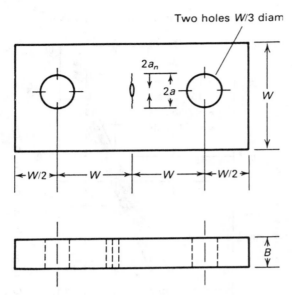

Fig. 16 Standard center-cracked tension specimen for fatigue crack propagation testing when the width (W) of the specimen ≤75 mm (3 in.)

$2a_n$ is the machined notch; a is the crack length; B is the specimen thickness.

$R = 0.1$ is commonly used for developing data for comparative purposes.

Testing often is performed in laboratory air at room temperature; however, any gaseous or liquid environment and temperature of interest may be used to determine the effect of temperature, corrosion, or other chemical reaction on cyclic loading.

Data Analysis. For constant-amplitude loading, a set of crack-length versus elapsed-cycle data (a versus N) is generated, with the specimen loading, P_{max} and P_{min}, generally held constant. Figure 18 illustrates a typical a versus N plot. The minimum crack-length interval, Δa, between data points (see Fig. 18) should be 0.25 mm (0.01 in.) or ten times the crack-length measurement precision, which is defined as the standard deviation on the mean value of crack length determined for a set of replicate measurements. This prevents the measurement of erroneous growth rates from a group of data points that are spaced too closely relative to the precision of data measurement and relative to the scatter of data.

Crack measurement intervals are recommended in ASTM E 647 according to specimen type. For compact-type specimens:

$$\Delta a \leq 0.04\,W \text{ for } 0.25 \leq \frac{a}{W} \leq 0.40$$

$$\Delta a \leq 0.02\,W \text{ for } 0.40 \leq \frac{a}{W} \leq 0.60$$

$$\Delta a \leq 0.01\,W \text{ for } \frac{a}{W} \geq 0.60$$

For center-cracked tension specimens:

$$\Delta a \leq 0.03\,W \text{ for } \frac{2a}{W} < 0.60$$

$$\Delta a \leq 0.02\,W \text{ for } \frac{2a}{W} > 0.60$$

Fatigue crack growth rate data can be calculated by several methods. The most commonly used methods, however, are the secant and incremental polynomial methods. The secant method consists of the slope of the straight line connecting two adjacent data points. This method, although simpler, results in more scatter in measured crack growth rate.

The incremental polynomial method fits a second-order polynomial expression (parabola) to typically five to seven adjacent data points, and the slope of this expression is the growth rate. The incremental polynomial method eliminates some of the scatter in growth rate that is inherent in fatigue testing.

Numerous relationships have been generated to correlate crack growth rate and stress-intensity data. The most widely accepted relationship is that proposed by Paris. This is a linear relationship when plotted on log-log coordinates and generally yields a reasonable fit to the data in Region II (see Fig. 14) of the crack growth regime.

Other relationships based on the Paris equation, such as the commonly used Forman equation, are used to represent the variation of da/dN with other key variables, including load ratio, R, and the critical K value, K_c, at which rapid fracture of the specimen occurs (Region III in Fig. 14). The Forman equation is:

$$\frac{da}{dN} = \frac{C(\Delta K)^n}{(1-R)(K_c - \Delta K)}$$

where C and n are material constants of the same types as those in the Paris equation, but of different values. An advantage of the Forman equation is that it describes the type of accelerated da/dN behavior that is often observed at high values of ΔK, which is not described by the Paris equation.

Additionally, the Forman equation describes the frequently observed increase in da/dN asso-

Fatigue Crack Propagation

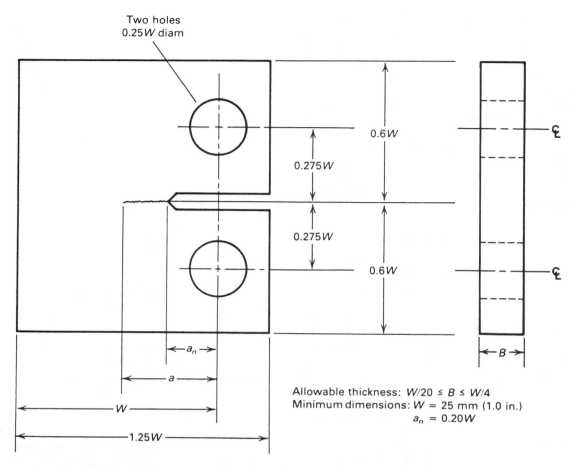

Fig. 17 Standard compact-type specimen for fatigue crack propagation testing (see Fig. 16 for explanation of symbols)

ciated with an increase in R from 0 toward 1. When it is necessary to describe the effect of K approaching K_c, or the effect of R on da/dN, the Forman equation can be used to represent the da/dN behavior. When only ΔK in Region II is involved, the less complex Paris equation may be used.

Cyclic Crack Growth Rate Testing in the Threshold Regime

Cyclic crack growth rate testing in the low-growth regime (Region I in Fig. 14) complicates acquisition of valid and consistent data, because the crack growth behavior becomes more sensitive to the material, environment, and testing procedures under this regime. Within this regime, the fatigue mechanisms of the material that slow the crack growth rates are more significant.

The precise definition of the cyclic crack growth rate threshold, ΔK_{th}, varies significantly.

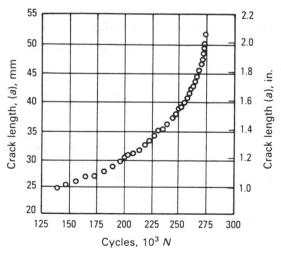

Fig. 18 Crack growth versus constant-amplitude stress cycles for a Fe-10Ni-8Co-1Mo high-strength steel

The most accurate definition would be the stress-intensity value below which fatigue crack growth will not occur. It is extremely expensive to obtain a true definition of ΔK_{th}, and in some materials a true threshold may be nonexistent. Generally, designers are more interested in the near-threshold regime, such as the ΔK that corresponds to a fatigue crack growth rate of 10^{-8} to 10^{-10} m/cycle (3.9×10^{-7} to 10^{-9} in./cycle). Because the duration of the tests increases greatly for each additional decade of near-threshold data (10^{-8} to 10^{-9} to 10^{-10}, etc., m/cycle), the precise design requirements should be determined in advance of the test.

Behavior of Short Cracks

Recently, it has been well documented that short cracks may behave differently from large cracks when plotted in the standard form of cyclic crack growth rate versus stress intensity.

A short crack is difficult to define. It may be small compared to the microstructure of the material to be studied (1 to 50 µm) when the concepts of continuum mechanics are of interest. It can also be small compared to the plastic zone size (10 to 1000 µm). In this situation, linear elastic fracture mechanics might be replaced with elastic-plastic fracture mechanics. The crack may also be physically small (500 to 1000 µm) when crack closure, crack tip shape, environment, and growth mechanisms are of concern. Figure 19 schematically illustrates the possible behavior of short cracks.

Selection of Test Specimens

Selection of a fatigue crack growth test specimen is usually based on the availability of the material and the types of test systems and crack-monitoring devices to be used. The two most widely used types of specimens are the center-cracked tension specimen and the compact-type specimen (see Fig. 16 and 17). However, any specimen configuration with a known stress-intensity factor solution can be used in fatigue crack growth testing, assuming that the appropriate equipment is available for controlling the test and measuring the crack dimensions. Stress-intensity factor solutions for center-cracked tension and compact-type specimens are given in Table 2.

Consideration of the range of application of the stress-intensity solution of a specimen configuration is very important. Many stress-inten-

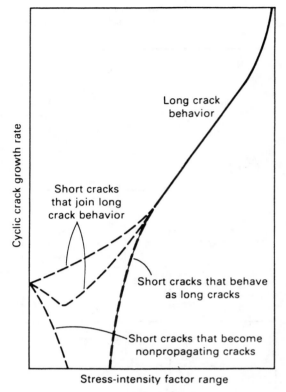

Fig. 19 Typical short crack behavior

sity expressions are valid only over a range of the ratio of crack length to specimen width (a/W). For example, the expression given in Table 2 for the compact-type specimen is valid for $a/W > 0.2$; the expression for the center-cracked tension specimen is valid for $2a/W < 0.95$. The use of stress-intensity expressions outside their applicable crack-length region can produce significant errors in data.

The size of the specimen must also be appropriate. To follow the rules of linear elastic fracture mechanics, the specimen must be predominantly elastic. However, unlike the requirements for plane-strain fracture toughness testing, the stresses at the crack tip do not have to be maintained in a plane-strain state. The stress state is considered to be a controlled test variable. The material characteristics, specimen size, crack length, and applied load will dictate whether the specimen is predominantly elastic. Because the loading mode of different specimens varies significantly, each specimen geometry must be considered separately.

Notch Preparation. The method by which a notch is machined depends on the specimen

Table 2 Stress-intensity factor solutions for standardized (ASTM E 647) fatigue crack growth specimen geometries

Center-cracked tension specimens (Fig. 16)

$$\Delta K = \frac{\Delta P}{B}\sqrt{\frac{\pi\alpha}{2W}\sec\frac{\pi\alpha}{2}}$$

where $\alpha = \frac{2a}{W}$; expression valid for $\frac{2a}{W} < 0.95$

Compact-type specimens (Fig. 17)

$$\Delta K = \frac{\Delta P(2+\alpha)}{B\sqrt{W}(1-\alpha)^{3/2}}(0.886 + 4.64\alpha - 13.32\alpha^2 + 14.72\alpha^3 - 5.6\alpha^4)$$

where $\alpha = \frac{a}{W}$; expression valid for $\frac{a}{W} \geqslant 0.2$

---

material and the desired notch root radius (ρ). Sawcutting is the easiest method, but is generally acceptable only for aluminum alloys. For a notch root radius of $\rho \leqslant 0.25$ mm (0.010 in.) in aluminum alloys, milling or broaching is required. A similar notch root radius in low- and medium-strength steels can be produced by grinding. For high-strength steel alloys, nickel-base superalloys, and titanium alloys, electrical discharge machining may be necessary to produce a notch root radius of $\rho \leqslant 0.25$ mm (0.010 in.).

Precracking of a specimen prior to testing is conducted at stress intensities sufficient to cause a crack to initiate from the starter notch and propagate to a length that will eliminate the effect of the notch. To decrease the amount of time needed for precracking to occur, common practice is to initiate the precracking at a load above that which will be used during testing and to subsequently reduce the load.

Load generally is reduced uniformly to avoid transient effects. Crack growth can be arrested above the threshold stress-intensity value due to formation of the increased plastic zone ahead of the tip of the advancing crack. Therefore, the step size of the load during precracking should be minimized. Reduction in the maximum load should not be greater than 20% of the previous load condition. As the crack approaches the final desired size, this percentage may be decreased.

Gripping of the specimen must be done in a manner that does not violate the stress-intensity solution requirements. For example, in a single-edge notched specimen, it is possible to produce a grip that permits rotation in the loading of the specimen, or it is possible to produce a rigid grip.

Each of these requires a different stress-intensity solution. In grips that are permitted to rotate, such as the compact-type specimen grip, the pin and the hole clearances must be designed to minimize friction. It is also advisable to consider lateral movement above and below the grips.

Gripping arrangements for compact-type and center-cracked tension specimens are described in ASTM E 647. For a center-cracked tension specimen less than 75 mm (3 in.) in width, a single pin grip is generally suitable. Wider specimens generally require additional pins, friction gripping, or some other method to provide sufficient strength in the specimen and grip to prohibit failure at undesirable locations, such as in the grips. Grips designed for compact-type specimens are illustrated in Fig. 20.

Crack-Length Measurement Techniques

Several different techniques have been developed to monitor the initiation, growth, and instability of cracks, including optical (visual and photographic), electrical (eddy current and resistance), compliance, ultrasonic, and acoustic emission monitoring techniques.

Optical Crack Measurement Techniques

Monitoring of fatigue crack length as a function of cycles is most commonly conducted visually by observing the crack at the specimen surfaces with a traveling low-power microscope at a magnification of 20 to 50×. Crack-length measurements are made at intervals such that a nearly even distribution of da/dN versus ΔK is achieved. The minimum amount of extension be-

Fig. 20 Grips designed for fatigue crack propagation testing of compact-type specimens (courtesy of MTS Systems Corp.)

tween readings is commonly about 0.25 mm (0.10 in.).

The optical technique is straightforward and, if the specimen is carefully polished and does not oxidize during the test, produces accurate results. However, the process is time consuming, subjective, and can be automated only with complicated and expensive video-digitizing equipment. In addition, many fatigue crack growth rate tests are conducted in simulated-service environments that obscure direct observation of the crack.

Compliance Method of Crack Extension Measurement

The compliance of an elastically strained specimen containing a crack of length a measured from the load line to the crack tip is usually expressed as the quotient of the displacement, δ, and the tensile load, P, with the displacement measured along, or parallel to, the load line. Fig-

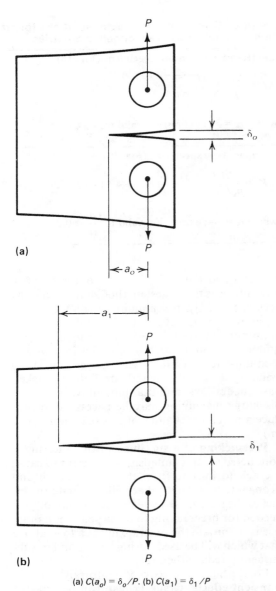

(a) $C(a_o) = \delta_o/P$. (b) $C(a_1) = \delta_1/P$

Fig. 21 Schematic of the relationship between compliance and crack length

ure 21 illustrates that the more deeply a specimen is cracked, the greater the amount of δ measured for a specific value of tensile load. Compliance can also be defined for shear and torsional loads applied to cracked specimens, and crack extension under these loading modes can be similarly determined.

Specimen load is simultaneously measured by an electronic load cell and conditioner/amplifier system, and the output is directed to the same data-acquisition system. A generalized schematic of the circuits involved is shown in Fig. 22.

Fatigue Crack Propagation

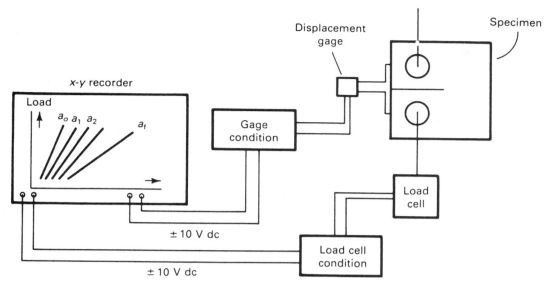

Fig. 22 Components of a compliance measurement system

The required sensitivity of the systems depends on specimen geometry and size; in general, noise-free, amplified output on the order of 1 V dc per 1 mm (0.04 in.) of deflection is satisfactory. Similarly, for the load range applied to the specimen, an approximately 1 V dc change in signal from the load cell is required for accurate calculation of the compliance.

Electric Potential Crack Monitoring Technique

The electrical potential, or potential drop, technique has gained increasingly wide acceptance in fracture research as one of the most accurate and efficient methods for monitoring the initiation and propagation of cracks. This method relies on the fact that there will be a disturbance in the electrical potential field about any discontinuity in a current-carrying body, the magnitude of the disturbance depending directly on the size and shape of the discontinuity.

For the application of crack growth monitoring, the electric potential method entails passing a constant current (maintained constant by external means) through a cracked test specimen and measuring the change in electrical potential across the crack as it propagates. With increasing crack length, the uncracked cross-sectional area of the test piece decreases, its electrical resistance increases, and thus the potential difference between two points spanning the crack rises. By monitoring this potential increase, V_a, and comparing it with some reference potential, V_o, the crack length to width ratio, a/W, can be determined through the use of the relevant calibration curve for the particular test piece geometry concerned.

Crack Growth Studies. By far the most useful application of the electrical potential method has been in measurements of crack length during crack propagation, where it has been utilized to monitor almost all mechanisms of subcritical crack growth and most notably to follow fatigue crack growth. Typical crack propagation rates derived from direct current potential measurements are shown in Fig. 23 for tests on a 2.25Cr-1Mo steel in air, gaseous hydrogen, and hydrogen sulfide environments.

Electromechanical Fatigue Testing Systems

The primary function of electromechanical fatigue testers is to apply millions of cycles to a test piece at oscillating loads up to 220 kN (50 000 lbf) to investigate fatigue life, or the number of cycles to failure under controlled cyclic loading conditions. Variables associated with fatigue-life tests are frequency of loading and unloading, amplitude of loading (maximum loads and minimum loads), and control capabilities. The fundamental data output requirement is the number of cycles to failure, as defined by the application.

A variety of electromechanical fatigue testers have been developed for different applications. Forced-displacement, forced-vibration, rota-

Fatigue Testing

Table 3 Comparison of electromechanical fatigue systems

Parameter	Forced displacement	Forced vibration
Tension	Yes	Yes
Compression	Yes	Yes
Reverse stress	Yes	Yes
Bending	Yes	Yes
Frequency range	Fixed	Fixed, 1800 rpm
Load range	Typically < 450 N (< 100 lbf)	Up to 220 kN (50 000 lbf)
Type:		
Control	Open-loop	Open-loop
Mode	Displacement	Load
Maximum deflection	—	25.4 mm (1.00 in.)
Advantages	Simple, straightforward	Versatile, efficient, durable
Disadvantages	No load control, very limited applications (soft samples)	Fixed frequency, limited control (open-loop)

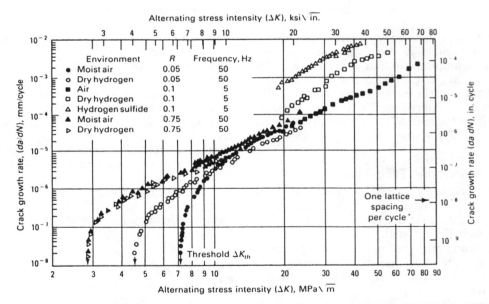

Fig. 23 Fatigue crack propagation data over a wide spectrum of growth rates. Data derived from direct current potential measurements in martensitic 2.25Cr-1Mo steel (SA542-C12) at $R = 0.05$ to 0.75 in air, hydrogen, and hydrogen sulfide at ambient temperature.

tional bending, resonance, and servomechanical systems are discussed in this article and are compared in Table 3. Other specialized electromechanical systems are available to perform specific tasks.

Forced-Displacement Systems

Forced-displacement motor-driven systems are the simplest type of electromechanical fatigue testers. They effectively reproduce service environments that impart fixed, reciprocating displacements to a component or test piece. An electric motor-driven flywheel is used to carry a loading arm at a variable distance from the center of rotation, much in the same manner as a connecting rod in an automotive engine. This rotational displacement is transformed into a guided, vertical displacement and is used to fatigue the specimen.

Although load can be monitored in such sys-

Rotational bending	Resonance	Servomechanical
No	Yes	Yes
No	Yes	Yes
Yes	Yes	Yes
Yes	Yes	Yes
0–10 000 rpm	40–300 Hz	0–1 Hz
—	Up to 180 kN (40 000 lbf)	Up to 90 kN (20 000 lbf)
Open-loop Rotation/bending	Closed-loop Load	Closed-loop Load, displacement, strain
—	1.0 mm (0.040 in.)	100 mm (4 in.)
Efficient, durable, simple Rotational bending only, limited applications	Fully closed-loop, extremely efficient Operating frequency directly proportional to sample stiffness	Fully closed-loop, high precision Low frequency only

tems, the fixed displacement precludes the ability to control load, which is a function of specimen characteristics. Therefore, the load generally drops as failure progresses. These systems typically are custom-built, inexpensive fatigue machines, used primarily for bend tests on soft samples in which load control, high frequencies, and large loads are not required.

Forced-Vibration Systems

Forced-vibration motor-driven systems were the first production fatigue testers in commercial use. The centrifugal forces of an imbalanced rotor is used to impart a cyclic load to the test piece.

In operation, an electric motor is used to rotate an eccentric mass via flexible couplings. The rotating mass is mounted in a frame that is guided by flexure plates to restrict movement to vertical motion only. The centrifugal force produced by the rotating eccentric mass (m) is transmitted through the vertically guided frame to the test piece. The horizontal component of the centrifugal force is absorbed by the restraining flexure plates.

Because the centrifugal force usually is totally absorbed by the mounting frame (of mass M), the inertial reaction is separated from the centrifugal force in such a way as to transmit only the centrifugal forces to the specimen. This technique involves the use of frame-support compensator springs; the natural frequency of the spring (K)/mass (M) system is tuned to the revolutions per minute of the motor. Thus, neither the specimen nor the rotating eccentric mass (m) "sees" an inertial reaction from the frame, because the inertial effects of the frame are totally compensated for by the frame support springs (not the specimen).

This technique has two requirements: the rotating frequency (ω) must be kept constant and the mass of mounting frame (M) must be kept constant. Consequently, the loading frequency of the device is fixed at 1800 rpm, and masses must be added or removed from the frame to compensate for fixturing to keep M constant.

The magnitude of the dynamic load is determined by placing the rotating mass at a known distance from the axis of rotation (r). Because ω, m, M, and K are known, the force on the specimen, F, is calibrated directly as a function of r as follows:

$$F = M\omega^2 r \text{ (centrifugal)} - Ma_z \text{ (inertial)} + Kz \text{ (spring compensated)}$$

where a_z is the acceleration of the frame in the z direction, and Kz is the spring-compensated displacement in the z direction. Because Ma_z is tuned to equal Kz, $F = M\omega^2 r$.

Thus, the forced-vibration rotating eccentric mass system is an open-loop, load-controlled system with the ability to accommodate up to 25 mm (1.0 in.) of total sample deflection at loads up to 220 kN (50 000 lbf) using special fixtures. The mean or static load, onto which the dynamic load is superimposed, is achieved by preloading the inertia compensator spring, K.

Through special fixturing, forced-vibration devices are capable of testing in tension, compression, bending, torsion, or reverse stress. Although servo-controlled, mean-load-maintenance systems are available, the open-loop nature of the system prevents direct load measurement or control, which is characteristic of closed-loop systems. The load applied to the specimen is assumed to be a function of r, and a graduated scale is provided to permit reasonably accurate setup.

Rotational Bending Systems

Rotational bending systems effectively apply reversed loading to the outer surface of rods or shafts. The basic operating principle of the rotating beam consists of the use of a motor to rotate a shaft of known dimensions around its longitudinal axis. By applying a known static force at the end of the shaft, a bending moment can be applied to the test section, the outer surface of which oscillates between tension and compression during each rotation.

The cantilevered specimen, however, is subjected to a nonuniform bending moment, which is large at the supported end of the specimen and zero at the free end. To produce a more meaningful, uniform bending moment throughout the test piece, a specially designed tapered specimen should be used or bending moments should be applied to each end of the specimen. Figure 24 illustrates the rotating-beam operating mechanism and the resulting stress distribution in the specimen.

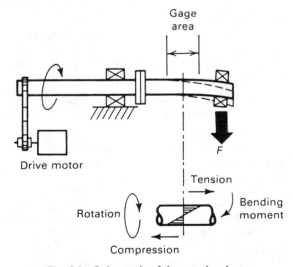

Fig. 24 Schematic of the rotating-beam operating mechanism and the resulting stress distribution in the specimen

Resonance Systems

A high-speed fatigue testing system was developed by Amsler that operated at 40 to 300 Hz, achieved high loads (up to 90 kN, or 20 000 lbf), and consumed minimal energy. It is based on a resonant spring/mass system, in which the specimen is used, like a spring, as an integral part of the oscillating mechanism.

The fatigue load, in the form of a sine wave, is achieved by preloading the sample in the frame via a complex optomechanical procedure and dynamically loading the sample at the natural oscillating frequency of the spring/mass system. The preload is maintained automatically during the test. The dynamic load is achieved by pulsing an electromagnet at the natural frequency of the spring/mass system. During resonance, the electromagnet restores any hysteresis energy lost during the previous cycle, thereby maintaining a constant, controllable dynamic load. Capable of tension, compression, bending, torsional, and reverse-stress fatigue tests, the Amsler resonant fatigue testers were instrumental in obtaining the vast amount of fatigue data currently available.

The resonant system is based on a similar principle, but incorporates solid-state technology to achieve fully closed-loop control of mean and dynamic loads. This system uses dual opposing masses (unlike the single oscillating mass/seismic base of earlier systems), linked by the specimen to achieve vibration-free resonance. A strain gage load cell, in series with the specimen, senses the load and automatically triggers the electromagnet to achieve self-tuning capability.

The mean load is achieved by physically moving the upper mass up or down to achieve tension or compression, respectively; the dynamic load is achieved by varying the width of the pulse to the magnet beneath the lower mass. The dynamic load, like the mean load, is electronically maintained at a preset command level through solid-state closed-loop circuitry. The remainder of the controls and mechanisms associated with the resonant fatigue system maintain a preset air gap between the magnet and the oscillating lower mass, maintain preset loading conditions (shutting down at preset load levels or frequencies), and power the electromagnet.

The high efficiency of resonant systems makes them well suited to high-cycle fatigue tests, in which closed-loop load control, high loads (up to 180 kN, or 40 000 lbf), low power consumption (around 750 W maximum for closed-loop sys-

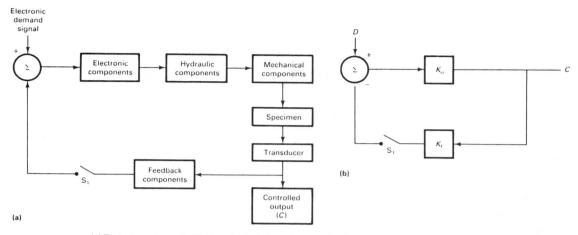

(a) Typical components. (b) Transfer functions. See text for details and explanation of symbols.

Fig. 25 Simplified block diagram for a negative-feedback closed-loop testing machine

tems), and high throughput are required. These systems tolerate minimal hysteresis and produce optimum testing results when used with stiff metallic samples.

Closed-Loop Servomechanical Systems

The most recent development in electromechanical fatigue testers is based on an electric actuator/load frame assembly. The system closely resembles its servohydraulic counterpart in that it consists of an actuator, a load frame, a load cell, a power supply, and a solid-state closed-loop electronic control console. Closed-loop systems compare live feedback signals to an input command signal to maintain accurate control of preset conditions. The closed-loop servomechanical system is, by virtue of its design, primarily intended for low-cycle and creep-fatigue studies.

Servohydraulic Fatigue Testing Systems

Servohydraulic testing machines are particularly well suited for providing the control capabilities required for fatigue testing. Extreme demands for sensitivity, resolution, stability, and reliability are imposed by fatigue evaluations. Displacements may have to be controlled (often for many days) to within a few microns, and forces can range from 100 kN to just a few newtons. This wide range of performance can be obtained with servomechanisms in general and, in particular, with the modular concept of servohydraulic systems.

Usually, the problem of selecting the appropriate system is simply a matter of optimizing the various components to form a system best suited to the given testing application. In this section, the principles underlying closed-loop servo systems are discussed briefly. In addition, the interaction between system components is illustrated, and a brief description of their operating principles and characteristics is provided.

With any type of control system, the objective is to obtain an output that relates as closely as possible to the programmed input. In a fatigue testing system, it may be desired to vary the force on a specimen in a sinusoidal manner, at a frequency of 1 Hz over a force range of 0 to 100 kN (0 to 22 000 lbf). The only practical means to accomplish this with precision is through the use of a negative-feedback closed-loop system. An overview of the basic principles of operation of negative-feedback systems is provided in Fig. 25. The blocks shown in Fig. 25(a) represent a group of typical components of a testing machine. The transfer functions of each of these blocks can be combined to produce the more simplified diagram shown in Fig. 25(b).

Placement of the switch, S_1, has been added to the diagram to permit analysis of the system when it is open (no feedback, or an open-loop condition) and when it is closed (providing feedback to the system). The equation governing this simplified open-loop system is:

$$C = K_o D$$

where C represents the controlled output, K_o represents the open-loop transfer function, and D represents the electronic demand signal. Therefore, the output is simply proportional to the system demand if K_o is a constant. Unfortunately, K_o is seldom a constant, because it can be influenced by several common system variations. The electronic components may drift slightly, or their gain may vary. The behavior of the hydraulic components may change with temperature, contamination, or wear, and the mechanical components may vary because of thermal effects or friction.

Servohydraulic System Components

Many commercially manufactured units are available for each component in a typical servohydraulic testing system.

The programmer supplies the command signal to the system, which is generally an analog of the desired behavior of the controlled parameter. For example, assume the same test conditions as previously discussed (control the force on the specimen in a sinusoidal manner at a frequency of 1 Hz and a force range of 0 to 100 kN). In this instance, the programmer might be set to produce an electronic signal with a sinusoidal waveform that has a frequency of 1 Hz and a voltage output of 0 to 10 V. The analog is: 1 V represents 1000 N. The system can then be adjusted to produce the correct output. Any change in the programmer signal will result in a corresponding change in the controlled parameter.

The servo-controller makes most of the adjustments necessary to optimize system performance. For example, it compares the command signal with a signal produced by the controlled parameter (stress or strain, for example) and relays a correction signal, if needed, to the control device in the system (usually a flow-control servo-valve). A servo-controller incorporates numerous other compensatory features, such as:

- Means to adjust the gain or proportional band of the system
- Controls to modify the feedback or correction signals for improved stability
- Controls to adjust the mean level and amplitude of the command signal(s)
- Controls to enhance and adjust servo-valve response
- Means to monitor the system error signal (a measure of how well the command and feedback signals agree)
- Capability to select various command and feedback signals
- Auxiliary functions such as recorder signal conditioning, calibration, and system start-up and shutdown

The servo-valve controls the volume and direction of flow of hydraulic fluid between the hydraulic power supply and the hydraulic ram. Within the control loop, it is the intermediary between the low-power servo-controller and the hydraulic ram, which can supply large forces and displacements to the specimen. Characteristics of the device are such that the output flow is approximately proportional to the input current when the output pressure is constant. Also, the output pressure is approximately proportional to the square of the input current when the flow is constant.

Hydraulic rams, or actuators or cylinders, furnish the forces and displacements required by the testing system. These rams usually are double ended to provide the greatest lateral rigidity and to produce the balanced flow and force characteristics desirable for push-pull testing. The effective area of the piston is therefore equal to the cross-sectional area of the piston minus the cross-sectional area of the piston rod. Under static conditions (very little flow), the maximum force capability of the ram will approach the hydraulic supply pressures multiplied by the effective area.

The force available during dynamic operation depends on the pressure drop and flow characteristics of the servo-valve. Reference should be made to the load/flow/pressure characteristics supplied by the servo-valve manufacturer.

Load Cells. The strain gage load cell is the most widely used force-measuring and feedback device in closed-loop fatigue machines. An external applied force causes the elastic deformation of an internal member to which a strain gage bridge has been attached. An electronic signal that is proportional to the resistance change in the bridge and to the applied force can thus be produced. Some load cells are designed specifically for fatigue evaluations. Variable features include sensitivity, natural resonant frequency, temperature stability, fatigue rating, linearity, hysteresis, deflection constant, load capacity, overload rating, resistance to extraneous loading, and compatibility with the testing machine and fixtures. Most commercially available cells are very competitive with respect to these features.

Load Frames. In a fatigue machine, the reaction forces to the specimen and to the housing of the ram are supplied by the load frame. Many styles of load frames are available, but for fatigue purposes the frames should be customized. The requirements of good high-frequency response demand that there be high axial stiffness in the load frame. When a deflection occurs in the load frame, additional flow is required from the servo-valve. Therefore, this deflection should be minimal in comparison with the deflection imparted to the specimen.

In addition, because fatigue specimens must be subjected to fully reversed loading (i.e., compressive as well as tensile forces), lateral rigidity must be increased to resist bending. This is generally considered necessary in the design of fatigue machines. The extra rigidity can be obtained by increasing the diameter of the support columns or by utilizing three- or four-column configurations.

Exceptional alignment is required of load frames used in fatigue evaluations to minimize undesirable bending forces. In addition, some means is usually provided to refine the alignment with manual adjustments when necessary. A strain-gaged specimen can be used to make this evaluation.

1-1. *S-N* Curves Typical for Steel

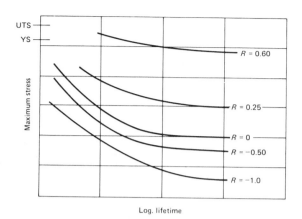

Schematic *S-N* curves for a material at various stress ratios. UTS and YS indicate ultimate tensile strength and yield strength, respectively, in uniaxial tensile testing.

The results of fatigue tests are usually plotted as maximum stress or stress amplitude to number of cycles, N, to fracture using a logarithmic scale for the number of cycles. Stress is plotted on either a linear or a logarithmic scale. The resulting curve of data points is called an *S-N* curve. A family of *S-N* curves for a material tested at various stress ratios is shown schematically in the above curves. Stress ratio is the algebraic ratio of two specified stress values in a stress cycle. Two commonly used stress ratios are the ratio, A, of the alternating stress amplitude to the mean stress ($A = S_a/S_m$) and the ratio, R, of the minimum stress to the maximum stress ($R = S_{min}/S_{max}$). If the stresses are fully reversed, the stress ratio R becomes -1; if the stresses are partially reversed, R becomes a negative number less than 1. If the stress is cycled between a maximum stress and no load, the stress ratio R becomes zero. If the stress is cycled between two tensile stresses, the stress ratio R becomes a positive number less than 1. A stress ratio R of 1 indicates no variation in stress, and the test would become a sustained-load creep test rather than a fatigue test. For carbon and low-alloy steels, *S-N* curves typically have a fairly straight slanting portion at low cycles changing into a straight, horizontal line at higher cycles, with a sharp transition between the two.

An *S-N* curve usually represents the median life for a given stress—the life that half the specimens attain. Scatter of fatigue lives can cover a very wide range.

Source: Metals Handbook, 9th Edition, Volume 1, Properties and Selection: Irons and Steels, American Society for Metals, Metals Park OH, 1978, p 667

1-2. S-N Curves Typical for Medium-Strength Steels

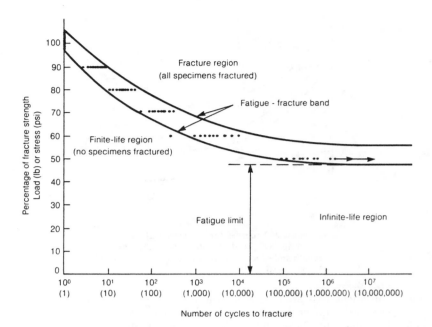

S-N curves that typify fatigue test results for testing of medium-strength steels.

As an explanation, if the single-load fracture strength of the specimens is considered to be 100 percent, for purposes of illustration this is the starting place, for the specimens can sustain no higher load without fracture. If ten specimens are fractured, the results are placed as points at the top of the left axis at one load application.

Intuitively it is known that if the maximum load (or stress) is lowered to 90 percent of the tensile strength, it will require more than one load application to fracture the specimens. The ten points shown in the diagram at 90 percent represent the possible life to fracture of each of the ten specimens. Because the scale is logarithmic, the points appear to be relatively close, but in fact the scatter in life from longest to shortest is on the order of more than 2 to 1. At this high stress, plastic deformation of the test specimen is likely to be great, such as in bending a paper clip or wire coat hanger to make it fracture. Actual parts are not intentionally designed to operate in this regime, and normal fatigue fractures have no obvious plastic deformation.

If the load is dropped to 80 percent of the single-load fracture strength and ten more specimens are tested, they will run longer with a fatigue life scatter of perhaps 3 to 1, which is not unusual, even for theoretically identical specimens (which, of course, they are not). When the load is dropped to 70 percent, the lives get longer and the scatter in fatigue life increases to perhaps 5 to 1. Again dropping the load, now to 60 percent of the single-load fracture strength, the fatigue lives again increase, as does the scatter from longest to shortest life. Invariably, in actual fatigue testing, there is at least one specimen that inexplicably fractures far earlier than any of the others in the same group. One such specimen is shown at the 60 percent level fracturing at about 150 cycles, while the other supposedly "identical" specimens or parts had lives of from about 1,000 to 10,000 cycles. The cause of such an early "anomaly" is often sought in vain, although it is possible that some metallurgical reason, such as a large inclusion on the surface, might be found. Frequently, this lone early fracture specimen is simply ignored.

Dropping now to 50 percent of the single-load fracture strength, the fatigue lives increase dramatically, as the S-N curve starts to flatten out. This flattening out is characteristic of ferrous metals of low and moderate hardness; many nonferrous metals and some very high-hardness ferrous metals tend to continue their downward path at very large numbers of cycles. Now, the problem is when to stop the tests. The test

1-2. S-N Curves Typical for Medium-Strength Steels (continued)

machine will be needed for another test specimen after a very long test time, depending upon the rate of loading, or cycles per minute. If ten million is selected as the end point, the test must be stopped at that figure even if a specimen is unbroken, and the point shown with an arrow pointing to higher values, for it did not actually fracture. Frequently, five million, or even one million, cycles is selected as the end point, depending upon the metal, purpose, and urgency of the tests. For example, five hundred million cycles is sometimes used in the aluminum industry.

The region below the lowest portion of the S-N curve is called the infinite-life region, because specimens that are tested at stresses below the curve should run indefinitely; that is, they should have infinite life. The leveling of the S-N curve is the fatigue limit, characteristic of ferrous metals but not of most nonferrous metals. However, the region to the left of the sloping part of the S-N curve is called the finite-life region, for at the higher stress levels the test specimens or parts will eventually fracture in fatigue. This is typical of certain structural parts in aircraft which have their histories carefully recorded so that they may be inspected and/or replaced as their fatigue lives are used up in service. Also, growing fatigue cracks must not be permitted to exceed the critical flaw size characteristic of the metal and the stress state.

Source: Donald J. Wulpi, Understanding How Components Fail, American Society for Metals, Metals Park OH, 1985, p 135-137

1-3. S-N Diagrams Comparing Endurance Limit for Seven Alloys

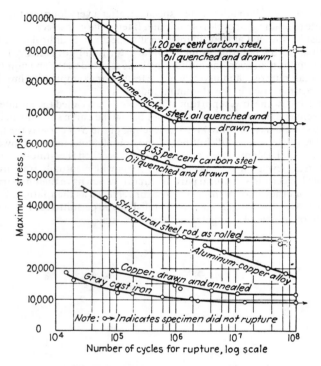

Typical S-N diagrams for determining endurance limit of metals under reversed flexural stress.

To determine the endurance limit of a metal, it is necessary to prepare a number of similar specimens that are metals tested, and for most nonferrous metals, the S-N diagrams become horizontal, as nearly as can be determined, for values of N ranging from 1,000,000 to 50,000,000 cycles, thus indicating a well-defined endurance limit. The S-N diagrams for duralumin and monel metal do not indicate well-defined endurance limits. The first specimen is tested at a relatively high stress so that failure will occur at a small number of applications of stress. Succeeding specimens are then tested, each one at a lower stress. The number of repetitions required to produce failure increases as the stress decreases. Specimens stressed below the endurance limit will not rupture. The results of fatigue tests are commonly plotted on diagrams in which values of stress are plotted as ordinates and values of number of cycles of stress for fracture are plotted as abscissas. Such diagrams are called S-N diagrams (S for stress, N for number of cycles). In general, the S-N diagrams are drawn using semilogarithmic plotting as shown in the above diagram, which presents the results for various typical materials.

Source: Fatigue and Creep Tests of Metals, p 220

1-4. Steel: Effect of Microstructure

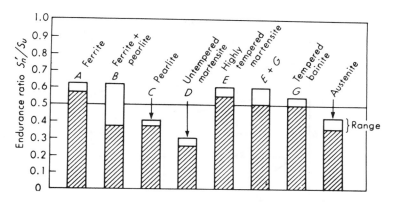

Effect of steel microstructure on endurance ratio.

One of the more extensive investigations on influence of microstructure was conducted by Cazaud. The results of some of his work are summarized in the above bar chart. His data confirm that 0.5 is a conservative number; he found ratios varying from 0.55 to 0.62 for highly tempered martensites. These data were also for steels in the 0.40% carbon range. When untempered martensite is included, the total ratio range is from 0.26 to 0.62. Untempered 0.40% carbon martensite is about 55 HRC. Above 40 HRC, factors other than microstructure become more significant, especially nonmetallic content and residual stress.

Many believe that tempered martensite gives optimum fatigue properties. However, much of the early work was with medium-carbon steels with intermediate hardnesses. Only limited data are available for other structures, including low-carbon martensites. Borik and Chapman determined the endurance limit of bainite and martensite in the range 36 to 61 HRC. They used 51100, a 1.00% carbon steel. They concluded that above 40 HRC, bainite had better fatigue properties at the same hardness than did martensite, whereas below 40 HRC the reverse was true. They explained the results in terms of carbide morphology and distribution. Below 40 HRC, the carbides in the martensite are spheroidal. Above 40 HRC, the carbide associated with the bainite was very fine and well-distributed, but below 40 HRC the carbides had a "pearlitic mode," which was less favorable in resisting fatigue.

Source: D. H. Breen and E. M. Wene, "Fatigue in Machines and Structures—Ground Vehicles," in Fatigue and Microstructure, American Society for Metals, Metals Park OH, 1979, p 77

1-5. Steel: Influence of Derating Factors on Fatigue Characteristics

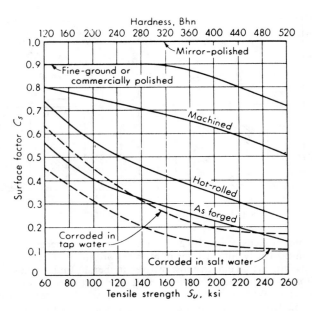

Derating factors for influence of surface condition on fatigue.

The graph above gives C_s factors for various surface conditions. It should be obvious that these factors are approximate, since it is impossible to represent such variable conditions by a single curve. C_o, the size factor, is significant. Early work by Horger firmly established that large-diameter samples of the same metallurgy were not as good in bending fatigue as were small samples. In the presence of a stress gradient, as in bending, a larger volume of metal is subject to high stress in a large part than in a small-diameter part. Since a large volume is subject to maximum stress, there is a higher probability of a critical-size nonmetallic inclusion to be in that volume. The fatigue properties established by testing large specimens are thought to represent the lower bound for a large number of small samples. Since axial tests, by their nature, test fairly large volumes at maximum stress, they also give lower-bound results. C_o is usually taken at 1.0 for diameters less than 0.4 inches and 0.9 for diameters between 0.4 and 2.0 inches. It must be borne in mind that this is a very rough estimate and that the curves shown in the above graph are thought to be toward the conservative side of scatter bands. The 0.5 relation for S_n and S_u is only reasonably accurate in the low and intermediate hardness ranges because of limitations related to microstructure, nonmetallic-inclusion content, and carbon content at higher hardnesses.

Source: D. H. Breen and E. M. Wene, "Fatigue in Machines and Structures—Ground Vehicles," in Fatigue and Microstructure, American Society for Metals, Metals Park OH, 1979, p 72

1-6. Steel: Correction Factors for Various Surface Conditions

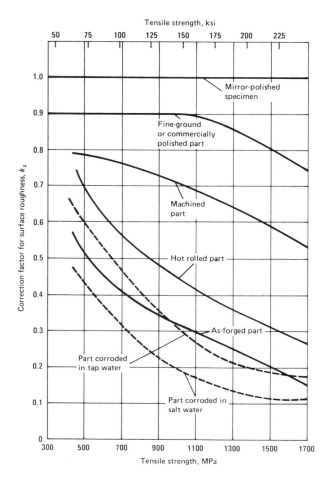

Factor	Value for loading in		
	Bending	Torsion	Tension
K_l	1.0	0.58	0.9(a)
K_d, where:			
$d \leq 0.4$ in.	1.0	1.0	1.0
0.4 in. $< d \leq 2$ in.	0.9	0.9	1.0
K_s	From chart above		

(a) A lower value (0.06 to 0.85) may be used to account for known or suspected undetermined bending because of load eccentricity.

Correction factors for surface roughness (K_s), type of loading (K_l), and part diameter (K_d), for fatigue life of steel parts.

Comparative effects of various surface conditions on fatigue limit of steels at various levels of tensile strength.

Source: Metals Handbook, 9th Edition, Volume 1, Properties and Selection: Irons and Steels, American Society for Metals, Metals Park OH, 1978, p 671

1-7. Fatigue Behavior: Ferrous vs Nonferrous Metals

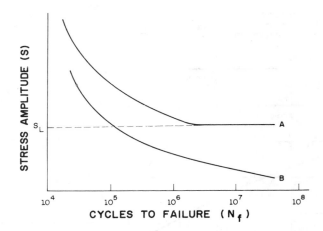

S (stress)—N (cycles to failure) curves. A, ferrous metals; B, nonferrous metals. S_L is the endurance limit.

Traditionally, the behavior of a material under conditions of fatigue has been studied by obtaining the S-N curves (see above), where S is the stress and N is the number of cycles to failure. For steels, in general, one observes a fatigue limit or endurance limit (curve A above) which represents a stress level below which the material does not fail and can be cycled infinitely. Such an endurance limit does not exist for nonferrous metals (curve B above). The relation between S and N, it must be pointed out, is not a single-value function but serves to indicate a statistical tendency.

Up until the 1960s, almost all fatigue failures, and consequently all the research in the field, was confined to moving mechanical components (e.g., axles, gears, etc.). Starting in the late 1950s, entire structures or very large structural elements (e.g., pressure vessels, rockets, airplane fuselages, etc.) have been studied and tested for fatigue. This can be attributed to the use of materials such as high-strength alloys, together with the advances in the fabrication technology, resulting in monolithic structures meant to undergo high cyclic stresses in service. It is this class of materials which has shown catastrophic failures in fatigue, and it is for this kind of material that fracture mechanics is being applied, with considerable success, to fatigue problems.

Source: Marc André Meyers and Krishan Kumar Chawla, Mechanical Metallurgy: Principles and Applications, Prentice-Hall, Inc., Englewood Cliffs NJ, 1984, p 689

1-8. Comparison of Fatigue Characteristics: Mild Steel vs Aluminum Alloy

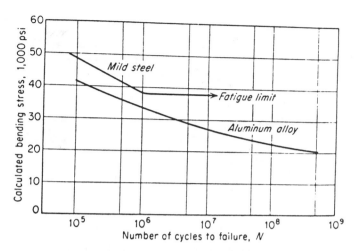

Typical bending ($R = -1$) fatigue curves for ferrous and nonferrous metals.

Here it is noted the lack of the "knee" for the aluminum alloy compared with steel; that is, the point on the curve where the number of cycles to failure becomes a straight line—essentially infinity.

Source: N. R. LaPointe, "Monotonic and Fatigue Characterizations of Metals," in Proceedings of the SAE Fatigue Conference P-109, Society of Automotive Engineers, Inc., Warrendale PA, 1982, p 32

1-9. Carbon Steel: Effect of Lead as an Additive

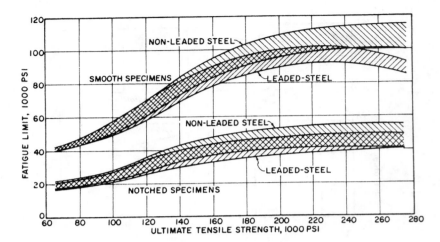

Fatigue limit of leaded and nonleaded alloy steels as a function of ultimate tensile strength.

Lead is often added to steels to improve machinability, although usually at the cost of a minor (usually) loss in mechanical properties,. The interrelationship of lead additions with tensile strength and fatigue limit is summarized in the above graph.

Source: George M. Sinclair, "Some Metallurgical Aspects of Fatigue," in Fatigue—An Interdisciplinary Approach, John J. Burke, Norman L. Reed and Volker Weiss, Eds., Syracuse University Press, Syracuse NY, 1964, p 68

1-10. Corrosion Fatigue: General Effect on Behavior

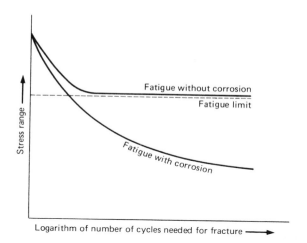

Effect of alternating stresses with and without corrosion.

If a specimen is subjected to alternating stress (tension and compression in turn) over a range insufficient to cause immediate fracture, gliding may occur within some of the grains, but when the dislocations reach a grain-boundary they are halted, retracing their movement along the gliding-plane when the stress is reversed. If the material were ideal, it might be hoped that the dislocations would merely move to and fro along the plane, and that no damage would result. In practice a large number of cycles can be withstood without apparent damage, but in material as we know it, slight irregularities will prevent smooth gliding indefinitely, and roughening along the original gliding-plane will make movement difficult, so that gliding will then start on another parallel plane. In the end, bands of material will have become disorganized, and ultimately one of two things must happen: (1) if the stress range is low, gliding will cease altogether, the only changes still produced by the alternating stress being elastic, (2) if it exceeds a certain level (the **fatigue limit**) the gliding will become so irregular, as to cause separation between the moving surfaces, first locally, producing gaps, which later will join up into cracks. Thus above the fatigue limit (after a time which is shorter at high stress ranges), there will be failure; below the fatigue limit, the life, in absence of corrosion, should be indefinitely long as shown above.

In the presence of a corrosive environment the situation will be different. Disorganized atoms along a gliding-plane may require less activation energy to pass into a liquid than more perfectly arrayed atoms elsewhere; certainly, while the atoms are in motion along a gliding-plane, preferential attack may reasonably be expected even below the fatigue limit. This means that there is no "safe stress range" within which the life should be infinite. It is, however, convenient to determine an **endurance limit**—namely, the stress range below which the material will endure some specified number of cycles (the number must be stated).

It should be noted that, although stress-corrosion cracking is often intergranular, corrosion-fatigue cracks are usually transgranular, following gliding-planes inclined at such an angle as to provide high resolved shear stress. There are exceptions to both rules. Whitwham, studying corrosion-fatigue cracks on steel, found that, although mainly transgranular, they followed grain-boundaries for short distances, where such boundaries chanced to run in a convenient direction.

Source: Ulick R. Evans, An Introduction to Metallic Corrosion, 3d Edition, Edward Arnold (Publishers) Ltd and American Society for Metals, Metals Park OH, 1982, p 160

1-11. Effect of Corrosion on Fatigue Characteristics of Several Steels

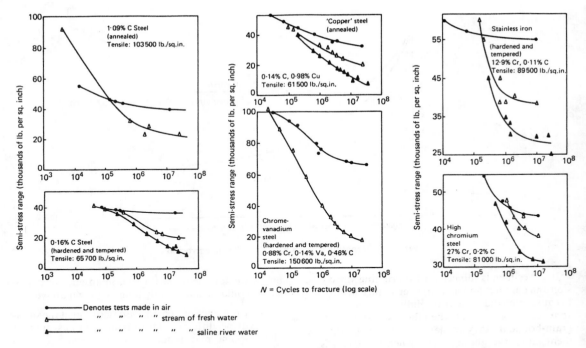

- ●——— Denotes tests made in air
- △——— "   "   "   " stream of fresh water
- ▲——— "   "   "   "   "   " saline river water

Typical curves showing the number of cycles needed to produce fracture at different stress ranges in absence and presence of corrosion.

Two main procedures are available for corrosion-fatigue tests:

One-stage tests. Here the corrosion fatigue is *continued until breakage*. The logarithm of the number of cycles needed to produce breakage is generally plotted against the stress range, as in the above curves selected by Gough from McAdam's experimental data.

Two-stage tests. Here the corrosion fatigue is interrupted after a definite number of cycles, and the residual strength is estimated by measuring either (a) the endurance limit in the absence of corrosive influences (i.e., the stress which can be withstood for some definite number of cycles, (b) the number of cycles needed to produce fracture in the absence of corrosive influences at some definite stress, (c) the tensile strength, or (d) the shock resistance (Izod number).

Source: Ulick R. Evans, An Introduction to Metallic Corrosion, 3d Edition, Edward Arnold (Publishers) Ltd and American Society for Metals, Metals Park OH, 1982, p 165

1-12. Steel: Effect of Hydrogen on Fatigue Crack Propagation

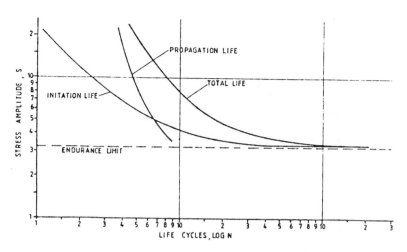

S-N type of fatigue curve.

In the majority of all cases, the external load changing with time, whereby low frequencies ($<10^{-2}$ Hz) have the highest practical importance. Under these circumstances a structural component can be subject to fatigue which is conventionally described by an *S-N* curve relating the cycle life, N, to applied stress, S, as in the above chart. In non-aggressive environments an endurance limit can be defined below which no fatigue failure occurs. A disadvantage of this approach is that *S-N* curves do not differentiate between crack initiation and crack propagation. The number of the cycles corresponding to the endurance limit presents initiation life primarily, whereas the number of cycles for crack initiation at a high value of applied stress is negligible. Consequently *S-N* type data do not necessarily provide information regarding safe-life predictions in structural components. Particularly, if the structure contains surface irregularities different from those of the test specimens, these are likely to reduce or even eliminate the crack initiation portion of the fatigue life.

Source: M. Kesten and K.-F. Windgassen, "Design of Equipment to Resist Hydrogen Fatigue Service," in Current Solutions to Hydrogen Problems in Steels, C. G. Interrante and G. M. Pressouyre, Eds., American Society for Metals, Metals Park OH, 1982, p 390

1-13. Relationship of Stress Amplitude and Cycles to Failure

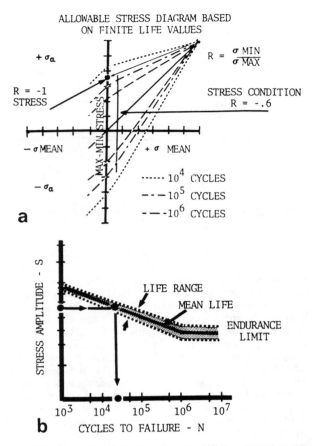

(a) Finite life ASR diagram, showing $R = -1$ equivalent stress for $R = 0.6$ loading. (b) S-N diagram, showing life prediction for $R = 0.6$ loading using $R = -1$ equivalent stress.

The ASR diagrams normally use the endurance-limit fatigue-strength value, but substitution of the fatigue strength at specific finite lives can also be used (see chart a above). Life estimations from the diagram can be done using such information as is shown in charts a and b above. Here the known stress range at some R value is converted to an equivalent completely reversed ($R=-1$) stress, and this equivalent stress is applied to the material's $R=-1$ S-N curve for the life estimate.

Designers have the ability to calculate the component's stresses using classical formulas or the computer-based finite-element-analysis (FEA) techniques. Both of these methods examine the elements for the maximum stresses that are normally in the areas of a discontinuity, or stress concentration.

Source: D. H. Breen and E. M. Wene, "Fatigue in Machines and Structures—Ground Vehicles," in Fatigue and Microstructure, American Society for Metals, Metals Park OH, 1979, p 67

1-14. Strain-Life and Stress-Life Curves

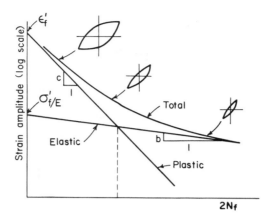

Reversals to failure (log scale)

Strain-life and stress-life curves.

Fatigue damage is caused by cyclic plastic strain, and consequently, the fatigue life should be related to the plastic-strain amplitude. Coffin and Manson independently proposed a relationship between the plastic-strain amplitude and the cycles to failure of the form:

$$\Delta\epsilon_p/2 = \epsilon_f' (2N_f)^c$$

where ϵ_f' is the fatigue-ductility coefficient, $2N_f$ is the number of reversals to failure, and c is the fatigue-ductility exponent. Their equation is very similar to the Basquin equation relating the elastic-strain or true-stress amplitude to the number of load reversals to failure:

$$\Delta\epsilon_e E/2 = \sigma_a = \sigma_f' (2N_f)^b$$

where $\Delta\epsilon_e/2$ is the elastic-strain amplitude, E is the modulus of elasticity, b is the fatigue-strength exponent, and σ_f' is the fatigue-strength coefficient. A schematic representation of these relationships and their superposition is shown in the above diagram. The summation curve is analogous to the stress-life, Wöhler diagram, if the strain amplitudes are replaced by their respective stress amplitudes. The intersection of the Basquin and Coffin-Manson plots is normally defined as the transition between high- and low-cycle fatigue. Consequently, the regime of low-cycle fatigue depends on the properties (for example, the ductility) of a particular material.

Source: Edgar A. Starke, Jr., and Gerd Lütjering, "Cyclic Plastic Deformation and Microstructure," in Fatigue and Microstructure, American Society for Metals, Metals Park OH, 1979, p 211

1-15. Fatigue Plot for Steel: Ultrasonic Attenuation vs Number of Cycles

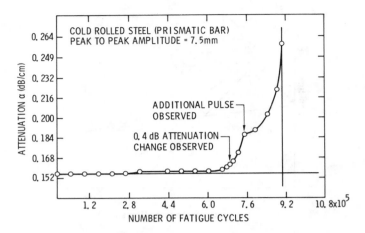

Typical plot of ultrasonic attenuation versus number of fatigue cycles for steel.

Joshi and Green determined the attenuation coefficient α for longitudinal bulk waves in aluminum and steel at 10 and 5 MHz, respectively. The measurements have been performed in a pulse-echo mode, with the acoustic pulse reflected at the back surface of the material. The above chart shows their results obtained on cold rolled steel bars. The attenuation started to increase at about 6×10^5 fatigue cycles (65% of fatigue life). At roughly 7.5×10^5 cycles (85%), an additional pulse was observed, arriving earlier than the one reflected from the back surface.

Results are interpreted in terms of a series of microcracks being formed, probably at the surface. As soon as the microcracks are sufficiently deep, they will change the bulk attenuation. As soon as a macrocrack has been formed (by coalescence of microcracks), it will reflect part of the pulse. After that, the attenuation is primarily determined by the transmission coefficient of this single crack. Thus, the attenuation curve (versus fatigue cycles) becomes discontinuous, as may be noticed in the above chart.

Source: O. Buck and G. A. Alers, "New Techniques for Detection and Monitoring of Fatigue Damage," in Fatigue and Microstructure, American Society for Metals, Metals Park OH, 1979, p 135

2-1. Typical S-N Curve for Low-Carbon Steel Under Axial Tension

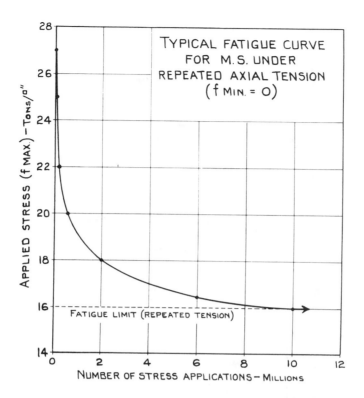

The term "fatigue" refers to the failure of metals from repetitions of stress rather than from a single application, as occurs for example in a simple tensile test or with a brittle failure. The value of the stress necessary to cause failure of a material from fatigue is lower than its nominal tensile strength. For example, a sample of mild steel may have a maximum stress of 27 t.p.s.i. when subjected to a single application of load as in an ordinary tensile test. If, however, a stress of say 25 t.p.s.i. is applied repeatedly to the same material, failure will not take place until this has been done a certain number of times, while at a lower stress still, the number of load cycles required to cause failure will be even greater. If testing is continued in this manner, a stress value will ultimately be found at which fracture will not occur, no matter how many stress repetitions are applied. This value is known as the fatigue limit of the material. If the results from such a series of tests are plotted, a graph such as the one above will be obtained, the curve tending to run parallel to the abscissa after approximately 10 million cycles (for steel), the corresponding value of the stress being known as the fatigue limit. Under conditions of repeated tension the value of the fatigue limit for the above mild steel which has a tensile strength of approximately 27 t.p.s.i. would be of the order of 16 t.p.s.i. If the same steel was tested under conditions of reversed bending stresses a value of the order of ± 12 t.p.s.i. may be found.

It must also be pointed out that where corrosive conditions operate in addition to fluctuating stresses, failure from "corrosion-fatigue" may occur and, in these circumstances, the concept of a fatigue limit does not apply, since if the stress applications are continued for a sufficient number of times, ultimate failure will occur. Further, most nonferrous metals and alloys do not possess a fatigue limit.

Source: F. R. Hutchings, "Fatigue Failure of Components of Lifting Machinery," in Failure Analysis: The British Engine Technical Reports, F. R. Hutchings and Paul Unterweiser, Eds., American Society for Metals, Metals Park OH, 1981, p 344

2-2. AISI 1006: Effects of Biaxial Stretching and Cold Rolling

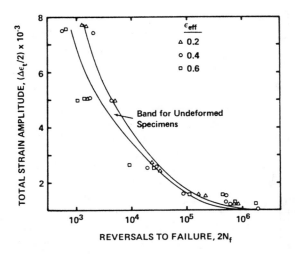

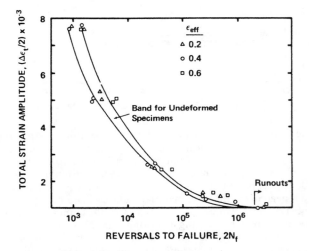

Strain-life plots for two modes of deformation for 1006 steel.

Plots in the top chart are for biaxial stretching; those in the bottom chart are for cold rolling. Included is the data band for the undeformed material. The effect of balanced biaxial stretching on fatigue life was as follows: at large strain amplitudes ($\Delta\epsilon_t/2 \geqslant \sim 2.5 \times 10^{-3}$), the fatigue life remained approximately the same or decreased slightly when compared to that of the undeformed material; in contrast, at small strain amplitudes the fatigue life increased as a result of the prior deformation.

After cold rolling, the fatigue life was approximately the same as in the undeformed material at large strain amplitudes (short lives) but it was longer at small strain amplitudes (long lives). Thus, unlike BBS, CR appeared to cause no reduction in fatigue life at short lives. Another difference between the two deformation modes was that the scatter of the data was larger after BBS than after CR. Thus, BBS was somewhat more detrimental to the fatigue life than CR.

Source: John M. Holt and Philippe L. Charpentier, "Effect of Cold Forming on the Strain-Controlled Fatigue Properties of HSLA Steel Sheets," in HSLA Steels—Technology & Applications, American Society for Metals, Metals Park OH, 1984, p 217

2-3. AISI 1006: Weldment; FCAW, TIG Dressed

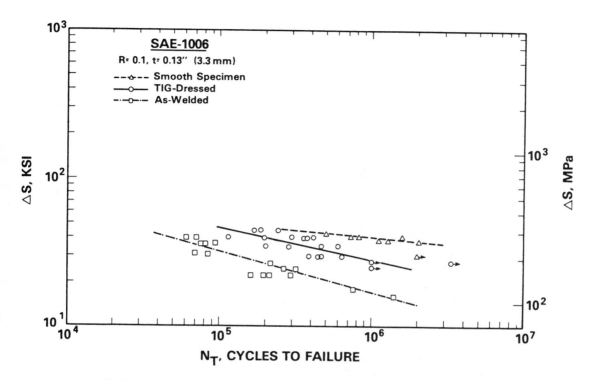

Fatigue strengths of FCAW/TIG- dressed joints compared to those without TIG dressing for AISI 1006 steel (unwelded). The improvement in fatigue provided by TIG dressing the welds is obvious.

Source: Kon-Mei Ewing, Pei-Chung Wang, Frederick V. Lawrence, Jr., and Albert F. Houchens, "Weld Fatigue of TIG-Dressed SAE-980X HSLA Steel," in HSLA Steels—Technology & Applications, American Society for Metals, Metals Park OH, 1984, p 557

2-4. AISI 1006: Weldment; Shear Joints

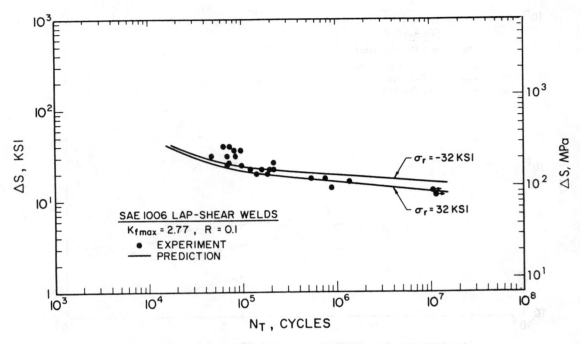

Total fatigue life predictions and experimental results for FCAW, AISI 1006 lap-shear joints. Note that the results and predictions compare closely.

Source: Kon-Mei Ewing, Pei-Chung Wang, Frederick V. Lawrence, Jr., and Albert F. Houchens, "Weld Fatigue of TIG-Dressed SAE-980X HSLA Steel," in HSLA Steels—Technology & Applications, American Society for Metals, Metals Park OH, 1984, p 562

2-5. AISI 1006: Weldment; Lap-Shear Joints

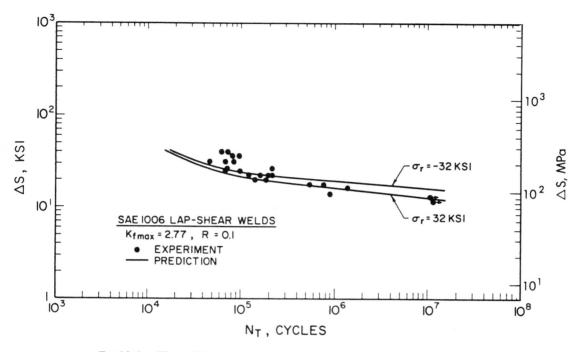

Total fatigue life predictions and experimental results for FCAW, AISI 1006 lap-shear joints. Here, the prediction and actual results are very close.

Source: Kon-Mei Ewing, Pei-Chung Wang, Frederick V. Lawrence, Jr., and Albert F. Houchens, "Weld Fatigue of TIG-Dressed SAE-980X HSLA Steel," in HSLA Steels—Technology & Applications, American Society for Metals, Metals Park OH, 1984, p 562

2-6. AISI 1015: Effect of Cold Working

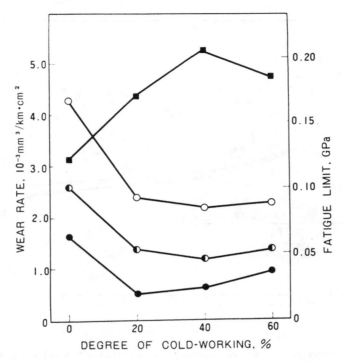

Comparison of effects of cold working on wear rate and fatigue limit of fully annealed 0.15%C mild steel. Wear was determined in sliding between the end surfaces of cylinders at a speed of 0.56 m/s under the loads ●:82 N, ◐:124 N and ○:147 N in machine oil. Fatigue limit (■) was determined by reversed bending fatigue tests of notched plate specimens 25 mm wide and 4 mm thick having a central hole 1.5 mm in diameter.

Attempts have been made to determine effects of cold-working on the resistance to wear and fatigue of a 0.15%C mild steel. Fully annealed material was then cold-worked to different degrees and the specimens were machined from it. Wear experiments were conducted in a rotating cylinder machine as described above with a machine oil as the lubricant. Care had been taken to avoid the effects of work hardening during machining by electrolytically polishing the sliding surface. Reversed bending fatigue tests were carried out by using notched test pieces of the same material. The wear rate and the fatigue limit are compared with the degree of cold-working in the above chart, which shows a definite correlation.

Source: Yoshitsugu Kimura, "The Role of Fatigue in Sliding Wear," in Fundamentals of Friction and Wear of Materials, David A. Rigney, Ed., American Society for Metals, Metals Park OH, 1981, p 215

2-7. A533 Steel Plate: Fatigue Crack Growth Rate

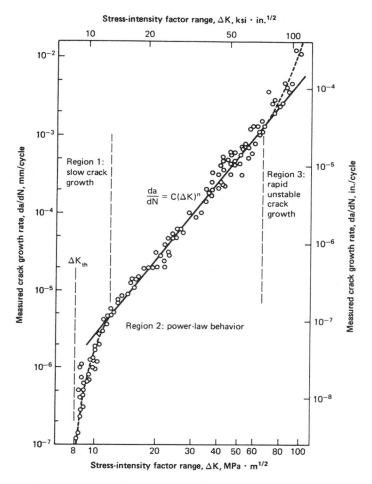

Fatigue crack growth behavior of A533 steel. The material was ASTM A533 B-1 steel, with a yield strength of 470 MPa (70 ksi). Test conditions: $R = 0.10$; ambient room air; 24 °C (75 °F).

The general nature of fatigue crack growth and its description using fracture mechanics can be briefly summarized by the example data shown in the above chart. This figure, based on the work of Paris et al, shows a logarithmic plot of the crack growth per cycle, da/dN, versus the stress-intensity factor range, ΔK, corresponding to the load cycle applied to a sample. The da/dN versus ΔK plot shown is from five specimens of ASTM A533 B-1 steel tested at 24 °C (75 °F). A plot of similar shape is expected with most structural alloys; the absolute values of da/dN and ΔK are dependent on the material. Results of fatigue crack growth rate tests for nearly all metallic structural materials have shown that the da/dN versus ΔK curves have the following characteristics: (a) a region at low values of da/dN and ΔK in which fatigue cracks grow extremely slowly or not at all below a lower limit of ΔK called the threshold of ΔK, ΔK_{th}; (b) an intermediate region of power-law behavior described by the Paris equation:

$$\frac{da}{dN} = C(\Delta K)^n$$

Source: J. H. Underwood and W. W. Gerberich, "Concepts of Fracture Mechanics," in Application of Fracture Mechanics for Selection of Metallic Structural Materials, James E. Campbell, William W. Gerberich and John H. Underwood, Eds., American Society for Metals, Metals Park OH, 1982, p 18

2-8. A514F Steel Plate: Fatigue Crack Growth Rates

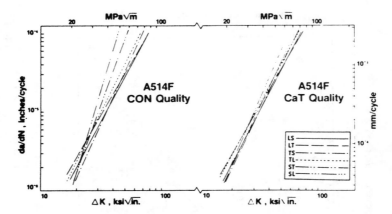

Plots of fatigue crack growth rate versus range of stress intensity factor (best fit lines) for A514F plates.

The increased isotropy in the CaT over the CON steels is evident with the through thickness (ST, SL) orientation having the fastest growth rate in the CON steel and showing the greatest improvement by CaT.

Source: Alexander D. Wilson, "The Effect of Inclusions on the Properties of Constructional Steels," in Wear and Fracture Prevention, American Society for Metals, Metals Park OH, 1981, p 196

2-9. A514F and A633C: Variation in Fatigue Crack Growth Rate With Orientation

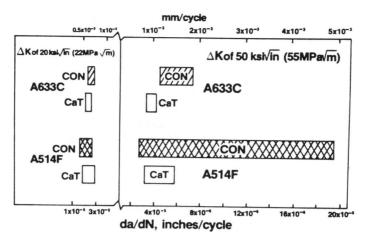

Comparison of fatigue crack growth rate variation with orientation for A633C and A514F plates at two ΔK levels.

These data show that the CaT improvement in FCP growth rate takes place only at higher ΔK levels. Additionally, this figure indicates that there is a more substantial enhancement in FCP behavior for A514F. Also there generally appears to be more anisotropy in the A514F steels of both quality levels. It has previously been shown that higher strength level steels tend to be more adversely affected by inclusions associated in groups, such as present in CON steels.

Source: Alexander D. Wilson, "The Effect of Inclusions on the Properties of Constructional Steels," in Wear and Fracture Prevention, American Society for Metals, Metals Park OH, 1981, p 197

2-10. A514F: Scatterbands of Fatigue Crack Growth Rate

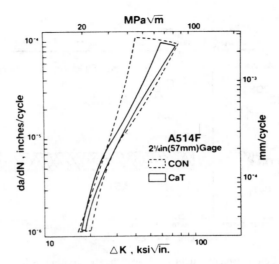

Summary scatterbands of fatigue crack growth rate versus range of stress intensity factor encompassing all data points in 6-orientation testing comparing CON and CaT quality A514F plates.

Source: Alexander D. Wilson, "The Effect of Inclusions on the Properties of Constructional Steels," in Wear and Fracture Prevention, American Society for Metals, Metals Park OH, 1981, p 197

2-11. A633C Steel Plate: Scatterbands of Fatigue Crack Growth Rates

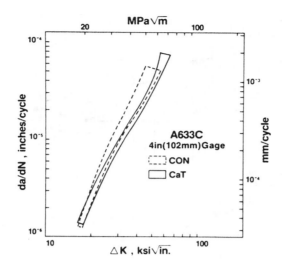

Summary scatterbands of fatigue crack growth rate versus range of stress intensity factor encompassing all data points in 6-orientation resting comparing CON and CaT quality A633C plates.

In this presentation the generally faster FCP growth rates for the CON steels at higher ΔK levels are displayed, as well as the improved isotropy of the CaT steels.

Source: Alexander D. Wilson, "The Effect of Inclusions of the Properties of Constructional Steels," in Wear and Fracture Prevention, American Society for Metals, Metals Park OH, 1981, p 196

2-12. Low-Carbon Steel Weldment: Effects of Various Weld Defects

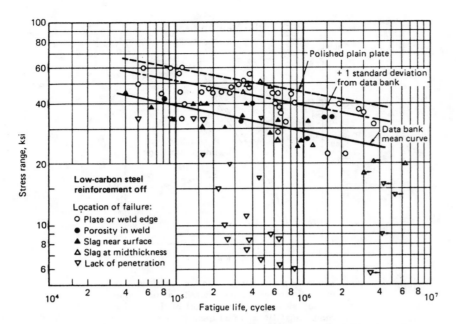

S-N curves showing effect of various weld defects on fatigue life of a low-carbon steel weldment, presented as a comparison with fatigue life of the plate.

Source: Metals Handbook, 9th Edition, Volume 6, Welding, Brazing, and Soldering, American Society for Metals, Metals Park OH, 1983, p 848

2-13. Low-Carbon Steel Weldment: Effect of Weld Reinforcement and Lack of Inclusions

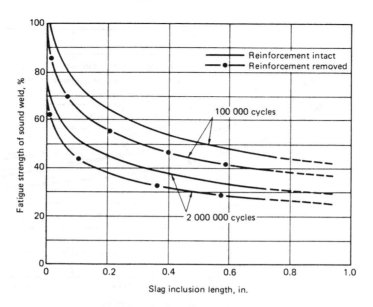

Fatigue strength of a weldment containing slag inclusions as a percentage of the mean fatigue strength of a sound low-carbon steel weld.

Source: Metals Handbook, 9th Edition, Volume 6, Welding, Brazing, and Soldering, American Society for Metals, Metals Park OH, 1983, p 850

2-14. Low-Carbon Steel Weldment: Effect of Weld Reinforcement and Lack of Penetration

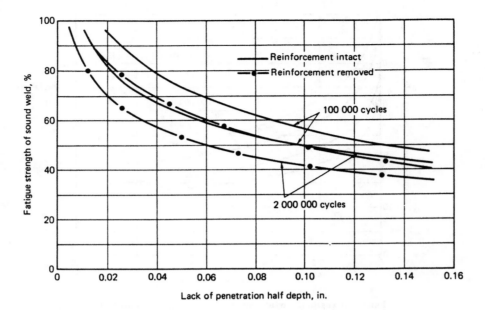

Fatigue strength of a weldment containing lack of penetration as a percentage of the mean fatigue strength of a sound low-carbon steel weld.

Source: Metals Handbook, 9th Edition, Volume 6, Welding, Brazing, and Soldering, American Society for Metals, Metals Park OH, 1983, p 849

2-15. Low-Carbon Steel Weldment: Computed Fatigue Strength; Weldment Contained Lack of Fusion

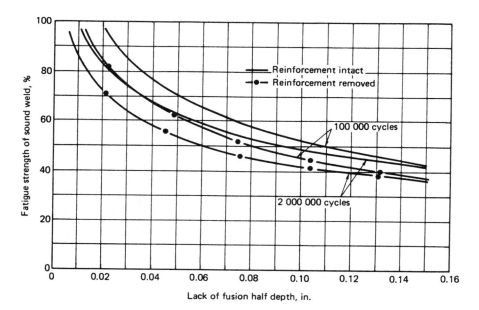

Computed fatigue strength of a weldment containing lack of fusion as a percentage of the mean fatigue strength of a sound low-carbon steel weld.

Source: Metals Handbook, 9th Edition, Volume 6, Welding, Brazing, and Soldering, American Society for Metals, Metals Park OH, 1983, p 849

2-16. Low-Carbon Steel Weldment: Effect of Reinforcement and Undercutting

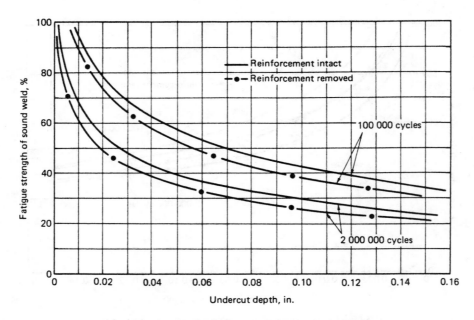

Effect of depth of undercut in terms of percentage of fatigue strength of a sound low-carbon steel weld.

Source: Metals Handbook, 9th Edition, Volume 6, Welding, Brazing, and Soldering, American Society for Metals, Metals Park OH, 1983, p 848

2-17. Low-Carbon Steel: Transverse Butt Welds; Effect of Reinforcement

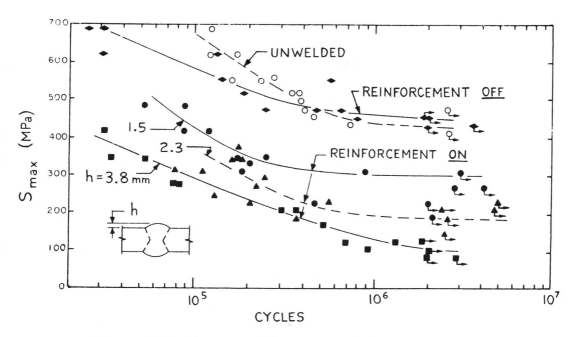

Influence of weld reinforcement on fatigue strength ($R=0$) of transverse butt welds of quenched and tempered carbon steels. From these data it is evident that removal of the reinforcement (weld dressing) improves fatigue strength and fatigue life.

Source: Drew V. Nelson, "Fatigue Considerations in Welded Structure," in Proceedings of the SAE Fatigue Conference P-109, Society of Automotive Engineers, Inc., Warrendale PA, 1982, p 206

2-18. A36/E60S-3 Steel Plate: Butt Welds

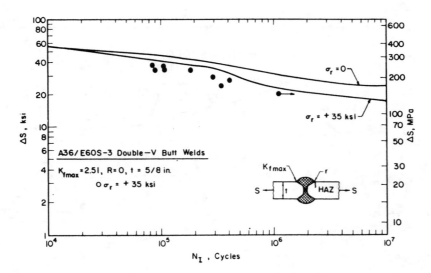

Fatigue crack initiation life predictions and experimental results for 5/8-in. (16-mm) A36/E60S-3 butt welds.

Source: F. V. Lawrence, "The Predicted Influence of Weld Residual Stresses on Fatigue Crack Initiation," in Residual Stress for Designers and Metallurgists, Larry J. Vande Walle, Ed., American Society for Metals, Metals Park OH, 1981, p 114

2-19. A514F/E110 Steel: Bead on Plate Weldment

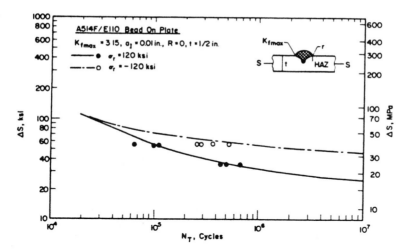

Total fatigue life predictions and experimental results for A514F/E110 weldments with tensile and compressive residual stresses.

Source: F. V. Lawrence, "The Predicted Influence of Weld Residual Stresses on Fatigue Crack Initiation," in Residual Stress for Designers and Metallurgists, Larry J. Vande Walle, Ed., American Society for Metals, Metals Park OH, 1981, p 113

2-20. A36 and A514 Steel Plates: Butt Welded

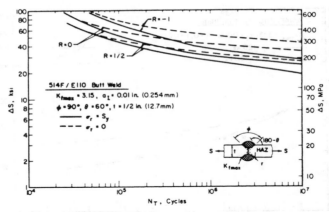

Predicted effect of stress relief and stress ratio on A514/E110 butt weld fatigue life.

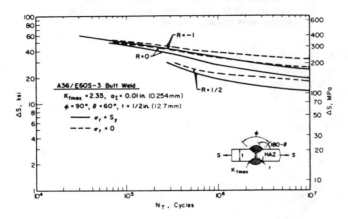

Predicted effect of stress relief and stress relief and stress ratio on A36/E60S-3 butt weld fatigue life.

The results for the high-strength, quenched-and-tempered steels (upper chart), indicate that such materials can sustain high residual stresses which do not relax. The total fatigue life of such materials is strongly influenced by both residual stress (σ_r) and stress ratio (R). Stress relief or mechanically induced compressive residuals should be highly effective. An intermediate case is mild steel as shown in the lower chart. Mild steels can have appreciable residual stresses; but, since the transition fatigue life (N_{tr}) is often very long ($\approx$ 500,000 cycles), there are large amounts of plasticity at the notch root even at long lives (10^6 cycles); this notch-root plasticity tends to relax rapidly the notch-root residual and mean stresses with the result that N_I is little affected for lives less than 10^6 cycles. The observed dependence of N_p on stress ratio does, however, result in a predicted variation of total fatigue life with stress ratio R.

Source: F. V. Lawrence, "The Predicted Influence of Weld Residual Stresses on Fatigue Crack Initiation," in Residual Stress for Designers and Metallurgists, Larry J. Vande Walle, Ed., American Society for Metals, Metals Park OH, 1981, p 112

2-21. A36 Plate Steel: Butt Welded

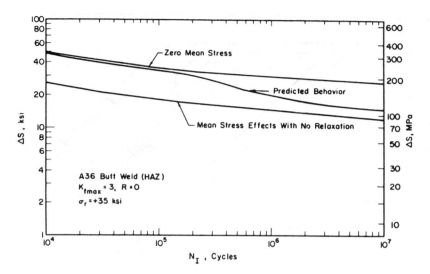

Mean stress relaxation behavior influence on fatigue crack initiation life (A36 HAZ material, $K_f = 3$, $R = 0$, $\sigma_r = +35$ ksi (242 MPa)).

Materials such as high-strength steels exhibit very little notch-root plasticity; consequently, σ_{os} may be larger than σ_r. The results obtained using the model agree with the experimentally observed behavior. The above chart shows the qualitative behavior of N_I predictions.

Source: F. V. Lawrence, "The Predicted Influence of Weld Residual Stresses on Fatigue Crack Initiation," in Residual Stress for Designers and Metallurgists, Larry J. Vande Walle, Ed., American Society for Metals, Metals Park OH, 1981, p 111

2-22. Low-Carbon Steel Tubes: Effect of Welding Technique

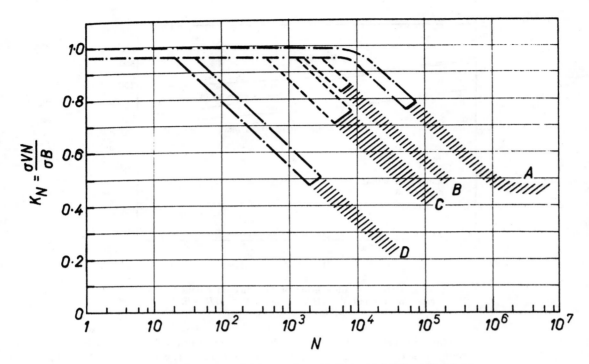

Fatigue strength of welded tubes: A — unwelded or welded without filler metal; B — helical welding (70° angle); C — longitudinal or helical welding (55° or 60° angle); D — helical welding (50° angle).

Source: R. V. Salkin, "Low Cycle Fatigue of Welded Structural Steels: A Material Manufacture and Design Approach," in Proceedings of the Conference of Fatigue of Welded Structures, Vol 2, The Welding Institute, Abington Cambridge, 1971, p 193

2-23. Low Carbon Steel: Effect of Applied Anodic Currents in 3% NaCl

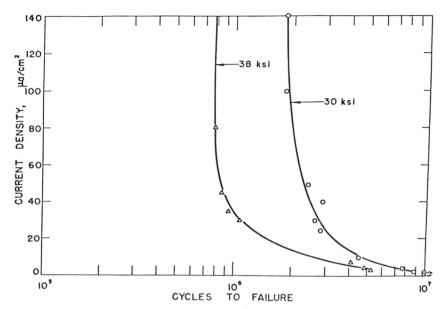

Effect of applied anodic currents on the fatigue lives of low-carbon steel in deaerated 3% NaCl solution. The corrosion rate of the steel in this solution is virtually zero in the absence of applied currents. Note the independence of fatigue life at currents greater than ~ 40 $\mu A/cm^2$, the absence of an applied stress effect and the reappearance of a fatigue limit at currents less than ~ 0.2 $\mu A/cm^2$.

The effects of salt concentration and temperature on the fatigue behavior of steels have been studied. Experiments performed on mild steel specimens in distilled water and in various concentrations of potassium chloride have shown that solutions ranging from 2 molal to 1/40 molal have virtually identical effects on corrosion-fatigue lives, but that at concentrations below 1/40 molal, the effect approaches that of distilled water, although corrosion rates increase in an almost linear manner with solution ion concentration. A similar result has been reported for deaerated 3% NaCl solution in which corrosion rates were controlled by applied anodic currents (see above chart). These observations indicate that a critical corrosion rate is a necessity to initiate corrosion-fatigue failures. Additionally, increasing over-all corrosion rates over a long range of rates has little effect on corrosion-fatigue resistance.

Source: D. J. Duquette, "Environmental Effects I: General Fatigue Resistance and Crack Nucleation in Metals and Alloys," in Fatigue and Microstructure, American Society for Metals, Metals Park OH, 1979, p 344

2-24. Low-Carbon Steel: Effect of pH in NaCl and NaOH

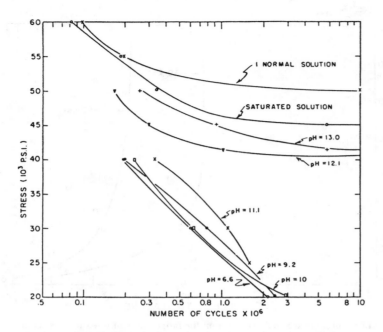

The effect of pH on the fatigue behavior of low-carbon steel in NaCl + NaOH.

The effect of stress frequency on corrosion fatigue has been studied by a number of investigators but is still not completely understood. For example, an early review of corrosion fatigue noted that it is difficult to compare the corrosion-fatigue properties of metals exposed to like environments because data reported are usually taken at different frequencies. In general, a given time was found to produce more damage at a higher frequency, but a given number of cycles was found to produce greater damage at low frequencies. For low-alloy steels in fresh water, a frequency of 1450 cycles/min produced failure in 10^6 cycles or 11½ hours, but at a frequency of 5 cycles/min, failure occurred in 0.11×10^6 cycles, or 400 hours.

To date, the effect of pH of aqueous solutions on corrosion-fatigue behavior has not received extensive study. A study of the effect of $0.1 N$ HCl on the fatigue life of steels showed greater damage in this medium than in neutral potassium chloride solutions. Tests conducted in alkaline media, at a pH above 12.1, showed that a fatigue limit is regained, this limit improving at still higher pH values (above chart). These investigators suggested that corrosion fatigue is a result of differential aeration cells, which produce pits in the metal surface, and that a high pH provides diffusion barriers (ferrous hydroxide) to oxygen on the surface. Higher fatigue limits at high pH are explained in terms of a "better and more perfect film barrier."

Source: D. J. Duquette, "Environmental Effects I: General Fatigue Resistance and Crack Nucleation in Metals and Alloys," in Fatigue and Microstructure, American Society for Metals, Metals Park OH, 1979, p 346

2-25. Low-Carbon Steel: Effect of Carburization and Decarburization

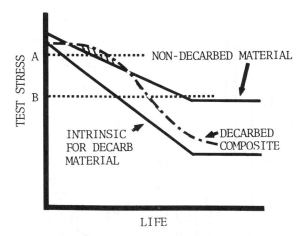

Influence on fatigue *S-N* curve of soft surface caused by decarburization.

Parts that were made from low-carbon steel, but have high-carbon surfaces resulting from carburizing, have special microstructural factors that must be considered. From the carburizing process an intergranular oxide network may develop. This oxide may be an alloy oxide which causes alloy depletion in grain-boundary areas. As a rule, this condition is thought to detract from fatigue properties. The two exceptions may be in combination rolling and sliding contact fatigue, where the oxide network may enhance low-cycle bending fatigue—somewhat the same as does decarburization. The effect on high-cycle bending fatigue is deleterious, as is decarburization. These concepts are shown schematically in the above chart.

Source: D. H. Breen and E. M. Wene, "Fatigue in Machines and Structures—Ground Vehicles," in Fatigue and Microstructure, American Society for Metals, Metals Park OH, 1979, p 80

2-26. A514B Steel: Effect of Various Gaseous Environments on Fatigue Crack Propagation

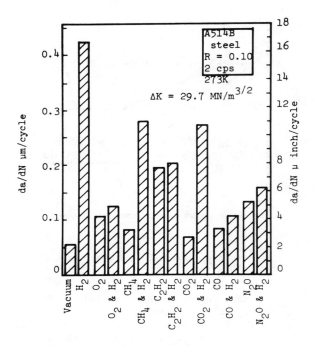

The fatigue-crack propagation of ASTM A514B steel in various gaseous environments.

The origin of the element (such as sulfur) on the surface could result from its presence in the gas phase (such as for hydrogen sulfide). It could also originate as an enriched sulfur layer associated with a propagating crack, as would be the case for sulfur segregated to a grain boundary. Oxygen alone on the surface tends to drive the hydrogen-dissociation reaction rates in the opposite direction from the sulfur. The above bar chart shows how a mixture of environments can influence the fatigue-crack growth of an alloy when all the loading factors are kept constant.

The main influence of the environment is to supply the active atoms to the vicinity of the crack tip. Subsequent interaction with the crack allows the degradation mechanism to take place. The next step in the environmental interaction is the transport of the active species to the location in the vicinity of the crack tip where the degradation mechanism takes place.

Source: H. L. Marcus, "Environmental Effects II: Fatigue-Crack Growth in Metals and Alloys," in *Fatigue and Microstructure*, American Society for Metals, Metals Park OH, 1979, p 371

2-27. Cast 1522 and 1541 Steels: Effect of Various Surface Conditions

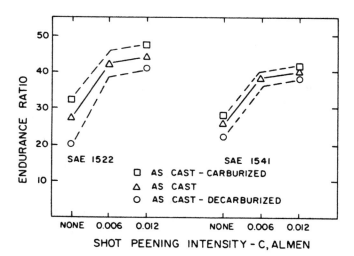

The effect of shot peening, carburization, and decarburization on the endurance ratio of normalized and tempered cast steel with cast surfaces. Plate bending fatigue specimens were used to secure these data.

Decarburization of the surface lowers fatigue resistance. This effect, along with the beneficial effects of carburization and shot peening on the endurance ratio of cast low alloy plate specimen in bending, is shown in the above diagram. The nominally 1.2% Mn steels with 0.22% C and 0.41% C, respectively, were normalized and tempered to 78 and 95 ksi (538 and 655 MPa) ultimate tensile strength, respectively. The depth of decarburization (0.05% C at the surface) was 0.06 in. (1.5 mm); that of carburization (1.15% C at the surface) was 0.08 in. (2 mm).

Source: Steel Castings Handbook, 5th Edition, Peter F. Weiser, Ed., Steel Founders' Society of America, Rocky River OH, 1980, p 15-29

2-28. Cast A216 (Grade WCC) Steel: Fatigue Crack Growth Rate

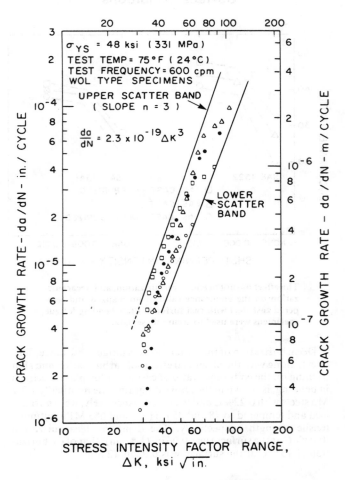

Fatigue crack growth rate as a function of ΔK for A216 (grade WCC) cast steel.

The equation $da/dN = C_0 \Delta K^n$ is sometimes referred to as the Paris law and predicts a linear plot of log da/dN versus log ΔK with slope n. This is observed for a wide variety of materials and is illustrated in the above diagram for an ASTM-A216, Grade WCC cast steel. Some materials show a significant influence of the mean load or K level on fatigue crack growth rates. The ratio of K_{min} to K_{max} is used to express the mean load conditions.

Source: Steel Castings Handbook, 5th Edition, Peter F. Weiser, Ed., Steel Founders' Society of America, Rocky River OH, 1980, p 4-16

3-1. AISI 1030 (Cast) Compared With AISI 1020 (Wrought)

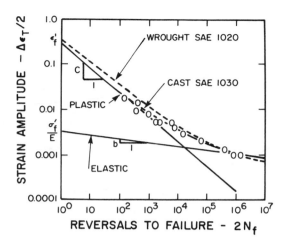

Low-cycle strain-control fatigue behavior of carbon steel.

A number of techniques are available for computing the low-cycle fatigue life, although a straightforward approach is simply to compute the fatigue life from the expected cyclic plastic strain amplitude in service. Errors in computing or estimating $\Delta\epsilon_p$ produce a smaller change in the computed cyclic life than similar errors in the elastic strain range. Note that there is a large difference in slopes "c" and "b" in the above diagram. Plastic strain ranges may be computed using sophisticated finite element techniques, estimated from simple approximations such as Neuber's rule or experimentally measured in component or model tests.

Source: Steel Castings Handbook, 5th Edition, Peter F. Weiser, Ed., Steel Founders' Society of America, Rocky River OH, 1980, p 4-13

3-2. AISI 1035: Effect of Gas and Salt Bath Nitriding

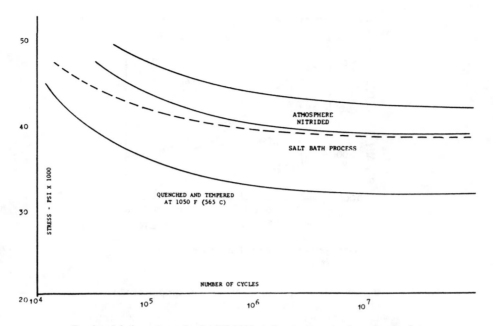

Torsional fatigue strength of AISI 1035 steel—stress vs number of cycles for completely reversing torsional fatigue, featuring the effects of gaseous atmosphere and salt bath nitriding on fatigue strength.

Source: J. A. Riopelle, "Short Cycle Atmosphere Nitriding," in Source Book on Nitriding, American Society for Metals, Metals Park OH, 1977, p 286

3-3. AISI 1040: Cast vs Wrought

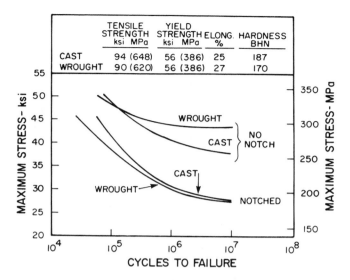

Fatigue characteristics (*S-N* curves) for cast and wrought 1040 steel in the normalized and tempered condition, both notched and unnotched. R. R. Moore rotating beam tests, $K_t = 2.2$.

Cast steel suffers less degradation of fatigue properties due to notches than equivalent wrought steel. When the ideal laboratory test conditions are replaced with more realistic service conditions, the cast steel shows much less notch sensitivity to variations in the values of the test parameters than wrought steel. Under the ideal laboratory test conditions and test preparation (uniform section size, polished and honed surfaces, etc.), the endurance limit of wrought steel is higher. The same fatigue characteristics as those of cast steel, however, are obtained when a notch is introduced, or when standard lathe-turned surfaces are employed in the rotating beam bending fatigue test. These effects are illustrated above.

Source: Steel Castings Handbook, 5th Edition, Peter F. Weiser, Ed., Steel Founders' Society of America, Rocky River OH, 1980, p 15-10

3-4. AISI 1045: Relationship of Hardness and Strain-Life Behavior

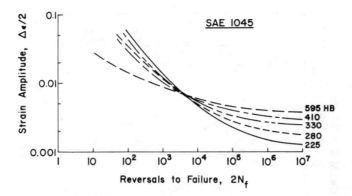

Strain-life behavior of medium-carbon steel as a function of hardness.

Strain-life curves at various hardnesses are presented in the diagram above to demonstrate the range of properties attainable by tempering. Such information, used in conjunction with life-prediction models, provides guidelines for optimizing material processing for specific situations.

Source: R. W. Landgraf, "Control of Fatigue Resistance Through Microstructure—Ferrous Alloys," in Fatigue and Microstructure, American Society for Metals, Metals Park OH, 1979, p 458

3-5. AISI 1141: Effect of Gas Nitriding

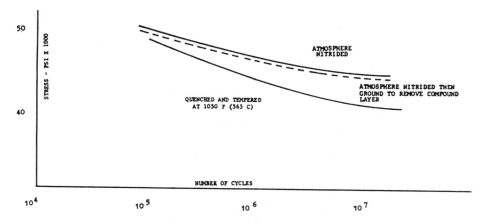

S-N curves for 1141 steel—gaseous-atmosphere nitrided vs not nitrided (quenched and tempered only)—showing stress vs number of cycles for completely reversing torsional fatigue.

Source: J. A. Riopelle, "Short Cycle Atmosphere Nitriding," in Source Book on Nitriding, American Society for Metals, Metals Park OH, 1977, p 287

3-6. Medium-Carbon Steels: Interrelationship of Hardness, Strain Life and Fatigue Life

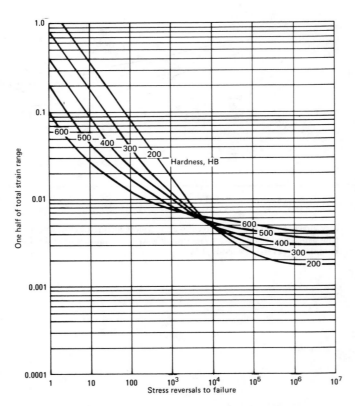

Predicted plots of strain versus fatigue life for typical medium-carbon steels at the hardness levels indicated above.

Source: Metals Handbook, 9th Edition, Volume 1, Properties and Selection: Irons and Steels, American Society for Metals, Metals Park OH, 1978, p 673

3-7. Medium-Carbon Steel: Effect of Fillet Radii

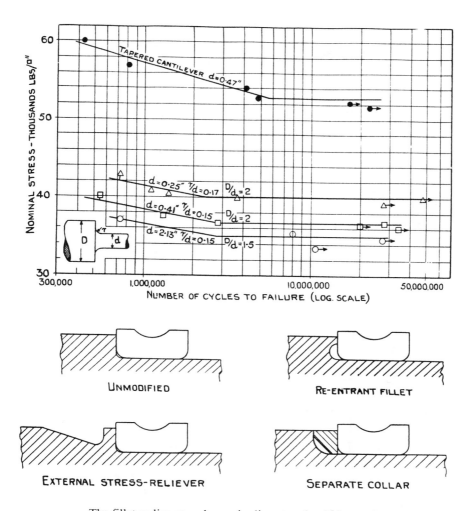

The fillet radius at a change in diameter should be made as great as possible. This cannot always be done; e.g., if the inner race of a rolling bearing must abut against a shoulder formed by the change in diameter. In such cases the stress-raising effect can be moderated very considerably by adopting one of the expedients illustrated above.

Source: G. A. Cottell, "Some Common Stress Raisers in Engineering Parts," in Failure Analysis: The British Engine Technical Reports, F. R. Hutchings and Paul Unterweiser, Eds., American Society for Metals, Metals Park OH, 1981, p 108

3-8. Medium-Carbon Steel: Effect of Keyway Design

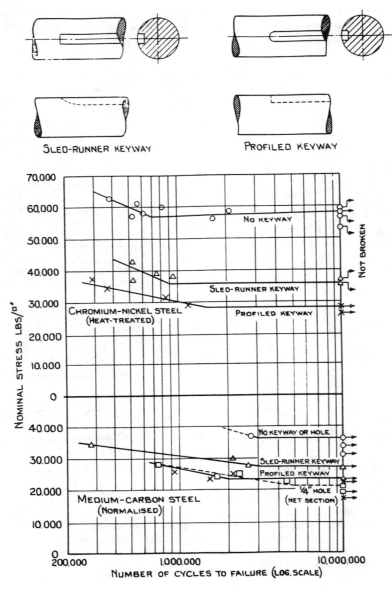

Keyways are severe stress raisers from which fatigue cracks are very liable to develop. Where bending stresses are predominant the cracks usually run transversely in the region of the keyway end, but where torsional stresses predominate they originate at the root at one side and may cause a portion of the shaft to peel off or they may lie diagonally across the bottom. Effects of various keyway designs on fatigue life are shown above.

Source: G. A. Cottell, "Some Common Stress Raisers in Engineering Parts," in Failure Analysis: The British Engine Technical Reports, F. R. Hutchings and Paul Unterweiser, Eds., American Society for Metals, Metals Park OH, 1981, p 109

3-9. Medium-Carbon Steel: Effect of Residual Stresses

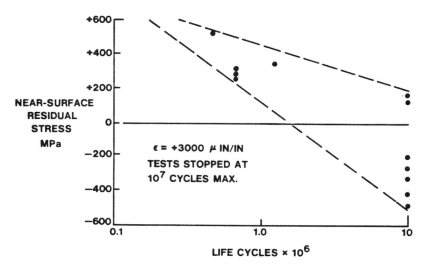

Fatigue life relationship to near-surface residual stress.

Fully reversed fatigue tests on smooth bar specimens in medium carbon steels fully hardened show, as expected, that fatigue life increases directly with surface and near-surface residual compressive stress (see above chart). Residual stress measurements are usually made in the direction of the applied stress.

The achievement of high residual compressive stress in a part requires a careful balance of the factors which affect this property and often involves a number of trade-offs which vary with the application.

Source: J. Alan Burnett, "Prediction of Stresses Generated During the Heat Treating of Case Carburized Parts," in Residual Stress for Designers and Metallurgists, Larry J. Vander Walle, Ed., American Society for Metals, Metals Park OH, 1981, p 44

3-10. Medium-Carbon Cast Steel: Effect of Changes in Residual Stress

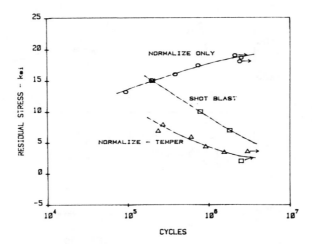

Residual stress at completion of testing.

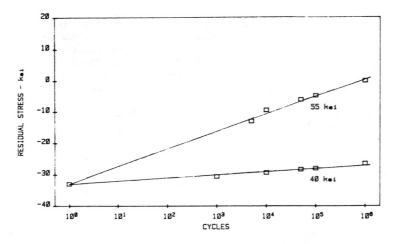

Change in residual stress with cycles at constant applied stress.

The upper chart shows residual stresses existing on the completion of individual tests. The similarity to *S-N* curves is apparent with the exception that the curve for normalized bars ($R_1 = +22$ ksi) is inverted. Since the initial residual stress was known, there was a question on the manner in which the residual stress changed during the progress of testing. To explore this point, two shot blasted bars were tested with applied stress levels of 40 and 55 ksi. The test on each specimen was interrupted periodically to measure the residual stress at that time. The results are shown in the lower chart. It is apparent that the change in residual stress is proportional to the number of cycles when the latter is represented on a logarithmic basis. The lower chart also points to the fact that the rate of change in residual stress is dependent on the level of applied stress. Since the initial and final residual stress values were known for all bars, the slope for each line could be determined.

Source: P. J. Neff, "A Quantitative Evaluation of Surface Residual Stress and Its Relation to Fatigue Performance," in Residual Stress for Designers and Metallurgists, Larry J. Vander Walle, Ed., American Society for Metals, Metals Park OH, 1981, pp 127-128

3-11. Medium-Carbon Cast Steel: S-N Projection (Effect of Applied Stress)

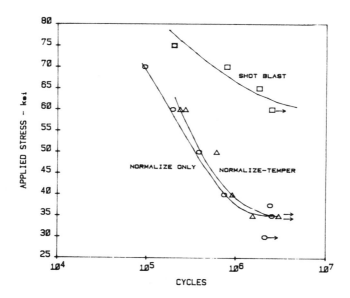

Observed *S-N* curves for bars having three treatments.

Test data for all three sets of bars representing treatments consisting of normalize, temper and shot blast; normalize, temper; and normalize only are shown above. Since no attempt was made to accurately determine endurance limits, the *S-N* curves are only approximate. Such curves merely estimate life at a given level of applied stress and give an indication of the stress level below which failure would not be expected. Attention is directed to the lower two curves where there is little or no difference in the estimated endurance limit in spite of the difference in initial residual stress (+2 and +22 ksi).

Source: P. J. Neff, "A Quantitative Evaluation of Surface Residual Stress and its Relation to Fatigue Performance," in Residual Stress for Designers and Metallurgists, Larry J. Vander Walle, E., American Society for Metals, Metals Park OH, 1981, p 126

3-12. Medium-Carbon Cast Steel: Effect of Applied Stress (Shot Blasting)

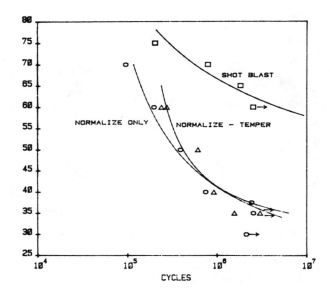

Computed *S-N* curves for bars having three treatments, showing the improvement in fatigue strength provided by shot blasting.

As expected, the shot blasted spot showed a comparatively large change. Another trial was attempted in which a small spot was pre-stressed by hammering the surface with a ball-peen hammer. This accomplished almost the same result as the shot blast operation. This technique obviously could be applied to a production part.

Since service stresses may be highly variable, various techniques have been made available for approximating the fatigue life. This requires a knowledge of all of the stress cycles. Since the fatigue life seems to be dependent on change in residual stress, which is, in turn, dependent on the particular applied stress, it is conceivable that the change in residual stress might represent an integral of the effect of variable stresses.

Source: P. J. Neff, "A Quantitative Evaluation of Surface Residual Stress and Its Relation to Fatigue Performance," in Residual Stress for Designers and Metallurgists, Larry J. Vander Walle, Ed., American Society for Metals, Metals Park OH, 1981, p 134

4-1. Medium-Carbon Alloy Steels, Five Grades: Effect of Martensite Content

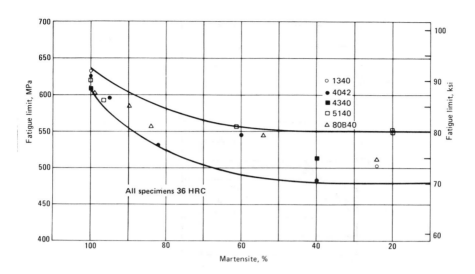

For specimens having comparable strength levels, resistance to fatigue depends somewhat on the microstructure. A tempered martensite structure provides the highest fatigue limit. However, if the structure as-quenched is not fully martensitic, the fatigue limit will be lower (see graph above). Pearlitic structures, particularly those with coarse pearlite, have poor resistance to fatigue. *S-N* curves for pearlitic and spheroidized structures in a eutectoid steel are provided in chart 4-40 (p 122).

Source: Metals Handbook, 9th Edition, Volume 1, Properties and Selection: Irons and Steels, American Society for Metals, Metals Park OH, 1978, p 676

4-2. Medium-Carbon Alloy Steels, Six Grades: Hardness vs Endurance Limit

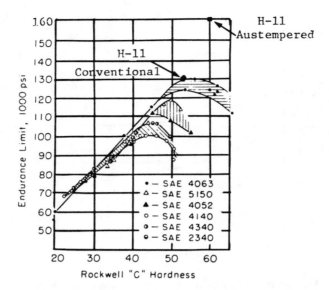

Relation of hardness and fatigue strength for several steels.

The above chart and other data can be used to show the importance of limiting the system to low and intermediate hardnesses as well as to point out the importance of residual stress in fatigue. These data from Garwood, Zurburg and Erickson show a very tight linear relation up to about 40 HRC. Above that hardness, the relation deviates from linearity, seemingly depending on carbon content. Carbon, however, is in an intermediate role here, because it affects temperability. Because response to tempering is dependent on carbon and alloy levels, it was necessary for samples of different grades to be tempered at different temperatures to achieve the same hardness; consequently, a variety of residual-stress conditions resulted. The tempering temperatures were necessarily sufficiently high to obtain 40 HRC; the residual stresses were reduced to a very low level, making all samples similar in that usually the tensile strength for small sections decreases with increasing section size and/or decreasing hardenability to compressive values. The sequence of transformation from surface to center, together with the temperature gradients, governs the outcome.

Source: D. H. Breen and E. M. Wene, "Fatigue in Machines and Structures—Ground Vehicles," in Fatigue and Microstructure, American Society for Metals, Metals Park OH, 1979, p 73

4-3. Medium-Carbon Alloy Steels: Effect of Specimen Orientation

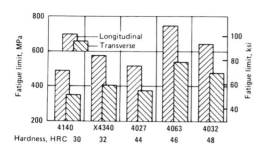

Steel	No. of tests(a)	Avg tensile strength MPa ksi		Hardness, HRC
Longitudinal Tests				
4027	11	1179	171	37 to 39
4063	12	1682	244	47 to 48
4032	11	1627	236	46 to 48
Transverse Tests				
4027	10	1130	164	34 to 39.5
4063	9	1682	244	47 to 48.5
4032	10	1254	182	47.5 to 48.5

(a) Number of fatigue specimens. For 4140 steel, 50 longitudinal and 50 transverse specimens were tested; for 4340 steel, 10 longitudinal and 10 transverse.

It must always be considered that in rolled steels fatigue behavior is affected significantly by specimen orientation. Shown above is the effect of orientation relative to fiber axis resulting from hot working on the fatigue limit of low-alloy steels. Through hardened and tempered specimens, 6.3 mm (0.250 in.) in diameter, were taken from production billets. Specimens for each grade were from the same heat of steel, but the tensile and fatigue specimens were heat trated separately, thus accounting for one discrepancy in hardness readings between the chart and the tabulation above. Fatigue limit is for 100 million cycles.

Source: Metals Handbook, 9th Edition, Volume 1, Properties and Selection: Irons and Steels, American Society for Metals, Metals Park OH, 1978, p 677

4-4. 4027 Steel: Carburized vs Uncarburized

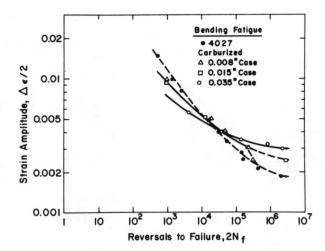

Bending-fatigue results for uncarburized and carburized 4027 steel.

Bending-fatigue results supporting the validity of the effect of carburizing are presented in the above curves. An uncarburized baseline curve is compared with curves for three case thicknesses. As predicted, all carburized specimens show inferior low-cycle resistance. At longer lives, the thinnest case offers some improvement but tends toward the baseline as a result of subsurface failure initiation. The thickest case, which shows the greatest life improvement and has been found to exhibit surface failure initiation, seems close to optimum.

Source: R. W. Landgraf, "Control of Fatigue Resistance Through Microstructure—Ferrous Alloys," in Fatigue and Microstructure, American Society for Metals, Metals Park OH, 1979, p 463

4-5. 4120 Steel: Effect of Surface Treatment in Hydrogen Environment

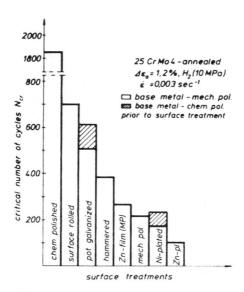

Effect of surface treatment on fatigue life in hydrogen environment for a 0.23C-0.98Cr-0.22Mo steel.

In the above bar chart, the effect of surface treatment on fatigue life is summarized. The base metal was mechanically polished before surface treatment. For comparison, pot galvanizing and Ni-plating have been performed after mechanical as well as after chemical polishing of the base metal. The results after chemical polishing are given above in the form of dashed columns.

The galvanizing such as Ni- and ZN-plating is by no means an appropriate method to increase the fatigue life in hydrogen in spite of the reduced surface roughness and protecting effect. This is because the galvanizing produces a relatively high tensile residual stress and the deposits possess generally poor ductility.

Source: Kyong-Tschong Rie and Werner Kohler, "Improvement of the Resistance of Metals to Cyclic Plastic Loading in High Pressure Hydrogen Environment," in Current Solutions to Hydrogen Problems in Steels, C. G. Interrante and G. M. Presouyre, Eds., American Society for Metals, Metals Park OH, 1982, p 380

4-6. 4120 Steel: Effect of Surface Treatment in Hydrogen Environment

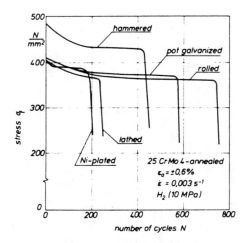

Effect of surface treatment on cyclic flow curve in hydrogen environment.

The above graph shows the cyclic strain hardening and softening curves for different surface treatments. It can be seen that the fatigue behavior in hydrogen environment can be improved by some surface treatments.

Source: Kyong-Tschong Rie and Werner Kohler, "Improvement of the Resistance of Metals to Cyclic Plastic Loading in High Pressure Hydrogen Environment," in Current Solutions to Hydrogen Problems in Steels, C. G. Interrante and G. M. Pressouyre, Eds., American Society for Metals, Metals Park OH, 1982, p 379

4-7. 4120 Steel: Effect of Various Surface Treatments on Fatigue Characteristics in Air vs Hydrogen

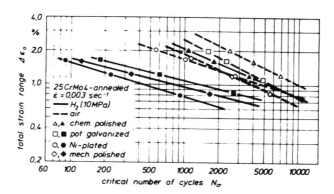

Fatigue life curve for various surface treatments of steel in hydrogen environment and in air. Steel contained 0.23 C, 0.98 Cr and 0.22 Mo (4120).

Source: Kyong-Tschong Rie and Werner Kohler, "Improvement of the Resistance of Metals to Cyclic Plastic Loading in High Pressure Hydrogen Environment," in Current Solutions to Hydrogen Problems in Steels, C. G. Interrante and G. M. Pressouyre, Eds., American Society for Metals, Metals Park OH, 1982, p 380

4-8. 4130 Steel: Fatigue Crack Growth Rate vs Temperature in Hydrogen

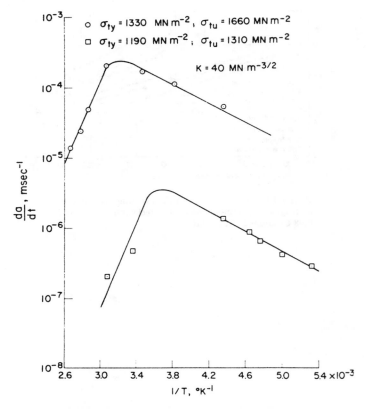

Crack growth rate versus temperature in hydrogen gas, for 4130 steel with yield strengths of 1330 and 1190 MN m^{-2}.

The striking characteristic of hydrogen which sets it apart from other causes of embrittlement is its large diffusivity. Although the diffusivity of hydrogen does vary significantly among metals and alloys, it is nevertheless always several orders of magnitude larger than the diffusivities of other species. Consequently, hydrogen transport is a prominent feature of discussions of hydrogen-induced crack growth kinetics, and of the unique strain rate and temperature dependence of hydrogen embrittlement.

Nelson and Williams reported the first complete investigation of the kinetics of crack growth in high strength steel exposed to hydrogen gas (see graph above).

Source: Herbert H. Johnson, "Keynote Lecture: Overview on Hydrogen Degradation Phenomena," in Hydrogen Embrittlement and Stress Corrosion Cracking, R. Gibala and R. F. Hehemann, Eds., American Society for Metals, Metals Park OH, 1984, p 17

4-9. 4135 and 4140 Steels: Cast vs Wrought

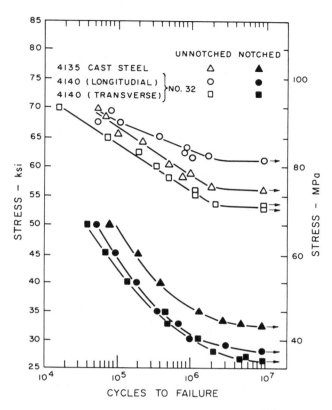

S-N curves of a normalized and tempered AISI 4140 wrought steel in the longitudinal and transverse direction and cast 4135 steel normalized and tempered. Tensile strength for wrought steel: longitudinal, 110.0 ksi (758 MPa); transverse, 110.7 ksi (763 MPa); cast steel: 112.7 ksi (770 MPa).

In general, if the longitudinal and transverse ductility, impact, or fatigue property values of rolled steel are averaged, they will be about the same as properties of cast steel. One example of this is shown in the *S-N* curves presented above. For these, a 4140 rolled steel was tested in fatigue in the longitudinal and transverse position and compared with a similar Cr-Mo cast steel.

Source: Steel Castings Handbook, 5th Edition, Peter F. Weiser, Ed., Steel Founders' Society of America, Rocky River OH, 1980, p 3-16

4-10. 4135 and 4140 Steels: Cast vs Wrought

		STRENGTH				
		TENSILE		YIELD		ELONG
		ksi (MPa)		ksi (MPa)		%	BHN
CAST	4135	113 (779)		87 (560)		43	223
WROUGHT	4140-L	110 (758)		80 (552)		61	217
	-T	111 (765)		81 (558)		30	217

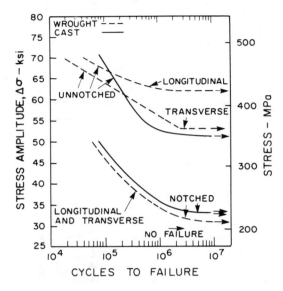

Fatigue characteristics (*S-N* curves) for cast and wrought 4100 series steels, quenched and tempered to the same hardness, both notched and unnotched.

The number of cycles to failure of a structure subjected to the above stress history can be expressed in terms of the *S-N* curve shown above. The fatigue life increases as the cyclic stress amplitude decreases. For ferrous alloys a true endurance or fatigue limit is reached below which fatigue failure is not observed.

The data presented in the above *S-N* curves illustrate several important points. First, a fatigue limit is evident. That is, below a certain cyclic stress amplitude, fatigue failure will not occur for any arbitrarily large number of cycles. Secondly, while the fatigue properties of cast steel are lower than those obtained with the wrought steel, it has less anisotropy. And, finally, the presence of a notch equalizes the fatigue properties of cast and wrought steels.

The above data also illustrate that the fatigue limit of notched test specimens is substantially below that of unnotched samples when the fatigue limit is computed on the basis of nominal stress.

Source: Steel Castings Handbook, 5th Edition, Peter F. Weiser, Ed., Steel Founders' Society of America, Rocky River OH, 1980, p 4-8

4-11. 4140, 4053 and 4063 Steels: Effect of Carbon Content and Hardness

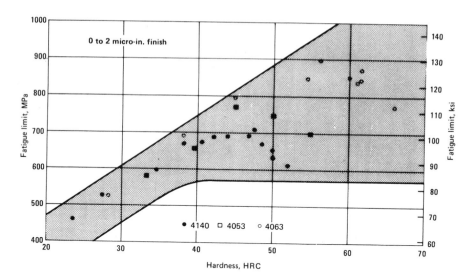

Effect of hardness and carbon level on fatigue limit of alloy steels.

As shown above, when steels are hardened to 45 HRC or higher an increase in carbon content can increase fatigue limit. Although other alloying elements may be required in order to attain desired hardenability, they have little effect on fatigue behavior.

Source: Metals Handbook, 9th Edition, Volume 1, Properties and Selection: Irons and Steels, American Society for Metals, Metals Park OH, 1978, p 676

4-12. 4140 Steel: Effect of Direction on Fatigue Crack Propagation

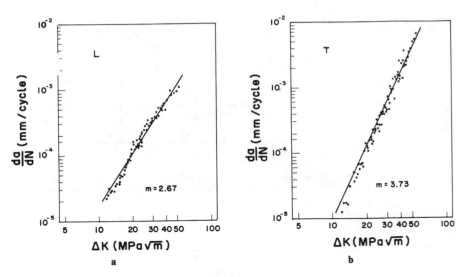

Fatigue crack propagation in an AISI 4140 steel: (a) longitudinal direction (parallel to rolling direction); (b) transverse direction (perpendicular to rolling direction).

The Paris power law, which describes the crack propagation rate in stage II for a series of metals, is very useful because of its extreme simplicity. For example, it has been observed experimentally that data points in the form of log (da/dN) versus log ΔK for a given material (constant metallurgical structure) from three different samples—edge crack in a compact tension sample, through-thickness central crack in a plate, and plate containing a partially through-thickness crack—all fall on the same line. Also, there is experimental evidence that shows that the stress level by itself does not influence the fatigue crack growth rate for stress levels below the general yielding. Thus, it can be considered that the parameter ΔK describes uniquely the crack growth rates for many engineering applications. However, the structure of material can influence fatigue crack growth rates drastically; the value of m can change a lot. The above charts illustrate the directionality in the fatigue crack propagation rate in an AISI 4140 steel. The exponent m has a much higher value in the transverse direction than in the longitudinal (rolling) direction, due to the presence of elongated inclusions.

Source: Marc André Meyers and Krishan Kumar Chawla, Mechanical Metallurgy: Principles and Applications, Prentice-Hall, Inc., Englewood Cliffs NJ, 1984, p 714

4-13. 4140 Steel: Effect of Cathodic Polarization

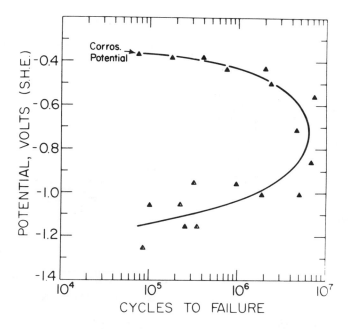

The effect of cathodic polarization on the fatigue behavior of 4140 steel (heat treated to HRC 52) in 3% NaCl solution at a stress level below the fatigue limit in air is shown above. The use of cathodic protection to prevent corrosion fatigue of steels depends sensitively on the hardness of the steel. For example, cathodic protection of a 4140 steel was shown to be feasible for hardness values of Rockwell C 40. At higher hardness values, an improvement in fatigue resistance is observed for moderate cathodic potentials, but complete protection is not possible. At potentials large enough to inhibit corrosion fatigue for softer steels, a decrease in fatigue resistance is observed, presumably due to hydrogen embrittlement (note above chart).

Source: D. J. Duquette, "Environmental Effects I: General Fatigue Resistance and Crack Nucleation in Metals and Alloys," in Fatigue and Microstructure, American Society for Metals, Metals Park OH, 1979, p 360

4-14. Cast 4330 Steel: Effects of Various Surface Conditions

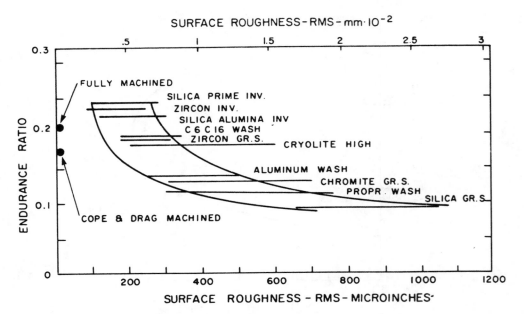

Relationship between surface roughness and endurance ratio (endurance limit divided by tensile strength) of quenched and tempered cast 4330 steel [UTS = 165-185 ksi (1138-1276 MPa)]. Fully reversed plate bending tests.

Plate bending tests for quenched and tempered low alloy cast 4330 steel indicate that investment cast surfaces, or conventional castings produced with special mold washes, performed equal to, or better than, fully machined and polished plate specimens. The data also suggest a tapering off of the surface effects on the endurance ratio at 600 or more RMS surface roughness as indicated in the above diagram.

Source: Steel Castings Handbook, 5th Edition, Peter F. Weiser, Ed., Steel Founders' Society of America. Rocky River OH, 1980, p 15-29

4-15. 4340 Steel: Scatter of Fatigue Limit Data

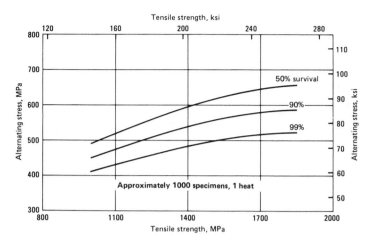

Interrelationships of alternating stress, tensile strength and expected percent survival for heat treated 4340 steel.

These data show survival after 10 million cycles of AISI-SAE 4340 steel with tensile strengths of 995, 1320, and 1840 MPa (144, 191, and 267 ksi). Rotating-beam fatigue specimens tested at 10 000 to 11 000 rpm. Coefficients of variation range from 0.17 to 0.20. From these data it is evident that scatter increases as strength level is increased.

Source: Metals Handbook, 9th Edition, Volume 1, Properties and Selection: Irons and Steels, American Society for Metals, Metals Park OH, 1978, p 678

4-16. 4340 Steel: Strength vs Fatigue Life

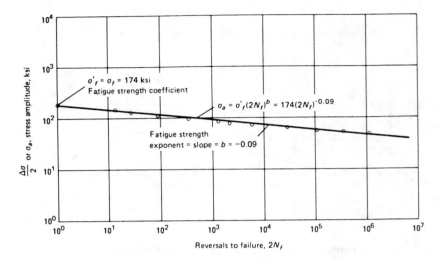

Typical data for strength versus fatigue life for annealed 4340 steel.

4-17. 4340 Steel: Total Strain vs Fatigue Life

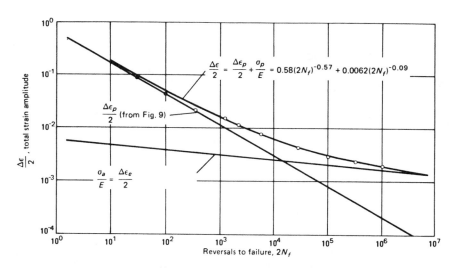

Typical data for total strain versus fatigue life for annealed 4340 steel.

Source: Metals Handbook, 9th Edition, Volume 1, Properties and Selection: Irons and Steels, American Society for Metals, Metals Park OH, 1978, p 672

4-18. 4340 Steel: Stress Amplitude vs Number of Reversals

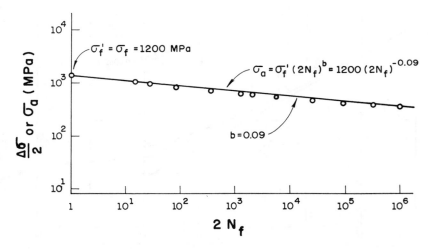

Stress amplitude ($\Delta\sigma/2$) versus number of reversals ($2 N_f$) for AISI 4340 steel.

It is convenient to consider separately the elastic and the plastic components of strain. The elastic component can be readily described by means of a relation between the true stress amplitude and the number of reversals (i.e., twice the number of cycles):

$$\frac{\Delta\epsilon_e}{2} = \frac{\sigma_a}{E} = \left(\frac{\sigma'_f}{E}\right)(2N_f)^b$$

where $\Delta\epsilon_e/2$ is the elastic strain amplitude, σ_a the true stress amplitude, σ'_f the fatigue strength coefficient (equal to stress intercept at $2N_f = 1$), N_f the number of cycles to failure, and b the fatigue strength exponent.

This relation is an empirical representation of the S-N curve above the fatigue limit. The above chart shows an application of this relation to SAE 4340 steel. It was observed that fatigue life increased with decreasing b. Morrow, based on energy considerations, showed that the fatigue strength exponent is given by:

$$b = -\frac{n'}{1 + 5n'}$$

where n' is the cyclic hardening coefficient.

Thus, the fatigue life under elastic cyclic conditions (whether stress- or strain-controlled) increases with a reduction in n'. Of course, the higher the material coefficient σ'_f, the better it is for fatigue. There is evidence that σ'_f is approximately equal to σ_f, the monotonic fracture strength.

The plastic strain component is better described by the Manson-Coffin relation:

$$\frac{\Delta\epsilon_p}{2} = \epsilon'_f(2N_f)^c$$

where $\Delta\epsilon_p/2$ is the plastic strain amplitude, ϵ'_f is the ductility coefficient in fatigue and is equal to strain intercept at $2N_f = 1$, $2N_f$ is the number of reversals to failure, and c is the ductility exponent in fatigue.

Source: Marc André Meyers and Krishan Kumar Chawla, Mechanical Metallurgy: Principles and Applications, Prentice-Hall, Inc., Englewood Cliffs NJ, 1984, p 697

4-19. 4340 Steel: Effect of Periodic Overstrain

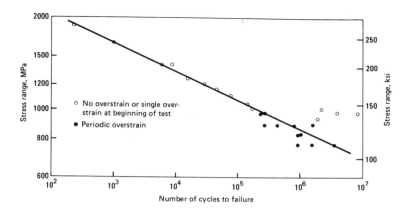

Overstrain superimposed on constant strain may have a significant effect on fatigue life. Shown above is the effect of periodic large strain cycles on fatigue life of AISI-SAE 4340 steel hardened and tempered to a yield strength of 1100 MPa (160 ksi).

Source: Metals Handbook, 9th Edition, Volume 1, Properties and Selection: Irons and Steels, American Society for Metals, Metals Park OH, 1978, p 681

4-20. 4340 Steel: Estimation of Constant Life

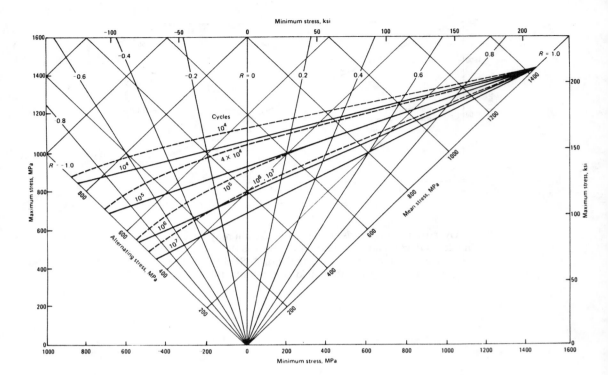

Potter has described a method for approximating a constant-lifetime fatigue diagram for unnotched specimens. Using this method, a series of points corresponding to different lifetimes are calculated and plotted along the diagonal line on the left side ($R = -1$). Each of these points is connected by a straight line to the point of the other diagonal ($R = 1.0$) that corresponds to the ultimate tensile strength. A comparison between the estimated constant-lifetime diagram and the experimentally determined diagram is given in the above illustration. Here is presented a comparison between a calculated constant-life fatigue diagram (solid lines) and experimentally determined data (dashed lines). The calculated lines correspond well with the experimental lines. Generally, the predicted lines represent lower stresses than the actual data. Estimating fatigue parameters from the Brinell hardness number provides more conservative estimates. These results are only approximations, and the methods may not apply for every material.

Source: Metals Handbook, 9th Edition, Volume 1, Properties and Selection: Irons and Steels, American Society for Metals, Metals Park OH, 1978, p 681

4-21. 4340 Steel: Effect of Strength Level on Constant-Life Behavior

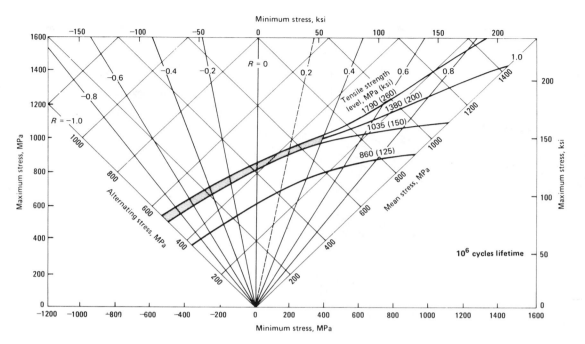

Constant-lifetime fatigue diagram for AISI-SAE 4340 alloy steel (bar), hardened and tempered to tensile strength levels of 860 MPa (125 ksi), 1035 MPa (150 ksi), 1380 MPa (200 ksi) and 1790 MPa (260 ksi). All lines represent fatigue lifetimes of one million cycles.

It may be noted that lives of the specimens at the three higher strength levels are about the same; the scatter in data is at least as great as any real differences in fatigue life among specimens.

Source: Metals Handbook, 9th Edition, Volume 1, Properties and Selection: Irons and Steels, American Society for Metals, Metals Park OH, 1978, p 669

4-22. 4340 Steel: Notched vs Unnotched Specimens

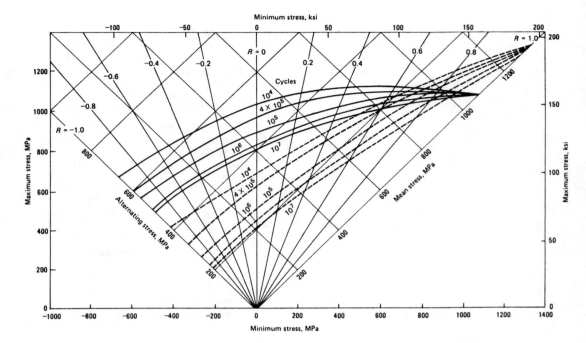

Constant-lifetime fatigue diagram for AISI-SAE 4340 alloy steel (bar), hardened and tempered to a tensile strength of 1035 MPa (150 ksi). Solid lines represent data obtained from unnotched specimens; dashed lines represent data from specimens having notches with $K_t = 3.3$.

Source: Metals Handbook, 9th Edition, Volume 1, Properties and Selection: Irons and Steels, American Society for Metals, Metals Park OH, 1978, p 667

4-23. 4340 Steel: Effect of Decarburization

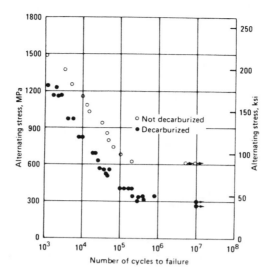

Decarburization is the removal of carbon from the surface of a steel part; as indicated in the above S-N curve, it significantly reduces the fatigue limits of steel. Decarburization of from 0.08 to 0.76 mm (0.003 to 0.030 in.) on AISI-SAE 4340 notched specimens that were heat treated to a strength level of 1860 MPa (270 ksi) reduces the fatigue limit almost as much as a notch with $K_t = 3$.

When subjected to the same heat treatment as the core of the part, the decarburized surface layer is weaker and therefore less resistant to fatigue than the core. Hardening a part with a decarburized surface can also introduce residual tensile stresses, which reduce the fatigue limit of the material. Results of research studies have indicated that fatigue properties lost through decarburization can be at least partially regained by recarburization (carbon restoration in the surfaces).

Source: Metals Handbook, 9th Edition, Volume 1, Properties and Selection: Irons and Steels, American Society for Metals, Metals Park OH, 1978, p 674

4-24. 4340H Steel: Effect of Inclusion Size

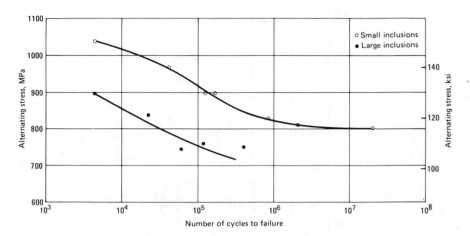

Fatigue life of two lots of AISI-SAE 4340H steel; one lot (lower curve) contained abnormally large inclusions; the other lot (upper curve) contained small inclusions.

Points on the lower curve represent the cycles to failure for a few specimens from one bar selected from a lot consisting of several bars of 4340H steel. Large spherical inclusions, about 0.13 mm (0.005 in.) in diameter, were observed in the fracture surfaces of these specimens; the inclusions were identified as corundum or silicate particles. No spherical inclusions larger than 0.02 mm (0.00075 in.) were detected in the other specimens.

Large nonmetallic inclusions can often be detected by nondestructive inspection; steels can be selected on the basis of such inspection. Vacuum melting, which reduces the number and size of nonmetallic inclusions, increases the fatigue limit of 4340.

Source: Metals Handbook, 9th Edition, Volume 1, Properties and Selection: Irons and Steels, American Society for Metals, Metals Park OH, 1978, p 673

4-25. 4340 Steel: Influence of Inclusion Size

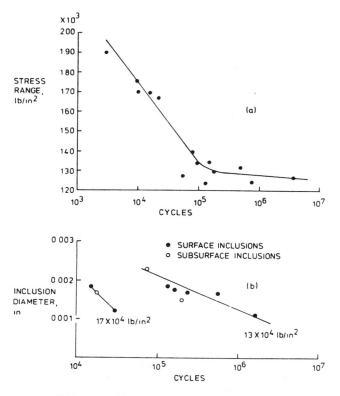

S-N curve and dependence of life on inclusion size for AISI 4340 steel.

Typical initiated crack sizes are 1-10 μm. As this is an order-of-magnitude greater than dislocation substructure sizes, such an initiated crack will behave as in a continuum. For materials with lower stacking fault energy cross-slip and PSB formation is difficult, thus inhibiting initiation. For such materials crack initiation can occupy a significant fraction of life. Other microstructural sites for initiation are discontinuities such as grain and twin boundaries, the latter being particularly operative in hcp metals. Usually, however, at ambient temperatures it is the dislocation substructures which dominate initiation.

For stronger, more complex alloys planar slip behavior dominates, making localized slip bands the initiation sites cf the random notch–peak topography generated by shearing a pack of cards. The interaction of slip bands with second-phase particles (inclusions, precipitates) can produce a local stress concentration which cracks the interface, producing a surface crack. The above S-N curves show the results of this process for a low alloy steel. Variations in fatigue life relate to variations in inclusion size. As well as debonding, oxide or carbide particles can crack under concentrated localized stresses.

Source: B. Tomkins, "Fatigue: Mechanisms," in Creep and Fatigue in High Temperature Alloys, J. Bressers, Ed., Applied Science Publishers Ltd, London, England, 1981, p 115

4-26. 4340 Steel: Effect of Hydrogenation; Static Fatigue

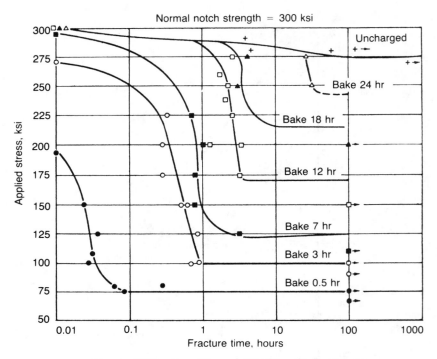

Static fatigue curves for quenched and tempered 4340 notched specimens charged with hydrogen and baked at 150 °C (300 °F) for the times shown.

There are many embrittling effects of hydrogen on steels: the ultimate strength of a steel may be reduced; ductility as measured by total elongation to fracture or reduction of area may be decreased; and crack growth may be significantly accelerated. The hydrogen responsible for these effects may be present in the environment external to the steel or may be present internally as a reslt of steelmaking or processing operations such as pickling or electroplating. Hydrogen may promote a transition from a ductile to brittle fracture mode or it may reduce ductility without a change in fracture mode.

The graph above shows the effects of baking at 150 °C (300 °F) on the static fatigue (sustained loading) of the hydrogen-charged specimens. Increasing baking time effectively lowers hydrogen content even in the plated specimens, and sufficient baking eventually restores the strength of charged specimens to that of uncharged specimens. The horizontal portions of the curves in the graph above are designated as static fatigue or endurance limits, i.e., the stress level below which failure would not occur no matter what the duration of stress application. As hydrogen content is decreased by baking, the static fatigue limit increases.

The specimens used to obtain the above data were notched and therefore the static fatigue limits hold for that particular notch geometry. In general, the sharper the notch, the lower the static fatigue limits, an indication that a critical combination of hydrogen concentration and triaxial stress state is required for crack initiation.

Source: George Krauss, Principles of Heat Treatment of Steel, American Society for Metals, Metals Park OH, 1980, p 223

4-27. 4340 Steel: Effect of Hydrogen

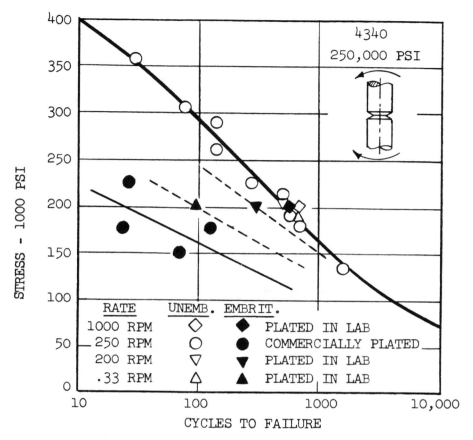

Schematic representation of the effect of cycling rate on the *S-N* curve of hydrogen-containing 4340 steel, heat treated to a strength level of 250,000 psi.

Source: George Sachs, "Test Methods for Evaluating Hydrogen Embrittlement," in Materials Evaluation in Relation to Component Behavior (Proceedings of the Third Sagamore Ordnance Materials Research Conference). Syracuse University, Syracuse NY, 1956, p 508

4-28. 4340 Steel: Effect of Nitriding

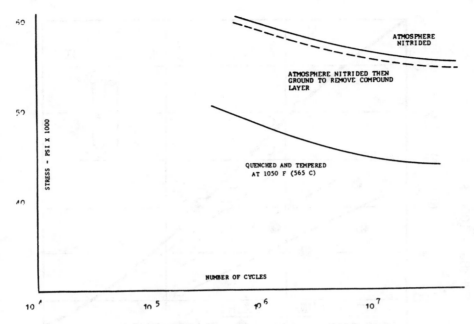

S-N curves for 4340 steel, gaseous atmosphere nitrided versus not nitrided (quenched and tempered only), showing stress versus number of cycles for completely reversing torsional fatigue.

Source: J. A. Riopelle, "Short Cycle Atmosphere Nitriding," in Source Book on Nitriding, American Society for Metals, Metals Park OH, 1977, p 287

4-29. 4340 Steel: Effect of Nitriding and Shot Peening

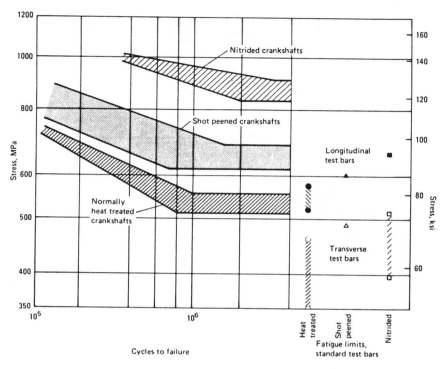

Comparison between fatigue limits of crankshafts (*S-N* bands) and fatigue limits for separate test bars, which are indicated by plotted points at right.

Mechanical working of the surface of a steel part effectively increases the resistance to fatigue. Shot peening and skin rolling are two methods for developing compressive residual stresses at the surface of the part. The improvement in fatigue life of a crankshaft that results from shot peening is illustrated in the above curves. Shot peening is useful in recovering the fatigue resistance lost through decarburization of the surface; decarburized specimens were shot peened, raising the fatigue limit from 275 MPa (40 ksi) after decarburizing to 655 MPa (95 ksi) after shot peening.

Source: Metals Handbook, 9th Edition, Volume 1, Properties and Selection: Irons and Steels, American Society for Metals, Metals Park OH, 1978, p 674

4-30. 4340 Steel: Effect of Induction Hardening and Nitriding

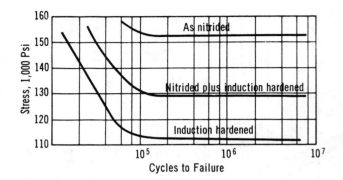

As demonstrated in the above *S-N* curves, fatigue tests of AISI 4340 steel in various surface hardened conditions show that combined treatments produce endurance limits between those developed by separate treatments.

Source: Sander A. Levy, Kenneth E. Barnes and Joseph F. Libsch, "Combining Nitriding With Induction Heating Pays a Bonus," in Source Book on Nitriding, American Society for Metals, Metals Park OH, 1977, p 241

4-31. 4340 Steel: Effect of Surface Coatings

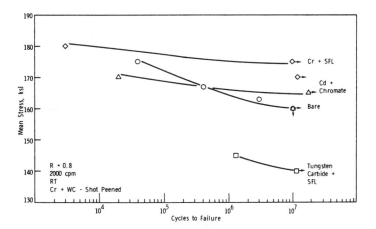

S-N curves (axial tension) of bare and coated 4340 steel in air environment.

Test	Condition	Air Stress ksi	Air Stress MN/m²	Change, %	3.5% NaCl Stress ksi	3.5% NaCl Stress MN/m²	Change, %
Rotating Bending R = -1	Bare	105	724	–	20	138	-81
	Cd + Chromate	105	724	0	80	552	-24
	Cr + Dry Film*	95	655	-9.5	90	621	-14.3
	WC + Dry Film*	90	621	-14.3	90	621	-14.3
	WC + Coricone + Dry Film*	90	621	-14.3	90	621	-14.3
Axial Tension R = 0.8	Bare	160	1103	–	110	758	-31.2
	Cd + Chromate	165	1138	+3.1	165	1138	0
	Cr + SFL*	175	1207	+9.4	90	621	-43.8† -48.6‡
	WC + SFL*	140	965	-12.5	60	414	-62.5† -57‡
	WC + Coricone + SFL*	140	965	-12.5	60	4.4	-62.5† -57‡

*Shot peened
†Compared to bare alloy air value
‡Compared to coated alloy air value

The above S-N curves, in conjunction with the table, contain the data obtained in air for 4340 steel, bare and coated. Fatigue data at 10^7 cycles showed that the cadmium and chromium electroplates, particularly the chromium, improved the fatigue strength. They were similar in both rotating bending and axial tension fatigue tests. But in NaCl solution, significantly greater reductions in axial fatigue strength of the coated alloys were observed due to environmental effects, which remains to be elucidated. Since the Cr and WC hard (brittle) coatings have a relatively low intrinsic fatigue strength in comparison with the steel, they will become discontinuous at a relatively low stress level owing to the development of fatigue cracks. (The Cr normally contains internal cracks.) These cracks will permit access of the corrosive NaCl solution to the steel base at the root of the fatigue crack. In the case of the axial tension test (high steady tensile load), it may be easier for the environment to reach the crack tip.

Source: M. Levy and C. E. Swindlehurst, Jr., "Corrosion Fatigue Behavior of Coated 4340 Steel for Blade Retention Bolts of the AH-1 Helicopter," in Risk and Failure Analysis for Improved Performance and Reliability, John J. Burke and Volker Weiss, Eds., Plenum Press, New York NY, 1980, p 275

4-32. 4340 Steel: Effect of Temperature on Constant-Lifetime Behavior

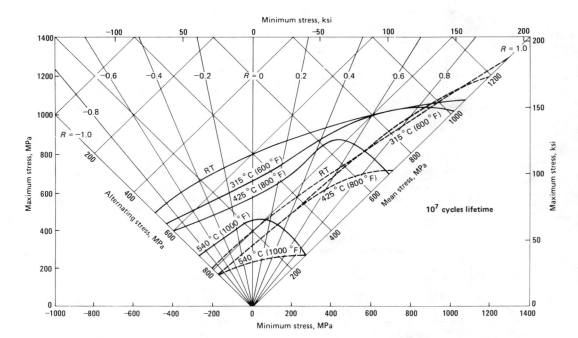

Constant-lifetime fatigue diagram for AISI-SAE 4340 alloy steel (bars) hardened and tempered to a tensile strength of 1035 MPa (150 ksi) and tested at the indicated temperatures. Solid lines represent data obtained from unnotched specimens; dashed lines represent data from specimens having notches with K_t = 3.3. All lines represent lifetimes of ten million cycles.

Source: Metals Handbook, 9th Edition, Volume 1, Properties and Selection: Irons and Steels, American Society for Metals, Metals Park OH, 1978, p 669

4-33. 4520H Steel: Effect of Type of Quench

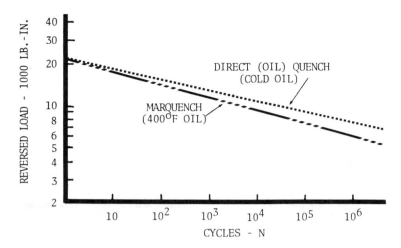

Effect of quench type on fatigue of carburized differential cross.

Since 4520H steel is relatively low in hardenability for the part involved, the depth-hardening characteristics of the two groups were significantly different. The marquenched group had shallower case depths, which resulted in fracture origins below the surface at the case-core junctures. However, when the comparison was made with higher-hardenability steels, with sufficient gradient strengths and thus all fracture origins at the surface, the difference was very slight, though still in favor of direct quenching. This is consistent with what is known concerning the differences in residual stress, which in this case would have been the only other contributing factor. In other instances—such as for gears, where distortion could be a factor—the results might turn out differently for the marquenching.

Source: D. H. Breen and E. M. Wene, "Fatigue in Machines and Structures—Ground Vehicles," in Fatigue and Microstructure, American Society for Metals, Metals Park OH, 1979, p 92

4-34. 4520H Steel: Effect of Shot Peening

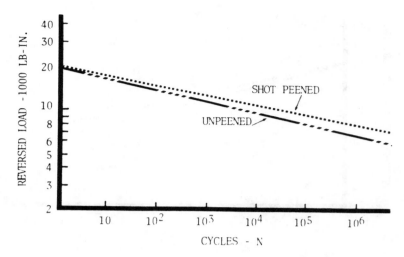

Effect of shot peening on fatigue of carburized differential cross.

Shot peening is known to increase fatigue strength; hence, tests were run to determine the amount of increase to be expected. The above chart shows some of the results. Shot peening was found to provide significant fatigue-strength improvements. Peening surfaces that had suffered grinding damage was found to be very beneficial, although not recommended because of the high risk of having grinding cracks. Peening parts that had marginal strength gradients improved the strength at the surface but moved the failure origin to a subsurface location. The net gain was small. It was also determined that to gain significant improvement the hardness of the shot used was very important. Since carburized surfaces are very hard, the shot must also be hard to be effective.

Source: D. H. Breen and E. M. Wene, "Fatigue in Machines and Structures—Ground Vehicles," in Fatigue and Microstructure, American Society for Metals, Metals Park OH, 1979, p 93

4-35. 4620 Steel: Effect of Nitriding

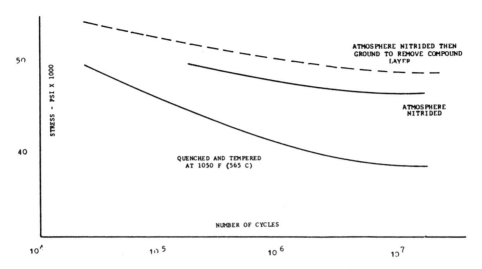

S-N curves for 4620 steel, nitrided versus not nitrided (quenched and tempered only), showing stress versus number of cycles for completely reversing torsional fatigue.

Source: J. A. Riopelle, "Short Cycle Atmosphere Nitriding," in Source Book on Nitriding, American Society for Metals, Metals Park OH, 1977, p 286

4-36. 4620 Steel: P/M-Forged

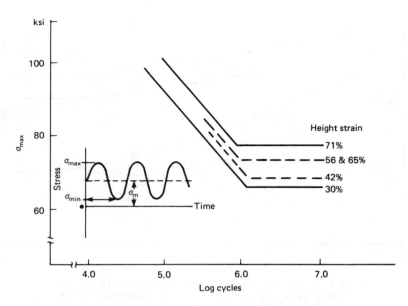

Axial fatigue of P/M-forged 4620 steel as a function of height strain during forging.

In general, sensitive properties improve as the level of upsetting is increased during the forging process. The diagram above shows the effect for fatigue resistance, although the cyclic stress state also influences fatigue behavior. An interesting feature of P/M-forged parts is the fact that deformation does not significantly affect through-thickness properties as it does detrimentally for wrought material. For re-pressed parts, through-thickness toughness is slightly lower than longitudinal toughness. Upsetting increases longitudinal toughness while toughness in the through-thickness direction remains at a relatively constant level.

Source: B. Lynn Ferguson, "Part II: Fully Dense Parts and Their Applications," in Powder Metallurgy: Applications, Advantages and Limitations, American Society for Metals, Erhard Klar, Ed., American Society for Metals, Metals Park OH, 1983, p 100

4-37. 4620 Steel: P/M-Forged at Different Levels

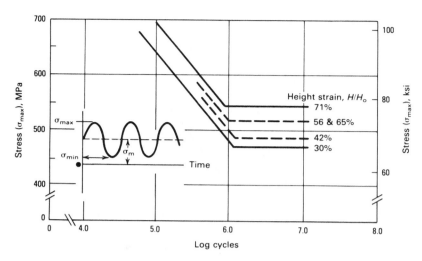

S-N curves for P/M-forged 4620 steel at various levels of forging deformation. As shown, fatigue limit increases as deformation (level of strain) increases.

Source: Metals Handbook, 9th Edition, Volume 7, Powder Metallurgy, American Society for Metals, Metals Park OH, 1984, p 416

4-38. 4625 Steel: P/M vs Ingot Forms

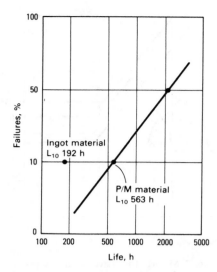

Fatigue life characteristics of P/M roller bearing cups, as shown by a typical Weibull plot. Shown is a 10% life (L_{10}) of 563 hr for P/M material compared with L_{10} life of 192 hr for ingot material.

Source: Metals Handbook, 9th Edition, Volume 7, Powder Metallurgy, American Society for Metals, Metals Park OH, 1984, p 620

4-39. 4640 Steel: P/M-Forged

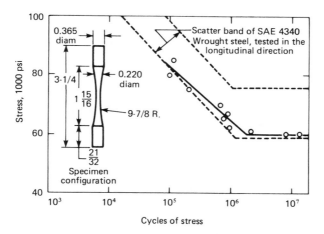

R. R. Moore fatigue curve for P/M-forged 4640 steel hardened and tempered to 33 HRC and a yield strength of 138,000 psi.

Water-atomized 4600 steel powder was blended with graphite and compacted in the split punch tooling. Fatigue data for P/M forged 4640 are shown above, and these data fall within the scatter band for 4340 steel. The most impressive statistic is that the P/M-forged parts passed the Army ambient and low-temperature firing endurance tests.

Source: B. Lynn Ferguson, "Part II: Fully Dense Parts and Their Applications," in Powder Metallurgy: Applications, Advantages and Limitations, American Society for Metals, Erhard Klar, Ed., American Society for Metals, Metals Park OH, 1983, p 103

4-40. High-Carbon Steel (Eutectoid Carbon): Pearlite vs Spheroidite

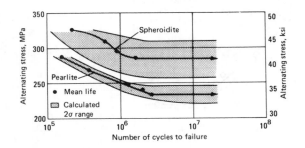

Property	Spheroidite	Pearlite
Tensile strength, MPa (ksi)	641(93)	676(98)
Yield strength, MPa (ksi)	490(71)(b)	248(36)(c)
Elongation in 2 in., %	28.9	17.8
Reduction in area, %	57.7	25.8
Hardness, HB	92	89

(a) Composition 0.78 C, 0.27 Mn, 0.22 Si, 0.016 S and 0.011 P. (b) Lower yield point. (c) 0.1% offset yield strength.

Both pearlitic and spheroidized structures have notably lower fatigue strength than martensitic structures (see 4-1, on p 83). As is shown above, the fatigue properties of spheroidized structures are superior to those of pearlitic structures for eutectoid steels.

Source: Metals Handbook, 9th Edition, Volume 1, Properties and Selection: Irons and Steels, American Society for Metals, Metals Park OH, 1978, p 677

4-41. 52100 EF Steel: Surface Fatigue; Effect of Finish and Additives

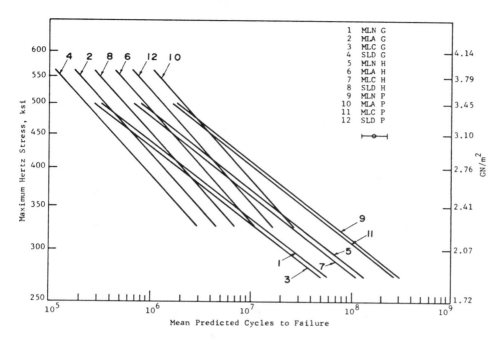

Effect of surface finish and additive on mean predicted surface fatigue life. 52100 EF steel, high slip, high speed.

The mean predicted fatigue life is highest with a polished surface and least with a ground finish (9 versus 1, etc.). Polished surface has about 6 times and honed surface about 3 times the fatigue life of ground finish. No interaction effect between additives and surface finish is revealed.

Source: S. Bhattacharyya, F. C. Bock, M. A. H. Howes and N. M. Parikh, "Chemical Effects of Lubrication in Contact Fatigue—Part II: The Statistical Analysis, Summary, and Conclusions," in Source Book on Gear Design, Technology and Performance, Maurice A. H. Howes, Ed., American Society for Metals, Metals Park OH, 1980, p 277

4-42. 52100 EF Steel: Surface Fatigue; Effect of Surface Finish and Speed

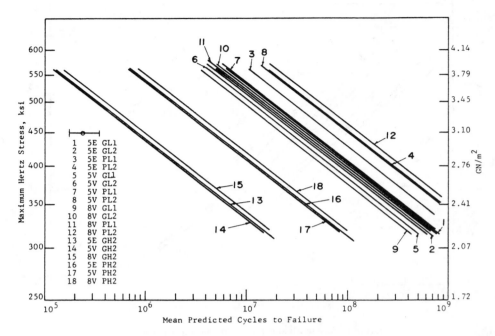

Effect of steel, surface finish and speed on mean predicted fatigue life. Low-viscosity mineral oil, no additive.

Interaction effects of steel with speed and of surface finish with slip and speed on fatigue life are shown in the above graph. The direct steel effects are nonsignificant. The effect of surface finish is shown in the difference between the two line groups 13, 14, 15 (ground) and 16, 17, 18 (polished). The difference in the line groups 4, 8, 12 (low slip) and 16, 17, 18 (high slip) again brings out the very large detrimental effect of high slip on life. Higher speed decreases life with the maximum effect observable on 8620 CV steel (compare lines 11 and 12), on polished specimens at low slip ratio.

Source: S. Bhattacharyya, F. C. Bock, M. A. H. Howes and N. M. Parikh, "Chemical Effects of Lubrication in Contact Fatigue—Part II: The Statistical Analysis, Summary, and Conclusions," in Source Book on Gear Design, Technology and Performance, Maurice A. H. Howes, Ed., American Society for Metals, Metals Park OH, 1980, p 277

4-43. 52100 EF Steel: Surface Fatigue; Effect of Lubricant Additives

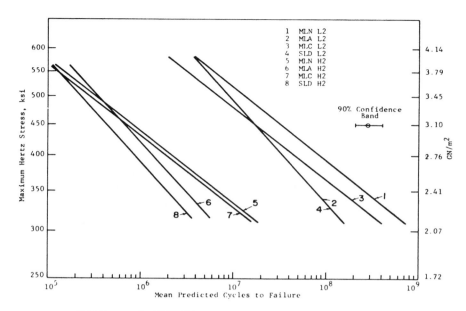

S-N diagram for 52100 EF steel showing the effect of lubricant additives on surface fatigue. The specimens had a ground finish, and a low-viscosity oil was used. Additives were used for 1, 2, 3 and 4; the favorable effect of the additives is obvious.

Source: S. Bhattacharyya, F. C. Bock, M. A. H. Howes and N. M. Parikh, "Chemical Effects of Lubrication in Contact Fatigue—Part II: The Statistical Analysis, Summary, and Conclusions," in Source Book on Gear Design, Technology and Performance, Maurice A. H. Howes, Ed., American Society for Metals, Metals Park OH, 1980, p 275

4-44. 52100 EF Steel: Surface Fatigue; Effect of Lubricant Viscosity, Slip Ratio and Speed

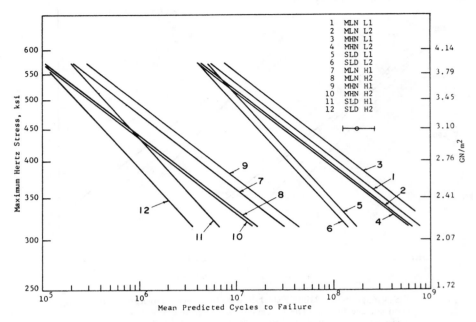

S-N diagram for 52100 EF steel.

The effects of lubricant viscosity, slip ratio, and speed on fatigue life are shown in the diagram. The 12 lines in the figure are separated in two distinct groups, low slip (lines 1 to 6) and high slip (lines 7 to 12). In each group the effects of viscosity and speed may be noted. Viscosity × speed interaction produces complex effects on mean predicted lives which under low slip conditions are not statistically significant in their differences. Only under high slip condition, lines 9 versus 10 indicate a small statistically significant lowering in mean fatigue life in high-viscosity oil under higher speed. A comparison of lines 11 and 12 shows that the lesser life in synthetic oil with additive is a statistically borderline case, though the trend is similar to that with mineral oil under the present operating conditions. The regression analysis shows that in the present tests both speed and viscosity have nonsignificant direct effect on life, and a few small interaction effects with steel, surface finish, viscosity, and slip were observed.

Source: S. Bhattacharyya, F. C. Bock, M. A. H. Howes and N. M. Parikh, "Chemical Effects of Lubrication in Contact Fatigue—Part II: The Statistical Analysis, Summary, and Conclusions," in Source Book on Gear Design, Technology and Performance, Maurice A. H. Howes, Ed., American Society for Metals, Metals Park OH, 1980, p 276

4-45. 52100 EF Steel: Rolling Ball Fatigue; Effect of Oil Additives

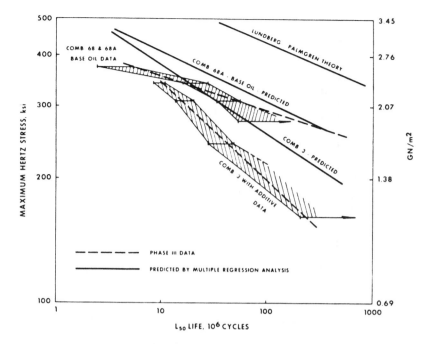

Comparison of stress/life data for the mineral oil with and without the ZnDDTP additive in surface fatigue; 85 percent confidence bands for the L_{50} life estimates are shown and compared with the stress/life relation predicted from regression analysis.

The synthetic and mineral oil no-additive conditions had about the same life. However, the life at all stress levels tested was significantly reduced for the mineral oil with additive below that without additive, by almost a factor of three at the L_{50} level, further indicating a detrimental effect of the ZnDDTP additive on life.

Both the synthetic and the mineral oil tests had lives almost two orders of magnitude below the standard Lundberg-Palmgren calculated life. A life reduction factor is used with the Lundberg-Palmgren theory when applied to rolling bearings having high contact angles and thus high slip; but rarely does the slip at bearing contacts approach that level used in these tests, so it is not surprising that the life reductions observed are much greater than the life reduction factors normally used for bearings.

The stress/life plot shown above is particularly revealing. There is no doubt that the stress/life slope for the additive oil is significantly steeper than for the base stock, which seems to approach the Lundberg-Palmgren theory in stress/life slope except for the highest stresses where it is even shallower.

Source: W. E. Littmann, B. W. Kelley, W. J. Anderson, R. S. Fein, E. E. Klaus, L. B. Sibley and W. O. Winer, "Chemical Effects of Lubrication in Contact Fatigue—Part III: Load-Life Exponent, Life Scatter, and Overall Analysis," in Source Book on Gear Design, Technology and Performance, Maurice A. H. Howes, Ed., American Society for Metals, Metals Park OH, 1980, p 285

4-46. 52100 Steel: Carburized vs Uncarburized

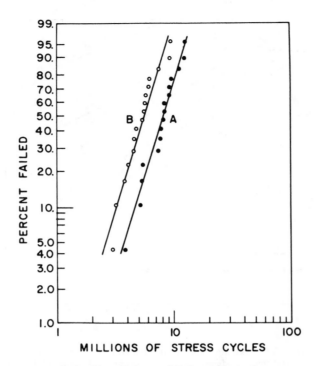

Rolling Contact Fatigue Tests

Bar specimens, 0.973 cm (0.383 in.) in diameter, about 8 cm long, were machined from spheroidize-annealed 52100 steel. Two pieces were copper plated to prevent carburizing, then, along with two unplated pieces, were austenitized at 815 °C (1500 °F) for two h in a carburizing atmosphere, oil-quenched and tempered for 1.5 h at 175 °C (350 °F). After finish grinding to 0.953 mm (0.375 in.), pieces were fatigue tested using a Polymet Model RCF-1 testing machine with a computed maximum hertzian contact stress of 503 MPa (729 ksi).

A Weibull plot, shown above, of the 16 tests on each type of specimen shows that pieces with a carburized surface had a fatigue life about 50% longer at all failure rate levels than pieces which were subjected to the same thermal cycle, but not carburized. The nonparametric Walsh test for statistical significance indicated at a 99.5% level of confidence, the two batches of fatigue test data came from different populations.

Source: C. A. Stickels and A. M. Janotik, "Controlling Residual Stresses in 52100 Bearing Steel by Heat Treatment," in Residual Stress for Designers and Metallurgists, Larry J. Vander Walle, Ed., American Society for Metals, 1981, p 34

4-47. 8620H Steel: Carburized; Results From Case and Core

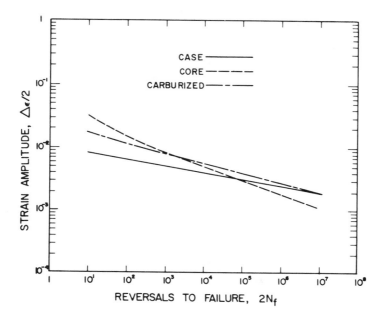

Summary plots of total strain amplitude: reversals-to-failure data for simulated case, simulated core, and carburized materials.

Carburized material is seen to have low-cycle fatigue resistance intermediate between the simulated case and core material, a common intersection with simulated core material at intermediate lives; and in the long-life regime, carburized material specimens are more fatigue resistant than either simulated core or case material specimens.

Plotting the strain-life curves for both case and core simulated materials on a common set of axes, as shown in the above chart, reveals an interesting feature. It has been observed that curves of these materials intersected at a life of approximately $2 \cdot N_f = 10^5$ reversals. This is in agreement with the results of this investigation. Intersection of the life curves for simulated case and core materials accounts for a shift of failure location in carburized components.

Source: J. M. Waraniak and D. F. Socie, "Cyclic Deformation and Fatigue Behavior of Carburized Steel," in Wear and Fracture Prevention, American Society for Metals, Metals Park OH, 1981, p 249

4-48. 8620H Steel: Effect of Variation in Carburizing Treatments

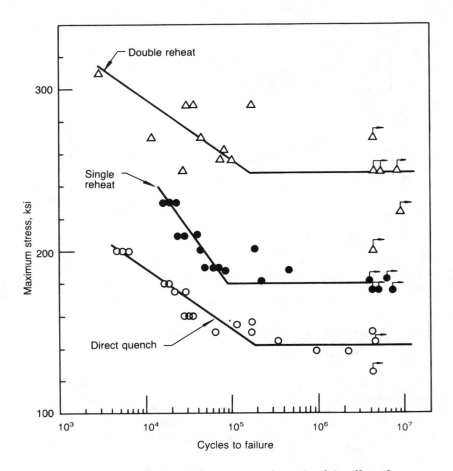

The above S-N curves show results of a study of the effect of martensite morphology, including the effects of microcracking on fatigue resistance of a carburized 8620 steel. These specimens, which were directly quenched from the carburizing temperature, had the coarsest structure and the highest density of microcracks, some of which were directly exposed on the specimen surfaces by chemical polishing. The single-reheat specimens had a finer austenite grain structure and therefore finer martensite plates and a lower density of microcracks. Since the retained austenite content and hardness profiles of the direct and single-reheat specimens were identical, the improved fatigue resistance of the single-reheat specimens is attributed to the smaller size of the microcracks and their lower density in the finer structure. The best fatigue resistance was shown by the double-reheat specimens.

Source: George Krauss, Principles of Heat Treatment of Steel, American Society for Metals, Metals Park OH, 1980, p 264

4-49. 8620 Steel: Effect of Nitriding

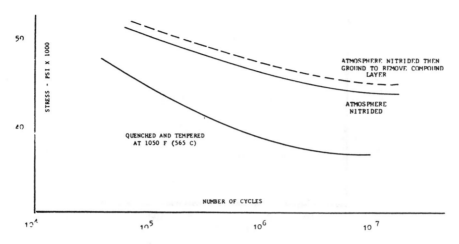

S-N curves for 8620 steel; nitrided versus not nitrided (quenched and tempered only), showing stress versus number of cycles for completely reversing torsional fatigue.

Source: J. A. Riopelle, "Short Cycle Atmosphere Nitriding," in Source Book on Nitriding, American Society for Metals, 1977, p 286

4-50. 8622 Steel: Effect of Grinding

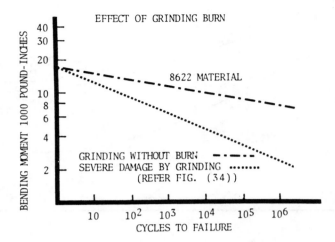

Influence of grinding quality on fatigue properties of carburized differential cross.

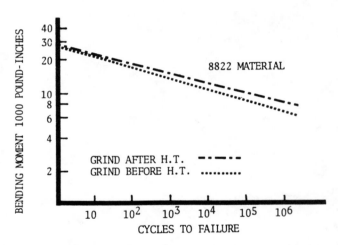

Influence of grinding sequence on fatigue of carburized differential cross.

As shown in the above charts, grinding has an important influence on fatigue. Elimination of grinding damage resulted in drastic improvement in fatigue performance (upper chart). However, it was also determined that a high-quality ground part gave better fatigue performance than when the carburized surface was left unground (lower chart).

Source: D. H. Breen and E. M. Wene, "Fatigue in Machines and Structures—Ground Vehicles," in Fatigue and Microstructure, American Society for Metals, Metals Park OH, 1979, pp 91-92

4-51. Cast 8630 Steel: Goodman Diagram for Bending Fatigue

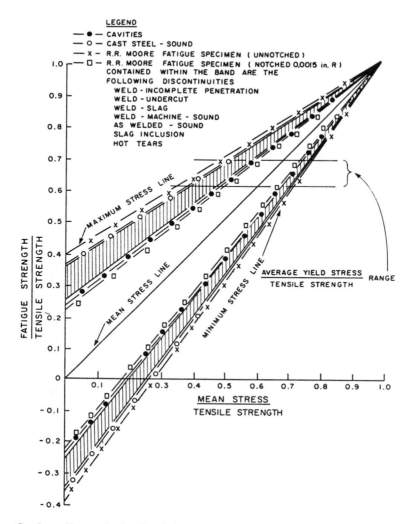

Goodman diagram for bending fatigue for normalized and tempered cast 8630 steel.

Data here show that severe discontinuities lower the fatigue strength of cast steel. However, it will be observed from the Goodman diagram above that the results of the notched [0.0015 in. (0.0381 mm) radius] R. R. Moore fatigue specimen fall below those of the other bending fatigue values. Goodman diagrams for torsion fatigue and for a quenched and tempered heat treatment show similar conditions with the notched fatigue values below the surface discontinuity values. In many cases, therefore, design, based upon notched R. R. Moore fatigue data, introduces a safety factor.

It must be remembered that the discontinuities were very severe and exceeded all ASTM classes of nondestructive inspection standards. The allowable discontinuities described in the ASTM standards are therefore expected to exert a somewhat less damaging effect on fatigue behavior.

Source: Steel Castings Handbook, 5th Edition, Peter F. Weiser, Ed., Steel Founders' Society of America, Rocky River OH, 1980, p 15-32

4-52. Cast 8630 Steel: Effect of Shrinkage

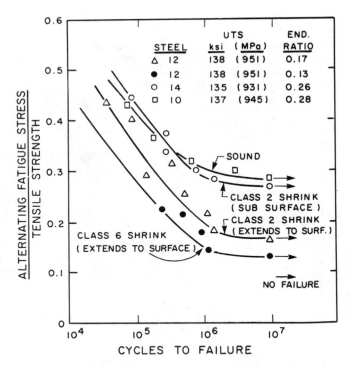

Effect of shrinkage on plate bending fatigue of quenched and tempered cast 8630 Ni-Cr-Mo steel.

As shown in the chart here, plate bending tests (completely reversed tension and compression with cast-to-size specimens) of low-alloy cast 8630 steel indicate only minor effects of Class 2 internal shrinkage.

Source: Steel Castings Handbook, 5th Edition, Peter F. Weiser, Ed., Steel Founders' Society of America, Rocky River OH, 1980, p 15-30

4-53. Cast 8630 Steel: Effect of Shrinkage on Torsion Fatigue

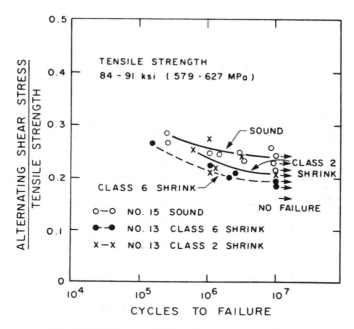

Effect of shrinkage on torsion fatigue properties of annealed cast 8630 steel.

Source: Steel Castings Handbook, 5th Edition, Peter F. Weiser, Ed., Steel Founders' Society of America, Rocky River OH, 1980, p 15-31

4-54. Cast 8630 Steel: Effect of Shrinkage on Torsion Fatigue

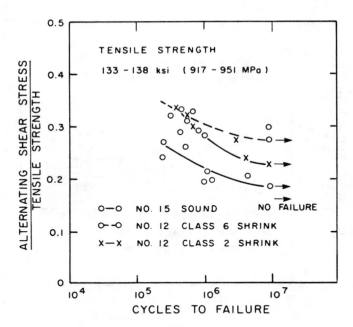

Effect of shrinkage on torsion fatigue properties of water quenched and tempered cast 8630 steel.

4-55. Cast 8630 Steel: Effect of Shrinkage on Plate Bending

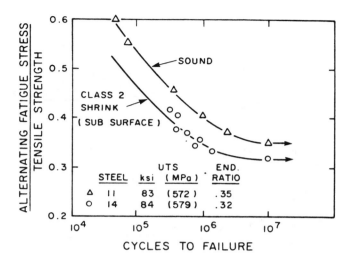

Effect of shrinkage on plate bending fatigue of normalized and tempered cast 8630 Ni-Cr-Mo steel.

Source: Steel Castings Handbook, 5th Edition, Peter F. Weiser, Ed., Steel Founders' Society of America, Rocky River OH, 1980, p 15-31

4-56. Cast 8630 vs Wrought 8640

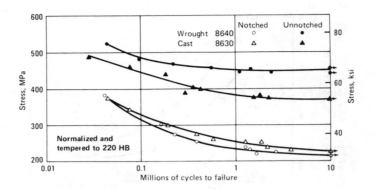

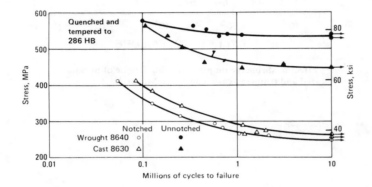

The fatigue limit for smooth-machined specimens is generally about one half the tensile strength, but is reduced considerably by notches or a rough cast surface. The S-N curves in the graphs above compare wrought 8640 and cast 8630 steel in two different conditions of heat treatment. In both of these comparisons, the wrought 8640 is superior, but the two steels are practically identical in the notched fatigue test. This is significant because most articles fabricated from either wrought or cast steel contain more than one notch and more than one type of notch.

Source: Metals Handbook, 9th Edition, Volume 1, Properties and Selection: Irons and Steels, American Society for Metals, Metals Park OH, 1978, p 397

4-57. 8630 and 8640 Steels: Effect of Notches on Cast and Wrought Specimens

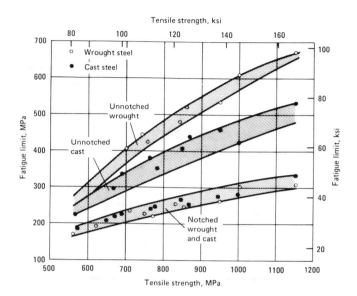

The effect of notches on fatigue limit is apparent when comparing similar wrought and cast steels with regard to fatigue limit at selected static tensile strength levels; note curves above.

Source: Metals Handbook, 9th Edition, Volume 1, Properties and Selection: Irons and Steels, American Society for Metals, Metals Park OH, 1978, p 397

4-58. Nitralloy 135 Steel: Effect of Nitriding

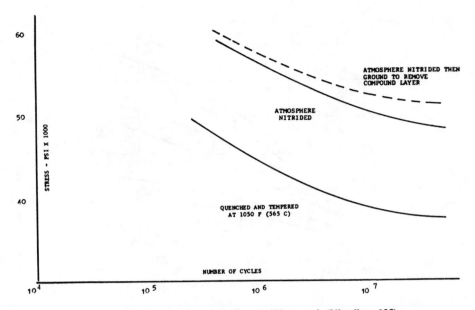

S-N curves for aluminum-bearing nitriding steel (Nitralloy 135), gaseous atmosphere nitrided versus not nitrided (quenched and tempered only), showing stress versus number of cycles for completely reversing torsional fatigue.

Source: J. A. Riopelle, "Short Cycle Atmosphere Nitriding," in Source Book on Nitriding, American Society for Metals, 1977, p 287

4-59. AMS 6475: Effects of Welding

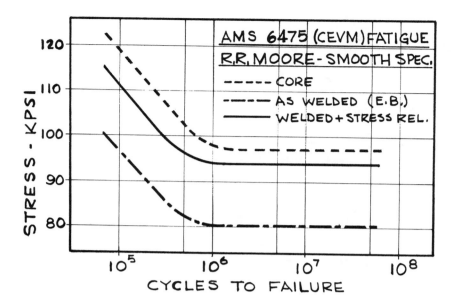

Fatigue strengths for case-hardened materials as well as through-hardened may be satisfactorily defined using the R. R. Moore rotating specimen test. The smooth unnotched Moore specimen is ideally suited for studying many of the effects of manufacturing and processing variables upon fatigue endurance. An example of the use of this testing technique in the evaluation of electron beam welding and postwelding aging effects upon the endurance limit of basic AMS 6475 material is shown in the above S-N diagram.

Source: Charles W. Bowen, "Review of Gear Testing Methods," in Source Book on Gear Design, Technology and Performance, Maurice A. H. Howes, Ed., American Society for Metals, Metals Park OH, 1980, p 346

4-60. Medium-Carbon, 1Cr-Mo-V Forging: Effect of Cycling Frequency

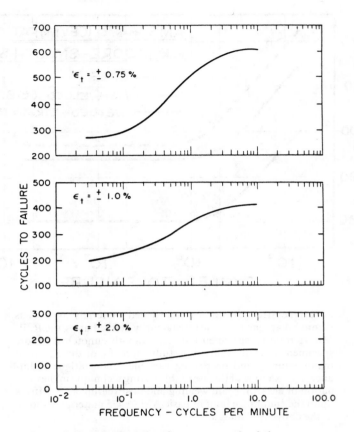

Influence of cycling frequency on the fatigue properties of forged 1Cr-Mo-V steel at 1049 °F (566 °C); no dwell period.

4-61. EM12 Steel: Effect of Temperature on Low-Cycle Fatigue

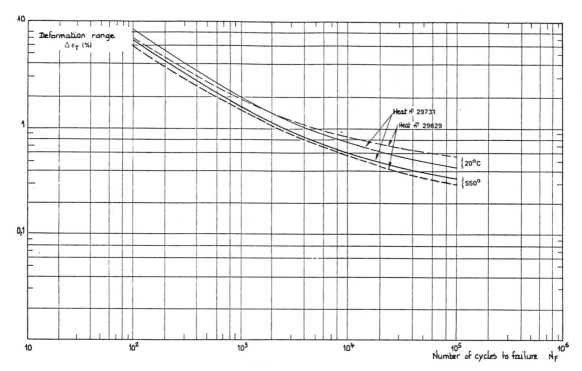

Low-cycle fatigue of EM12 at 20 and 550 °C (68 and 1020 °F).

As holds true for other ferritic steels, the effect of hold time in compression is slightly detrimental to fatigue life.

Source: Philippe Berge, Jean-Roger Donati, Félix Pellicani and Michel Weisz, "Properties of EM 12," in Ferritic Steels for High-Temperature Applications, Ashok K. Khare, Ed., American Society for Metals, 1983, p 114

4-62. Cast 0.5Cr-Mo-V Steel: Effects of Dwell Time in Elevated-Temperature Testing

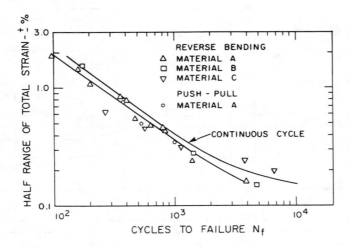

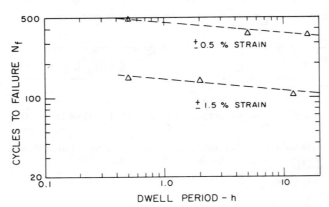

Effect of dwell periods on fatigue characteristics of low-alloy cast steel.

As the upper diagram shows, when a 0.5Cr-Mo-V steel was tested at 1022 °F (550 °C), a 20% drop in fatigue life in reverse bending resulted when a 0.5-hour dwell was added to each cycle. The lower diagram shows that extended dwell periods, up to 10 hours, have relatively little additional effect beyond that induced by the 0.5-hour dwell.

Source: Steel Castings Handbook, 5th Edition, Peter F. Weiser, Ed., Steel Founders' Society of America, Rocky River OH, 1980, pp 15-56 - 15-57

4-63. Cast 0.5Cr-Mo-V Steel: Effect of Environment at 550 °C (1022 °F)

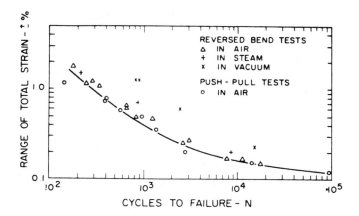

Fatigue endurance behavior of cast 0.5Cr-Mo-V steel at 1022 °F (550 °C) in air, steam, and vacuum (no dwell period).

4-64. Cast C-0.5Mo Steel: Effect of Temperature and Dwell Period on Cyclic Endurance at Various Strain Amplitudes

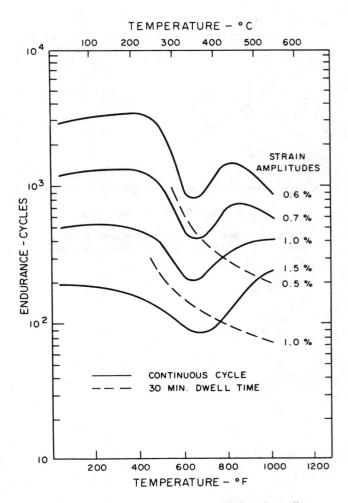

Influence of temperature and dwell period on the cyclic endurance of C-0.5Mo steel at various strain amplitudes.

Source: Steel Castings Handbook, 5th Edition, Peter F. Weiser, Ed., Steel Founders' Society of America, Rocky River OH, 1980 p 15-55

5-1. HI-FORM 50 Steel vs 1006

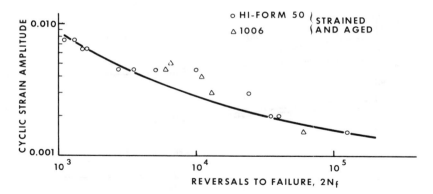

Strain-life data for AISI 1006 and HI-FORM 50 (a columbium-bearing HSLA steel) in the strained-and-aged condition.

Source: N. Lazaridis and S. P. Bhat, "Fatigue Behavior of Cold Rolled Dual Phase Steels," in Wear and Fracture Prevention, American Society for Metals, Metals Park OH, 1981, p 214

5-2. HI-FORM 50 Steel vs 1006: Stress Response

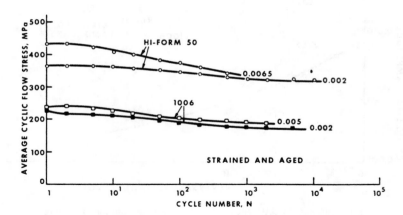

Stress response of strained-and-aged AISI 1006 and HI-FORM 50 steels.

The imposed constant total strain amplitudes are indicated on the graph. The degree of softening of these two steels is less compared with that of dual-phase steels, which simply reflects the significantly lesser degree of strain hardening of the 1006 and HI-FORM 50 compared with the dual-phase steels.

Source: N. Lazaridis and S. P. Bhat, "Fatigue Behavior of Cold Rolled Dual Phase Steels," in Wear and Fracture Prevention, American Society for Metals, Metals Park OH, 1981, p 209

5-3. HI-FORM 50 Steel Compared With 1006, DP1 and DP2

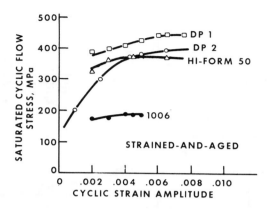

Comparison of four steels: AISI 1006, HI-FORM 50 (a columbium-bearing HSLA steel), a lean phosphorus-bearing dual-phase HSLA steel (DP1), and a carbon-manganese dual-phase HSLA steel (DP2).

Here it can be seen that all three high-strength steels offer substantial increase in load carrying capacity at the same gauge when compared to the plain low-carbon steel. This confirms the potential for gauge, and consequently weight, reduction that can be realized from the use of higher-strength steels.

Source: N. Lazaridis and S. P. Bhat, "Fatigue Behavior of Cold Rolled Dual Phase Steels," in Wear and Fracture Prevention, American Society for Metals, Metals Park OH, 1981, p 212

5-4. HSLA vs Mild Steel: Torsional Fatigue

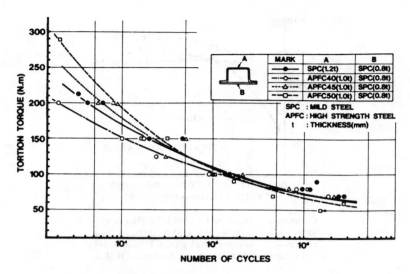

S-N curves showing torsional fatigue of automobile frame steels.

To determine whether the foregoing basic test results apply to the frame models, experiments were conducted. The above chart presents the torsional fatigue behavior of the frame models fabricated with the mild steel (0.8 mm) and each of the three high-strength steels. In the high-stress, low-cycle range, fatigue strength differs with the class of high-strength steel but virtually no differences of that nature are seen in the low-stress, high-cycle range. The three high-strength steel combinations showed virtually the same torsional fatigue strength values as those of the mild steel (1.2 mm) combination, indicating the possibility of gauge reduction.

Source: M. Takahashi, "Criteria of High Strength Steels for Applying to Automobile Frame Components," in HSLA Steels—Technology & Applications, American Society for Metals, Metals Park OH, 1984, p 498

5-5. Proprietary HSLA Steel vs ASTM A440

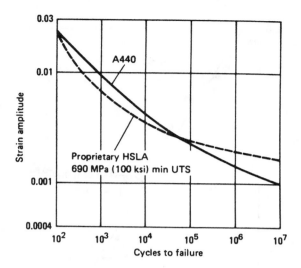

Total strain versus fatigue life for ASTM A440 having a yield strength of about 345 MPa (50 ksi) and for a proprietary quenched and tempered HSLA steel having a yield strength of about 750 MPa (110 ksi).

Source: Metals Handbook, 9th Edition, Volume 1, Properties and Selection: Irons and Steels, American Society for Metals, Metals Park OH, 1978, p 672

5-6. Comparison of HSLA Steel Grades BE, JF and KF for Plastic Strain Amplitude vs Reversals to Failure

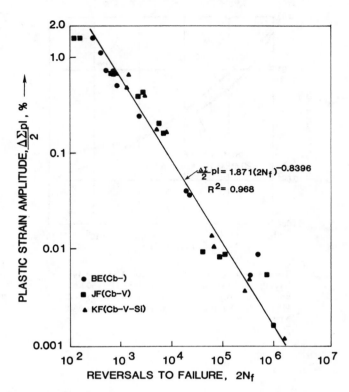

Plastic strain amplitude vs reversals to failure for Cb (BE), Cb-V (JF) and Cb-V-Si (KF) steels.

For plastic strain-life relationship the statistical analysis indicates that there are no significant differences between the three steels (F-ratio is not significant). This is further illustrated in the above chart, where all the plastic strain data are plotted as a function of reversals to failure. It is clear that a single straight line can adequately describe all the data. Such a regression line is drawn as the solid line in this chart.

Source: Shrikant P. Bhat, "Influence of Composition Within a Grade on the Fatigue Properties of HSLA Steels," in HSLA Steels—Technology & Applications, American Society for Metals, Metals Park OH, 1984, p 588

5-7. Comparison of HSLA Steel Grades BE, JF and KF for Total Strain Amplitude vs Reversals to Failure

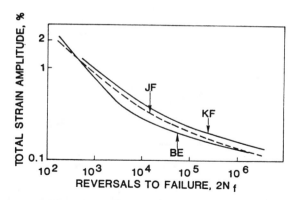

Total strain amplitude vs reversals to failure for Cb (BE), Cb-V (JF) and Cb-V-Si (KF) steels.

Strain-life behavior: The strain-life curves for the three steels are compared in this graph. It is clear that when plotted as total strain versus reversals to failure, the three steels behave similarly and the differences between them are minor.

Source: Shrikant P. Bhat, "Influence of Composition Within a Grade on the Fatigue Properties of HSLA Steel," in HSLA Steels—Technology & Applications, American Society for Metals, Metals Park OH, 1984, p 587

5-8. Comparison of a Dual-Phase HSLA Steel Grade With HI-FORM 50: Total Strain Amplitude vs Reversals to Failure

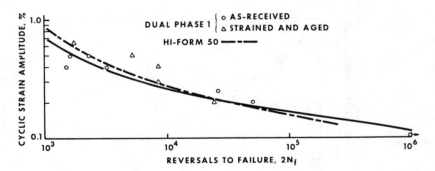

Total strain amplitude versus life data for DP1 (a lean-phosphorus HSLA steel) in the as-received and strained-and-aged conditions. Data for HI-FORM 50 (a columbium-bearing HSLA steel) are included for comparison.

Source: N. Lazaridis and S. P. Bhat, "Fatigue Behavior of Cold Rolled Dual Phase Steels," in Wear and Fracture Prevention, American Society for Metals, Metals Park OH, 1981, p 213

5-9. AISI 50 XF Steel: Effects of Cold Deformation

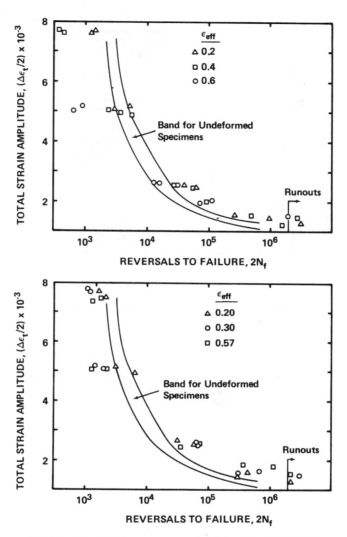

Total strain amplitude versus reversals to failure for AISI 50 XF HSLA steel. Upper chart: after balanced biaxial stretching; lower chart: after cold rolling.

Although the effects of prior deformation by BBS or CR on the strain-life behavior of 50 XF were generally similar to those in 1006, some specific differences were apparent; for example, the effect of prior deformation was stronger for 50 XF than for 1006 in that both the decrease in life at large strain amplitudes and the increase in life at small strain amplitudes were greater in 50 XF than in 1006.

Source: John M. Holt and Philippe L. Charpentier, "Effect of Cold Forming on the Strain-Controlled Fatigue Properties of HSLA Steel Sheets," in HSLA Steels—Technology & Applications, American Society for Metals, Metals Park OH, 1984, p 218

5-10. AISI 80 DF Steel: Effects of Cold Deformation

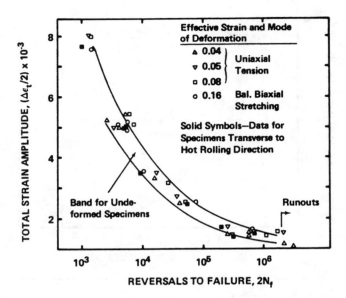

Strain-life curves after deformation for AISI 80 DF HSLA steel.

In this steel, the fatigue life appeared to remain unchanged or to increase very slightly as a result of deformation, at least for the effective strain levels investigated (see graph). Also, the fatigue life appeared to be unaffected by the mode of deformation and the specimen orientation.

Source: John M. Holt and Philippe L. Charpentier, "Effect of Cold Forming on the Strain-Controlled Fatigue Properties of HSLA Steel Sheets," in HSLA Steels—Technology & Applications, American Society for Metals, Metals Park OH, 1984, p 218

5-11. Comparison of Three HSLA Steel Grades, Cb, Cb-V and Cb-V-Si: Strain Life From Constant Amplitude

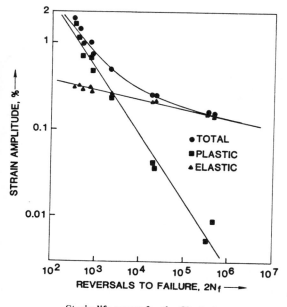

Strain-life curves for the Cb steel.

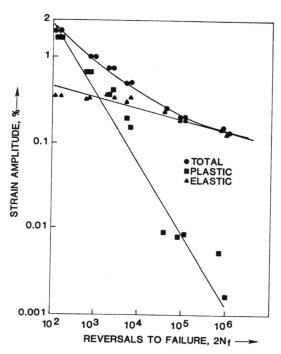

Strain-life curves for the Cb-V steel.

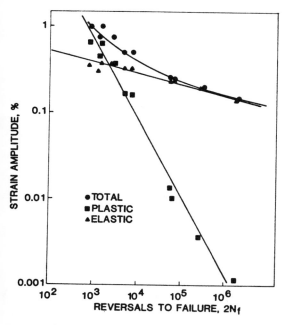

Strain-life curves for the Cb-V-Si steel.

Strain-life curves: Strain-life data from constant-amplitude tests for the three steels are plotted in the three charts here respectively in the form of total strain amplitude versus the number of complete reversals to failure.

Source: Shrikant P. Bhat, "Influence of Composition Within a Grade on the Fatigue Properties of HSLA Steels," in HSLA Steels—Technology & Applications, American Society for Metals, Metals Park OH, 1984, p 583

5-12. Comparison of Stress Responses: DP1 vs DP2 Dual-Phase HSLA Steels

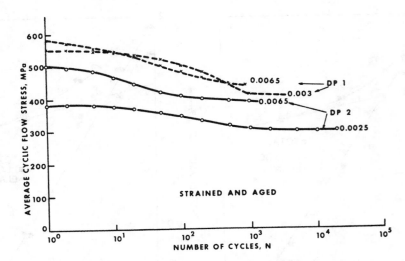

Comparison of stress response of strained-and-aged DP1 (a lean phosphorus HSLA steel) with that of DP2 (a carbon-manganese HSLA steel) for the total strain amplitudes indicated.

Source: N. Lazaridis and S. P. Bhat, "Fatigue Behavior of Cold Rolled Dual Phase Steels," in Wear and Fracture Prevention, American Society for Metals, Metals Park OH, 1981, p 209

5-13. Dual-Phase HSLA Steel Grade: Stress Response for As-Received vs Water-Quenched

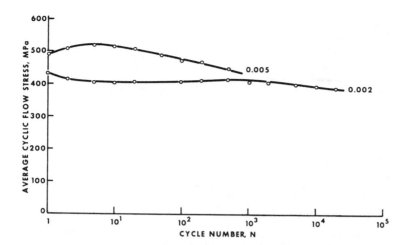

Stress response of a water-quenched dual-phase steel in the as-received condition for total strain amplitudes of 0.002 and 0.005.

Source: N. Lazaridis and S. P. Bhat, "Fatigue Behavior of Cold Rolled Dual Phase Steels," in Wear and Fracture Prevention, American Society for Metals, Metals Park OH, 1981, p 208

5-14. Dual-Phase HSLA Steel Grade: Stress Response for As-Received vs Gas-Jet-Cooled

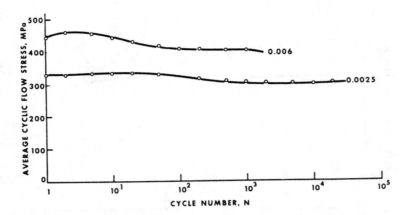

Stress response of the gas-jet-cooled dual-phase steel in the as-received condition for total strain amplitudes of 0.0025 and 0.006.

Source: N. Lazaridis and S. P. Bhat, "Fatigue Behavior of Cold Rolled Dual Phase Steels," in Wear and Fracture Prevention, American Society for Metals, Metals Park OH, 1981, p 208

5-15. S-N Comparison of Dual-Phase HSLA Steel Grades DP1 and DP2 With 1006

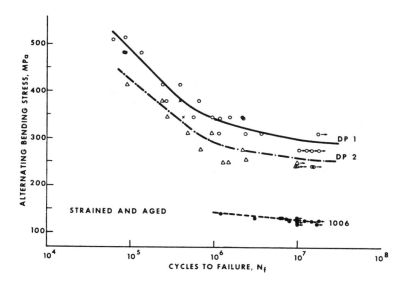

S-N curves for strained-and-aged DP1 (a lean-phosphorus HSLA steel), DP2 (a carbon-manganese HSLA steel), and AISI 1006. All tests were conducted in fully reversed bending.

Source: N. Lazaridis and S. P. Bhat, "Fatigue Behavior of Cold Rolled Dual Phase Steels," in Wear and Fracture Prevention, American Society for Metals, Metals Park OH, 1981, p 215

5-16. Comparison of Dual-Phase HSLA Steel DP2 With HI-FORM 50

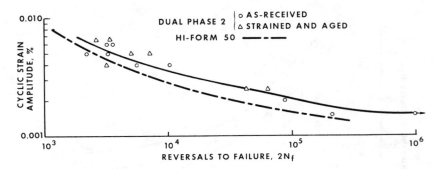

Strain-life curves for DP2 (a carbon-manganese HSLA steel) in two conditions compared with HI-FORM 50 (a columbium-bearing HSLA steel).

Source: N. Lazaridis and S. P. Bhat, "Fatigue Behavior of Cold Rolled Dual Phase Steels," in Wear and Fracture Prevention, American Society for Metals, Metals Park OH, 1981, p 214

5-17. Comparison of Cyclic Strain Response Curves for Cb, Cb-V, and Cb-V-Si Grades of HSLA Steel

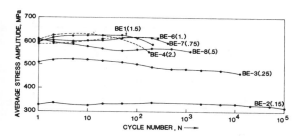

The cyclic response curves for the Cb steel (BE) at several constant strain amplitudes. The numbers in parentheses indicate the total strain amplitudes.

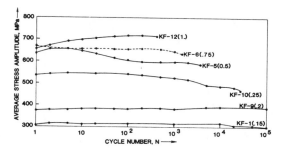

The cyclic response curves for the Cb-V-Si steel (KF) at several constant strain amplitudes. The numbers in parentheses indicate the total strain amplitudes.

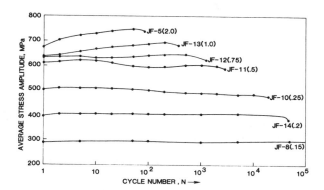

The cyclic response curves for the Cb-V steel (JF) at several constant strain amplitudes. The numbers in parentheses indicate the total strain amplitudes.

In a constant strain amplitude test, cyclic stability of a steel may be represented as a plot of the stress needed to impose the strain limits versus the number of cycles elapsed throughout the life. Cyclic response curves for the three steels are shown in the above charts. Here resistance to strain cycling is represented as the averages of the tensile and compressive maximum flow stress values (without regard to sign) plotted as a function of cycles. It is clear that the cyclic modification in stress response is a continuous process throughout fatigue life at most strain amplitudes. In a qualitative sense, the behavior of the three steels can be classed into three groups: (i) Low strain amplitude regime (up to 0.2% total strain); this region is characterized by very small or no cyclic hardening and the steels exhibit a well-defined saturation stage. (ii) Intermediate strain amplitude region ($0.2\% < \frac{\Delta\epsilon}{2}T < 0.75\%$); this region shows mixed behavior. After a brief hardening (lasting less than 0.1% of life in cycles), the specimen exhibits gradual softening and only a weak indication of saturation can be detected late in life. (iii) High strain amplitude region ($\frac{\Delta\epsilon}{2}T > 0.75\%$); typically the specimen in this region shows substantial hardening and again a well-defined saturation is not observed.

Source: Shrikant P. Bhat, "Influence of Composition Within a Grade on the Fatigue Properties of HSLA Steels," in HSLA Steels—Technology & Applications, American Society for Metals, Metals Park OH, 1984, p 581

5-18. Fatigue Crack Propagation Rate: Effect of Temperature for Two HSLA Steel Grades

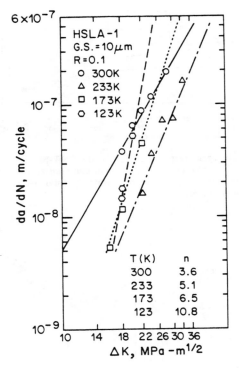
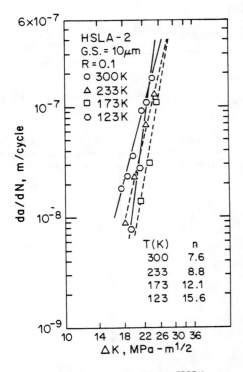

The effect of test temperature on the fatigue crack propagation rates in the Paris law regime for two HSLA steels in the as-received condition.

The only significant difference between HSLA-1 and HSLA-2 is that HSLA-2 contains double the amount of Nb that HSLA-1 contains (see compositions on p 165).

The effect of temperature is seen to decrease the crack propagation rate with decreasing temperature at low values of ΔK. However, as the stress intensity increases, a crossover occurs wherein higher growth rates were observed, as shown in the above charts. This crossover is further reflected in the increase in the Paris law exponent, n, where it ranged from 3.6 at room temperature to 10.8 at 123K for HSLA-1. The large increase is a result of the change in the fracture mechanism from ductile transgranular fracture to cleavage. This behavior has also been seen in iron binary alloys where n increased from 3.5 at room temperature to 20.9 at 123K.

Source: Khlefa A. Esaklul, William W. Gerberich and James P. Lucas, "Near-Threshold Behavior of HSLA Steels," in HSLA Steels—Technology & Applications, American Society for Metals, Metals Park OH, 1984, p 569

5-19. Effect of *R*-Ratio and Test Temperature on Crack Propagation of HSLA Steel Grade 1

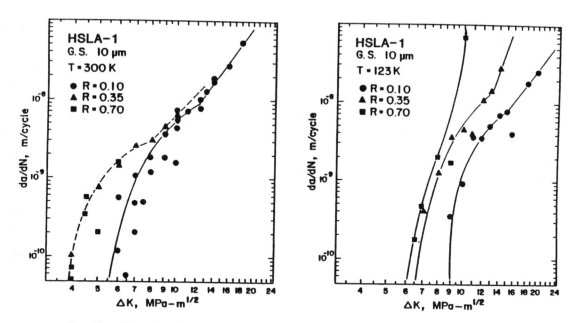

The effect of *R*-ratio on fatigue crack propagation behavior of HSLA-1 at test temperature of 300 and 123K in the as-received condition.

Compositions of HSLA-1 and HSLA-2

Alloy	C	Mn	Nb	SI	P	S	Al	Ni	Cr	Fe
HSLA-1	0.07	0.51	0.014	0.03	<0.005	0.005	—	0.01	0.01	Rem
HSLA-2	0.06	0.35	0.03	0.03	0.01	0.01	0.01	—	—	Rem

Source: Khlefa A. Esaklul, William W. Gerberich and James P. Lucas, "Near-Threshold Behavior of HSLA Steels," in HSLA Steels—Technology & Applications, American Society for Metals, Metals Park OH, 1984, p 571

5-20. Effect of Test Temperature on Fatigue Crack Propagation Behavior for Two HSLA Steel Grades

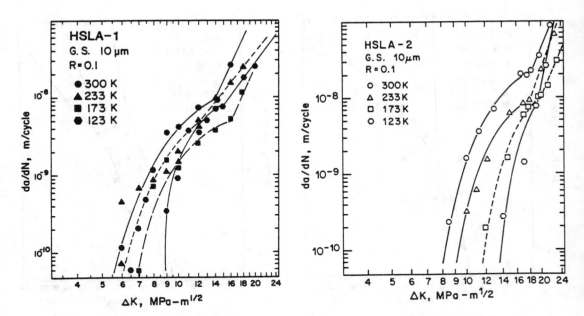

Fatigue crack propagation behavior of two HSLA steels tested at temperatures of 300, 233, 173 and 123K in the as-received condition.

The only significant difference between HSLA-1 and HSLA-2 is that HSLA-2 contains twice as much Nb as HSLA-1 (for compositions of the steels, see p 165).

Near-threshold crack growth and threshold stress intensities for both steels in the as-received condition are depicted in the above charts for all test temperatures. Comparison of crack growth rates and threshold stress intensities at room temperature indicate that HSLA-2 has a higher resistance to fatigue crack propagation than HSLA-1. The stress intensities amplitude, ΔK, for constant growth rates of 10^{-8} and 10^{-9} m/cycle are 2.0-2.5 MPa-m$^{1/2}$ higher in HSLA-2 than in HSLA-1. The threshold stress intensity ΔK_{th}, is also higher for HSLA-2 (8.0 MPa-m$^{1/2}$) compared to HSLA-1 (5.5 MPa-m$^{1/2}$). The 2.0-2.5 MPa-m$^{1/2}$ difference in threshold and for the two growth rates clearly demonstrates that there is an inherent difference in the fatigue crack propagation behavior of these two HSLA steels. This difference is also reflected at low temperatures, where HSLA-2 showed lower crack propagation rates and higher threshold stress intensities than HSLA-1. Furthermore, by comparison of threshold stress intensities for these two steels in relation to the effect of decreasing temperature on increasing ΔK_{th}, it was found that the ratios of $\Delta K_{th}(T)/\Delta K_{th}(300K)$ are the same for both steels.

Source: Khlefa A. Esaklul, William W. Gerberich and James P. Lucas, "Near-Threshold Behavior of HSLA Steels," in HSLA Steels—Technology & Applications, American Society for Metals, Metals Park OH, 1984, p 569

5-21. Stress-Cycle Curves for Weldments of Different HSLA Steel Grades

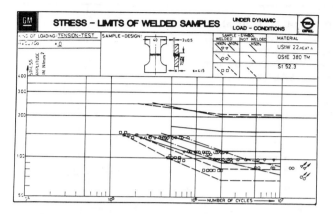

Stress-cycles curves of welded samples of different materials under tension load.

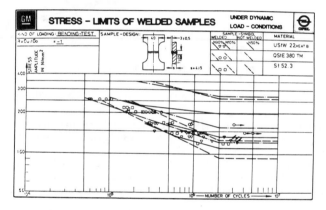

Stress-cycles curves of welded samples of different materials under alternating bending load.

Fatigue data were derived from testing welded samples under tension and bending loads. It was surprising that under both types of load the HSLA steel and the soft unalloyed steel hardly differed in fatigue strength; thus it can be said that the use of HSLA steels is not justified if a component has a weld in the highest-stressed area. An explanation for this is the loss of the thermal-mechanical effect, which is responsible for increased strength, by the heat influence during the welding operation; and it is thought that a higher-strength manganese-alloyed steel, such as St 52.3 (according to DIN), the strength of which results from the chemical composition, would be more favorable in this respect.

Source: Klaus E. Richter, "Cold and Hot-Rolled Microalloyed Steel Sheets in Opel Cars—Experience and Applications," in HSLA Steels—Technology & Applications, American Society for Metals, Metals Park OH, 1984, p 487

5-22. Weldments (FCAW): SAE 980 X Steel vs 1006

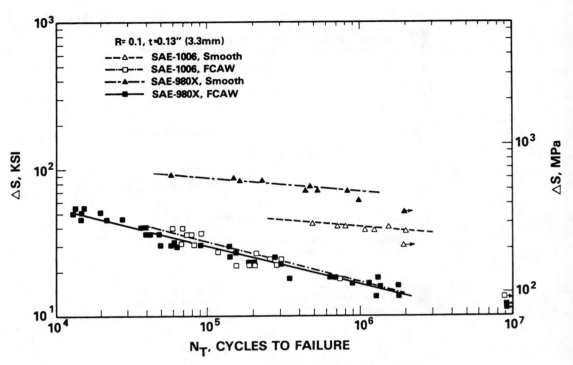

Fatigue properties of smooth and FCAW SAE 1006 and SAE 980 X steels.

The fatigue strengths of the smooth HSLA steel were higher than that of the low-carbon steel. The 10^6-cycle fatigue limit stress of the smooth SAE 980 X steel was 469 MPa (68 ksi) and that for the SAE 1006 steel was 283 MPa (41 ksi).

However, after welding, SAE 1006 and SAE 980 X steels exhibited similar fatigue properties over the 10^4-10^6-cycle life range studied. The 10^6-cycle fatigue limit stresses for FCAW SAE 1006 and SAE 980 X steels were between 114 MPa (16.5 ksi) and 117 MPa (17 ksi).

Source: Kon-Mei W. Ewing, Pei-Chung Wang, Frederick V. Lawrence, Jr., and Albert F. Houchens, "Weld Fatigue of TIG-Dressed SAE-980X HSLA Steel," in HSLA Steels—Technology & Applications, American Society for Metals, Metals Park OH, 1984, p 556

5-23. Weldments (TIG): DOMEX 640 XP Steel Welded Joints vs Parent Metal

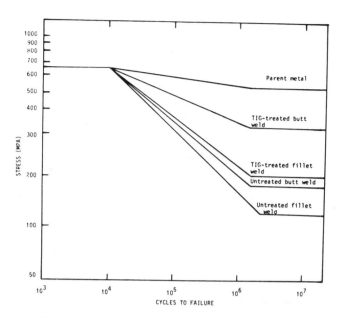

Fatigue strength for DOMEX 640 XP. Standard-Wöhler-diagram (log-log scale) with pulsating load (R=min stress/max stress=0). Sheet thickness 5 mm and ultimate tensile strength 767 MPa.

For unwelded parent metal the fatigue strength of a steel is improved with increasing static strength. For welded joints the fatigue strength in the endurance range 10^5-2×10^6 is mainly dependent upon the weld geometry and is therefore roughly the same irrespective of the static strength of the steels. For making full use of an increased static strength for a steel subjected to severe fatigue, special attention must be paid to the configuration of the welds. After welding, grinding or TIG-treatment can be used to improve the weld geometry. The notch effect at the weld toe is decreased and the fatigue properties can be improved. Another solution is to place the welds in areas where the stresses are low.

Source: Tony Nilsson, "Formable Hot-Rolled Steel With Increased Strength," in HSLA Steels—Technology & Applications, American Society for Metals, Metals Park OH, 1984, p 259

5-24. Weldments (FCAW Dressed by TIG): Fatigue Life Estimates Compared With Experimental Data for SAE 980 X Steel

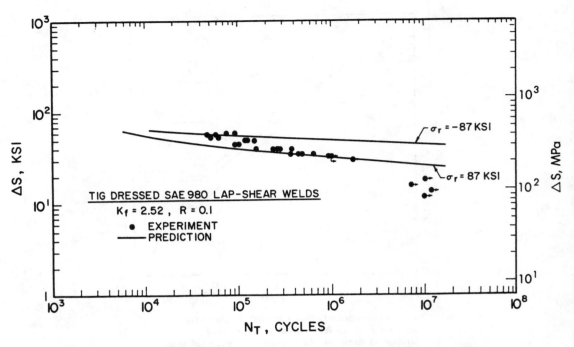

Total fatigue life estimates compared to the experimental data for the FCAW/TIG-dressed SAE 980 X steel.

It should be emphasized that life estimates made on the FCAW/TIG-dressed welds were based on geometry changes brought about by TIG-dressing. The other beneficial effects such as removal of slag intrusions and inclusions were not considered. The close agreement between the calculated and observed long-life fatigue properties suggested that the majority of fatigue improvement seen in TIG-dressed joints was attributable to the geometry change. The smaller flank angle contributed significantly to the increased fatigue strengths of TIG-dressed weldments.

Source: Kon-Mei W. Ewing, Pei-Chung Wang, Frederick V. Lawrence, Jr., and Albert F. Houchens, "Weld Fatigue of TIG-Dressed SAE-980X HSLA Steel," in HSLA Steels—Technology & Applications, American Society for Metals, Metals Park OH, 1984, p 563

5-25. SAE 980 X Steel Weldment (FCAW): Smooth Specimen vs TIG-Dressed vs As-Welded

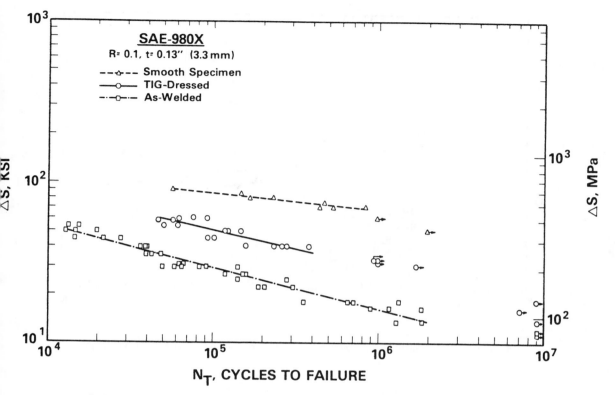

Fatigue properties of FCAW/TIG-dressed SAE 980 X steel compared to the smooth specimen and as-welded data. From these data, a significant improvement in fatigue characteristics can be obtained by TIG-dressing the welds.

Source: Kon-Mei W. Ewing, Pei-Chung Wang, Frederick V. Lawrence, Jr., and Albert F. Houchens, "Weld Fatigue of TIG-Dressed SAE-980X HSLA Steel," in HSLA Steels—Technology & Applications, American Society for Metals, Metals Park OH, 1984, p 558

5-26. SAE 980 X Steel Weldment (FCAW): Lap-Shear Joints

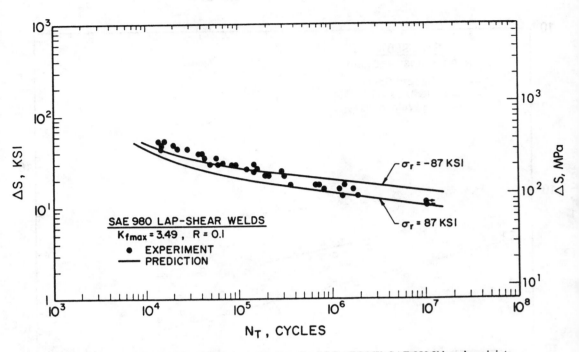

Total fatigue life predictions and experimental results for the FCAW, SAE 980 X lap-shear joints.

Source: Kon-Mei W. Ewing, Pei-Chung Wang, Frederick V. Lawrence, Jr., and Albert F. Houchens, "Weld Fatigue of TIG-Dressed SAE-980X HSLA Steel," in HSLA Steels—Technology and Applications, American Society for Metals, Metals Park OH, 1984, p 562

5-27. Microalloyed HSLA Steels: Properties of Fusion Welds

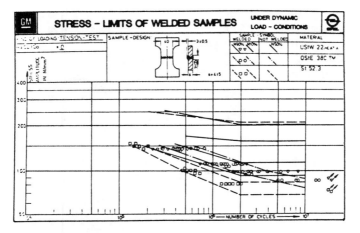

Stress-cycles curves of fusion welded samples of different materials under tension load.

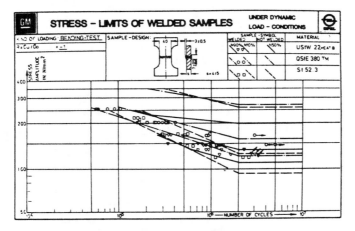

Stress-cycles curves of fusion welded samples of different materials under alternating bending load.

Due to the preferred crack location in the welded areas, it seemed necessary to examine the dynamic strength of fusion welded joints of HSLA steels in more detail, testing the steel used for the crossmember at a minimum yield strength of 380 N/mm^2, in comparison to a soft unalloyed, hot rolled steel sheet. Tensile load and alternating bending load were the selected types for dynamic test. The above charts show the respective stress-cycle curves.

Source: Klaus E. Richter, "Cold and Hot-Rolled Microalloyed Steel Sheets in Opel Cars—Experience and Applications," in HSLA Steels—Technology & Applications, American Society for Metals, Metals Park OH, 1984, p 487

5-28. Microalloyed HSLA Steels: Properties of Spot Welds

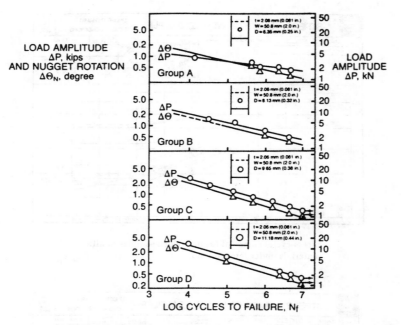

Fatigue test results for the 2.06 mm (0.081 inch) thick sheet with various weld diameters.

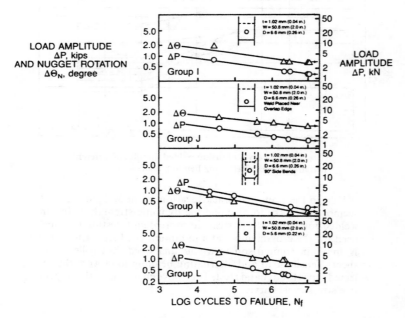

Fatigue test results for 1.02 mm (0.040 inch) thick sheet of different stiffnesses.

Results of spot-weld fatigue tests are presented in the four plots (above and on the facing page) for the stated conditions. Each curve shows the load amplitude, ΔP, and nugget rotation values, $\Delta \theta_N$, for each test as a function of cyclic life. Straight lines were fitted through the data.

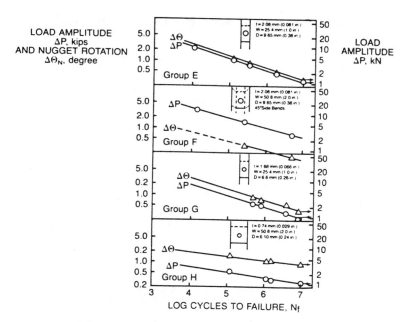

Fatigue test results for variations in specimen width and thickness.

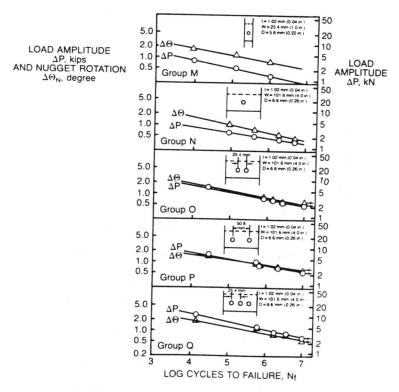

Fatigue test results for 1.02 mm (0.040 inch) thick sheet with single and multiple welds.

Source: James A. Davidson, "Design-Related Methodology to Determine the Fatigue Life and Related Failure Mode of Spot-Welded Sheet Steels," in HSLA Steels—Technology & Applications, American Society for Metals, Metals Park OH, 1984, p 542

6-1. HY-130 Steel: Effect of Notch Radii

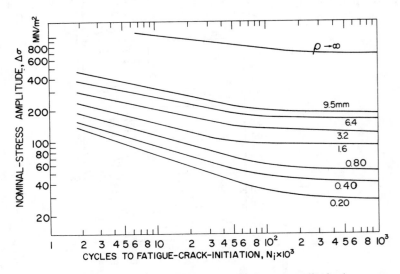

Cycles to fatigue-crack initiation versus nominal stress amplitude, $\Delta\sigma$, for notched specimens with various radii of curvature.

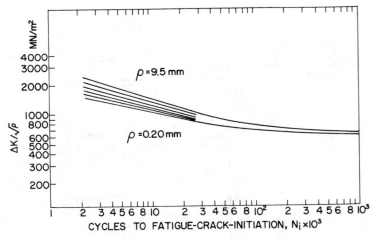

Same data as in upper graph but plotted versus $\Delta K/\sqrt{\rho}$ rather than $\Delta\sigma$.

Curvature of the notch and ΔK is the stress-intensity amplitude computed for an imaginary crack whose length is the same as the notch depth, a. Barsom and McNicol used this parameter to compare N_i, the cycles to fatigue-crack initiation, in HY-130 steel for notches of constant depth but various radii of curvature. The results are shown in the above graphs. In the upper graph, N_i is plotted versus $\Delta\sigma$, where N_i is defined as the number of cycles to give a 50-μm side notch. There is a wide spread in the curves. As expected, the sharpest notch, lowest ρ, gave the most rapid initiation at a given stress. The lower graph shows $\Delta K/\sqrt{\rho}$ plotted versus N_i. A narrow-spread family of curves results; these converge as the value of $\Delta K/\sqrt{\rho}$ is decreased to a threshold value $\Delta K/\sqrt{\rho}|_{th}$, the minimum value to initiate fatigue cracks in notches.

Source: M. E. Fine and R. O. Ritchie, "Fatigue-Crack Initiation and Near-Threshold Crack Growth," in Fatigue and Microstructure, American Society for Metals, Metals Park OH, 1979, pp 256-257

6-2. 300 M Steel: Effect of Notch Severity on Constant-Lifetime Behavior

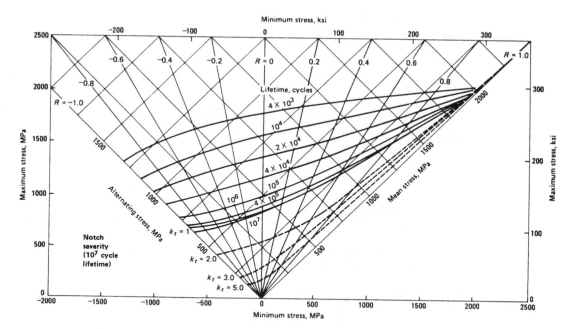

Constant-lifetime fatigue diagram for 300 M alloy steel, hardened and tempered to a tensile strength of 1930 MPa (280 ksi). Solid lines represent lifetimes obtained from unnotched specimens. Dashed lines represent lifetime of ten million cycles for specimens having the indicated notch severity.

Source: Metals Handbook, 9th Edition, Volume 1, Properties and Selection: Irons and Steels, American Society for Metals, Metals Park OH, 1978, p 670

6-3. TRIP Steels Compared With Other High-Strength Grades

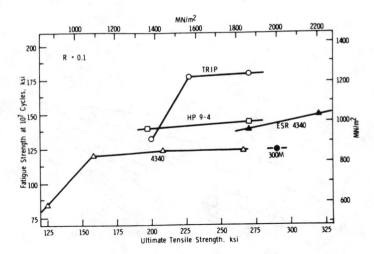

Fatigue strength at 10^7 cycles ($R = 0.1$) vs ultimate tensile strength for TRIP steels compared with other high-strength steels.

Studies on fatigue-crack propagation (FCP) conducted under controlled stress-intensity amplitude (ΔK) conditions indicate that deformation-induced transformation retards crack growth in lower-strength metastable austenites, particularly at low ΔK, and also exerts a beneficial influence in high-strength TRIP steels, although to a much lesser extent. This growth retardation may be due to crack-closure effects arising from the transformation volume change, which may be particularly effective in the fatigue-threshold regime. Smooth-bar fatigue properties appear to be dominated by transformation hardening, which is desirable under stress-control conditions (reducing strain amplitude) but generally undesirable under strain-control conditions (increasing stress amplitude). In lower-strength austenites, transformation reduces fatigue life under conditions of controlled *plastic* strain amplitude; under controlled *total* strain amplitude, transformation is detrimental to low-cycle fatigue life, but a small amount of transformation may be beneficial at high cycles. Similarly, the low-cycle fatigue properties of high-strength TRIP steels are found to be degraded by transformation under controlled total strain amplitude. Under *stress* control, the fatigue life of lower-strength austenites is greatly enhanced by transformation; for a stress ratio ($R = \sigma_{min}/\sigma_{max}$) of 0, fatigue limits in excess of the yield strength are observed. Investigation of the smooth-bar fatigue properties of high-strength TRIP steels at $R = 0.1$, in which thermodynamic stability was varied by heat treatment, also revealed transformation enhancement of fatigue life. Such enhancement allows the achievement of exceptional fatigue strength at high ultimate strength levels, as illustrated by comparison with other high-strength steels in the above graph.

Source: G. B. Olson, "Transformation Plasticity and the Stability of Plastic Flow," in Deformation, Processing, and Structure, George Krauss, Ed., American Society for Metals, Metals Park OH, 1984, p 419

6-4. Corrosion Fatigue: Special High-Strength Sucker-Rod Material

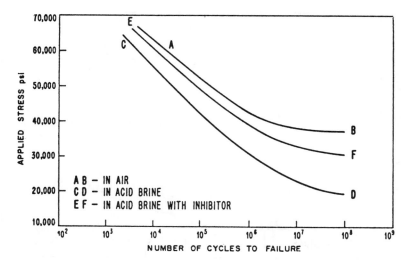

Effect of corrosion and corrosion inhibitors on the *S-N* curve for high-strength steel (sucker-rod material).

After the first brittle crack is initiated, No. 2 is the slow step in the process and electrochemical action is the slowest part of this step. Thus, the effect of corrosion can be illustrated with curves of stress vs logarithm-of-number-of-reversals-to-failure for sucker-rod steel. Corrosion accelerates cracks propagation, so the fatigue curve drops from AB to CD, as shown in the graph. Deceleration of the slow stage with a corrosion inhibitor will raise the *S-N* fatigue curve from CD to EF.

Source: Joseph F. Chittum, "Corrosion Fatigue Cracking of Oil Well Sucker Rods," in Corrosion: Source Book, Seymour K. Coburn, Ed., American Society for Metals, Metals Park, OH, 1984, p 380

6-5. Corrosion Fatigue Cracking of Sucker-Rod Material

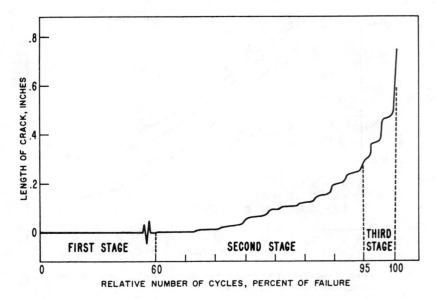

Corrosion fatigue cracking of sucker rods.

This graph shows typical progress of a crack at high stress plotted against number of cycles, showing stages in the fatigue process. Observations of sucker-rod crack penetration as a function of reversal accumulation are possible using a bending apparatus and a magnetic fluorescent powder technique. Penetration vs reversal curves resemble the one shown above when the stress is well in excess of the endurance limit.

During bending, no penetration is apparent in the first 40-60% of the specimen's fatigue life, even though intrusions and extrusions may form earlier. A crack eventually appears and progresses through the specimen. When the penetration reaches a certain percentage of the cross section, the cracking accelerates until catastrophic failure occurs.

Source: Joseph F. Chittum, "Corrosion Fatigue Cracking of Oil Well Sucker Rods," in Corrosion: Source Book, Seymour K. Coburn, Ed., American Society of Metals, Metals Park OH, 1984, p 378

6-6. Hydrogenated Steel: Effect of Baking Time on Hydrogen Concentration

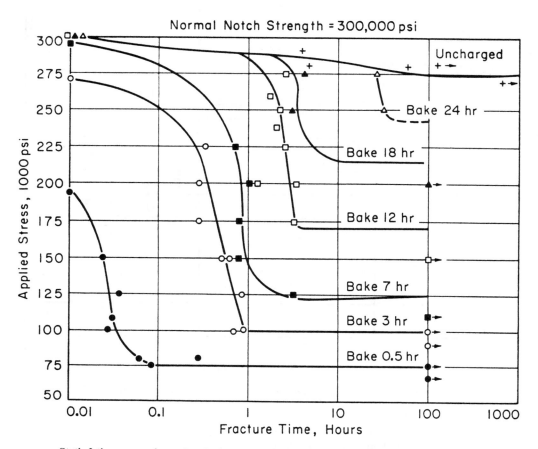

Static fatigue curves for various hydrogen concentrations obtained by baking different times at 150 °C (300 °F). Sharp-notch specimens. 230,000 psi strength level.

These are, in essence, static fatigue curves, and the lower critical stress may be considered a static endurance limit—that is, a stress below which failure will not occur for an indefinite period of time. This behavior is sensitive to hydrogen concentration as shown above, where it may be seen that all delayed-failure parameters—notch strength, rupture time, and static fatigue limit increase with decreasing hydrogen concentration. Also, even after 24 hours at 150 °C (300 °F), there is still a substantial stress range, of the order of 60,000 psi, over which delayed failure will occur. In an unnotched specimen, full recovery of the ductility as measured by the reduction of area can be attained in less than 20 hours at 150 °C (300 °F), yet delayed failure will occur after 24 hours or longer of baking time at 150 °C (300 °F).

Source: Alexander R. Troiano, "The Role of Hydrogen and Other Interstitials in the Mechanical Behavior of Metals," in Hydrogen Damage Source Book, Cedric D. Beachem, Ed., American Society for Metals, Metals Park OH, 1977, p 154

6-7. Hydrogenated Steel: Effect of Notch Sharpness

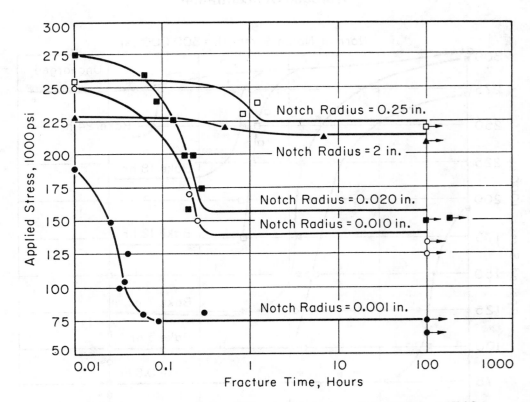

Static fatigue curves for specimens of different notch sharpness. Baked 0.5 hour at 150 °C (300 °F).

The variation of lower critical stress with notch severity is shown in this diagram. It is evident that the static fatigue limit rises as notch severity (radius) decreases for hydrogen-charged high-strength steels (using the same baking time).

Source: Alexander R. Troiano, "The Role of Hydrogen and Other Interstitials in the Mechanical Behavior of Metals," in Hydrogen Damage Source Book, Cedric D. Beachem, Ed., American Society for Metals, Metals Park OH, 1977, p 155

7-1. 0.5%Mo Steel: Effect of Hold Time in Air and Vacuum at Different Temperatures

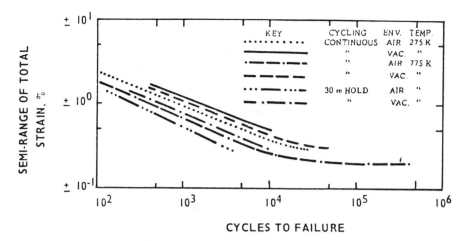

Effect of hold time in air and vacuum upon the fatigue endurances of a 0.5%Mo steel at 275 and 775 K.

Source: R. H. Cook and R. P. Skelton, "Environment-Dependence of the Mechanical Properties of Metals at High Temperature," in Source Book on Materials for Elevated-Temperature Applications, Elihu F. Bradley, Ed., American Society for Metals, Metals Park OH, 1979, p 83

7.2 DIN 14 Steel (1.5 Cr, 0.90 Mo, 0.25 V): Effect of Liquid Nitriding

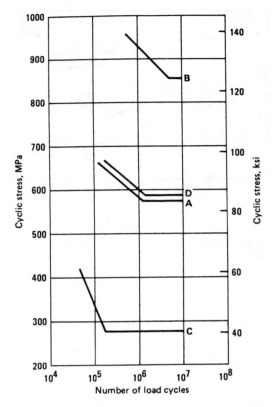

Effect of nitriding on fatigue behavior of DIN 14 CrMoV 69 steel (0.14 C, 1.5 Cr, 0.90 Mo, 0.25 V). Curves A and C are for hardened and tempered (not nitrided) specimens; B and D are for liquid nitrided specimens. A and B are for smooth specimens; C and D are for notched specimens $K_t = 2$.

Nitriding introduces residual compressive stresses at the surface of steel parts; these residual stresses, together with the increased strength of the nitrided layer, increase the fatigue resistance of the part. The increase in fatigue strength that results from nitriding is illustrated in these *S-N* curves.

Source: Metals Handbook, 9th Edition, Volume 1, Properties and Selection: Irons and Steels, American Society for Metals, Metals Park OH, 1978, p 541

7-3. 2.25Cr-1.0Mo Steel: Influence of Cyclic Strain Range on Endurance Limit in Various Environments

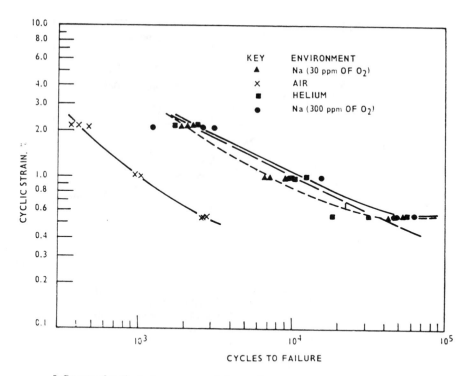

Influence of cyclic strain range upon fatigue endurance of 2.25Cr-1.0Mo steel in sodium, air, and helium at 865 K. (Cycle used was approximately up 5 s, hold 5 s, down 5 s, hold 5 s.)

Source: R. H. Cook and R. P. Skelton, "Environment-Dependence of the Mechanical Properties of Metals at High Temperature," in Source Book on Materials for Elevated-Temperature Applications, Elihu F. Bradley, Ed., American Society for Metals, Metals Park OH, 1979, p 83

7-4. 2.25Cr-1.0Mo Steel: Effect of Elevated Temperature

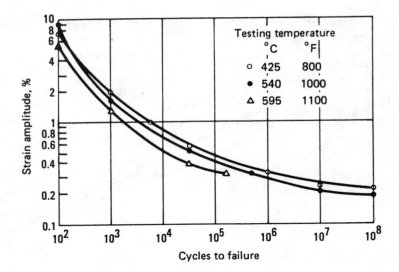

The results of strain-controlled fatigue tests of 2.25Cr-1.0Mo steel at 425, 540 and 595 °C (800, 1000 and 1100 °F) on specimens of annealed 2.25Cr-1.0Mo steel are presented in these *S-N* curves. Within this range, test temperature had relatively little effect on number of cycles to failure.

Source: Metals Handbook, 9th Edition, Volume 1, Properties and Selection: Irons and Steels, American Society for Metals, Metals Park OH, 1978, p 659

7-5. 2.25Cr-1.0Mo Steel: Effect of Elevated Temperature and Strain Rate

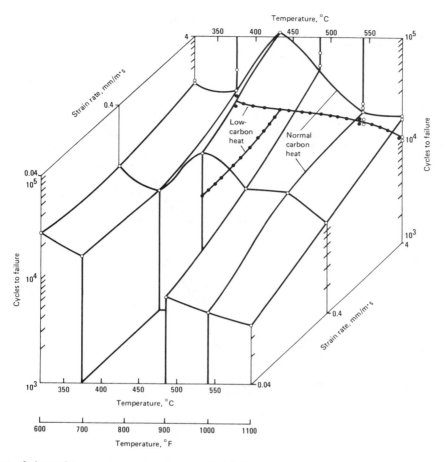

Effect of elevated temperature on strain-controlled fatigue behavior of annealed 2.25Cr-1.0Mo steel.

Strain-controlled fatigue tests have also shown (note above) that reducing carbon content to 0.03% results in a reduction in fatigue strength. Furthermore, because of variations in strain aging effect, specimens from one heat with a higher carbon content ran longer at 427 °C (800 °F) than at 316 °C (600 °F).

Source: Metals Handbook, 9th Edition, Volume 1, Properties and Selection: Irons and Steels, American Society for Metals, Metals Park OH, 1978, p 659

7-6. 2.25Cr-1.0Mo Steel: Effect of Temperature on Fatigue Crack Growth Rate

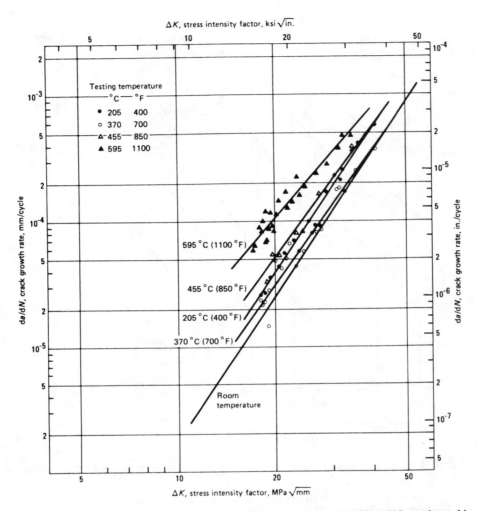

Variations in fatigue crack growth rate with test temperature for specimens of 2.25Cr-1.0Mo steel tested in air.

Specimens were subjected to cyclic loading at a constant maximum load. Stress ratio was 0.05; cyclic frequency was 400 per minute. As shown, the stress-intensity factor range increased as the crack length was increased.

Source: Metals Handbook, 9th Edition, Volume 1, Properties and Selection: Irons and Steels, American Society for Metals, Metals Park OH, 1978, p 660

7-7. 2.25Cr-1.0Mo Steel: Effect of Cyclic Frequency on Fatigue Crack Growth Rate

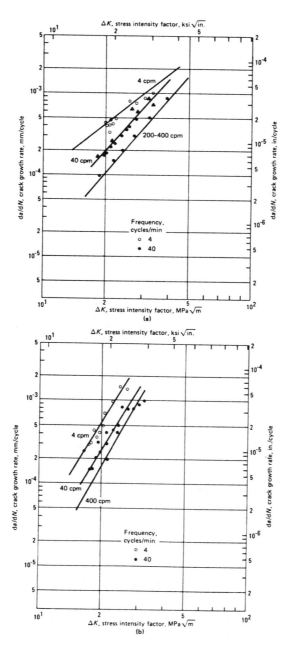

Data shown above indicate that in elevated temperature tests at a given stress-intensity factor range, crack growth rate increases as cyclic frequency is decreased. These fracture mechanics data may be applied to the design of structural components that may contain undetected discontinuities, or that may develop cracks in service. Stress ratio was 0.05. (a) Tested at 510 °C (950 °F); (b) tested at 595 °C (1100 °F).

Source: Metals Handbook, 9th Edition, Volume 1, Properties and Selection: Irons and Steels, American Society for Metals, Metals Park OH, 1978, p 661

7-8. 2.25Cr-1.0Mo Steel: Fatigue Crack Growth Rates in Air and Hydrogen

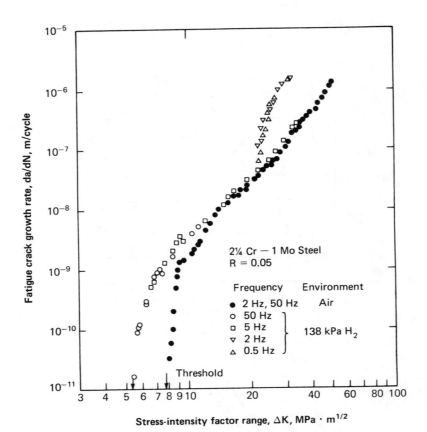

Fatigue crack growth rates in 2.25Cr-1.0Mo steel in air and in hydrogen.

Corrosion fatigue descriptions are further complicated by the fact that the environment may produce multiple effects. For example, Suresh *et al* demonstrated that dry hydrogen may produce a *frequency-sensitive* environmental effect analogous to SCC at intermediate ΔK values and a *frequency-insensitive* environmental effect near the threshold. This is illustrated in the above graph for 2.25Cr-1Mo steel tested in air and in 138-kPa hydrogen gas. Because the sustained-load threshold for this steel is on the order of 90 MPa · m$^{1/2}$ (82 ksi · in.$^{1/2}$), the $K_{th(f)}$ of about 22 MPa · m$^{1/2}$ (20 ksi · in.$^{1/2}$) gives $K_{th(f)} \ll K_{th}$.

It can be seen for ΔK values greater than $K_{th(f)}$ that there is a large increase in growth rate for the low test frequencies but not for the higher ones. Therefore, this regime may be considered to be one where superposition might apply. In addition, however, there is a true threshold, ΔK_{th}, which appears to be frequency-insensitive but which nevertheless decreased by about 30% to 5.4 MPa · m$^{1/2}$ (4.9 ksi · in.$^{1/2}$) because of the hydrogen environment. Such mulitple effects are poorly understood and are clearly possible in a large number of material/environment systems.

Source: W. W. Gerberich and A. W. Gunderson, "Design, Materials Selection and Failure Analysis," in Application of Fracture Mechanics for Selection of Metallic Structural Materials, James E. Campbell, William W. Gerberich and John H. Underwood, Eds., American Society for Metals, Metals Park OH, 1982, p 333

7-9. 2.25Cr-1.0Mo Steel: Effect of Holding Time

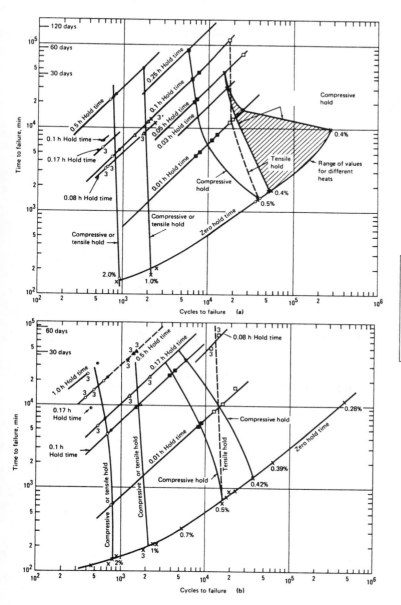

Time-to-failure/cycles-to-failure diagrams for annealed 2.25Cr-1.0Mo steel tested in strain-controlled cyclic loading at (a) 480 °C (900 °F) and (b) 540 °C (1000 °F). Hold time indicated on graph is length of time that specimens were held (during each cycle) in the state of maximum tensile strain (open symbols) or compressive strain (filled symbols). Strain amplitude indicated by shape of symbols and figures along zero-hold-time line.

In these "time-to-failure/cycles-to-failure" diagrams, the lowest curve (zero hold time) indicates the corresponding time period and number of cycles to failure for continuous strain-controlled fatigue tests over the strain range from 0.4 to 2.0% with no holding period at maximum strain. The other curves, which are approximately parallel, are for increasing periods of holding time at maximum strain levels in either tension or compression. The vertical curves are drawn through the number of cycles to failure for each particular cyclic strain. For all tests at 2% strain, failure occurred in less than 1000 cycles regardless of holding time or whether the stress was tensile or compressive. The effect of reducing the strain increment and increasing the holding time on number of cycles to failure can be determined from the appropriate curves in the figures.

Source: Metals Handbook, 9th Edition, Volume 1, Properties and Selection: Irons and Steels, American Society for Metals, Metals Park OH, 1978, pp 662-663

7-10. Cast 2.25Cr-1.0Mo Steel, Centrifugally Cast: Fatigue Properties at 540 °C (1000 °F)

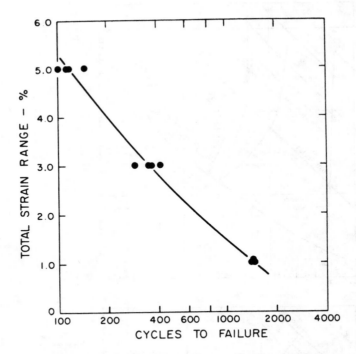

Fatigue properties of 2.25Cr-1.0Mo centrifugally cast pipe, A217, Grade WC9, at 540 °C (1000 °F).

Source: Steel Castings Handbook, 5th Edition, Peter F. Weiser, Ed., Steel Founders' Society of America, Rocky River OH, 1980, p 15-55

7-11. H11 Steel: Crack Growth Rate in Water and in Water Vapor

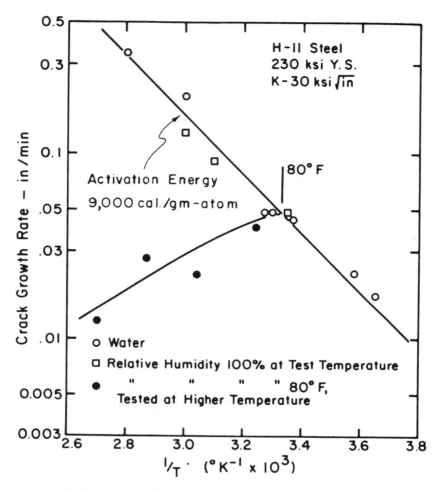

Crack growth rate versus temperature for an H11 steel in water and water vapor.

It is of considerable interest that the strain rate and temperature dependence of hydrogen embrittlement, as determined by ductility measurements after rising load tests on hydrogen-charged materials, show a characteristic behavior that resembles closely that seen with crack growth rate measurements and external hydrogen environments.

Source: Herbert H. Johnson, "Keynote Lecture: Overview on Hydrogen Degradation Phenomena," in Hydrogen Embrittlement and Stress Corrosion Cracking, R. Gibala and R. F. Hehemann, Eds., American Society for Metals, Metals Park OH, 1984, p 18

7-12. 9.0Cr-1.0Mo Steel: Creep-Fatigue Characteristics

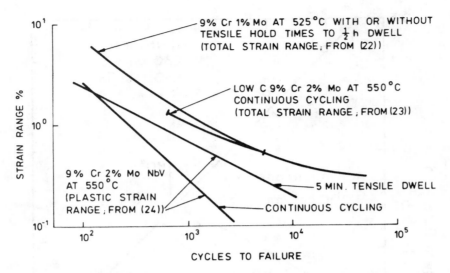

Illustrating the elevated temperature low-cycle fatigue and creep-fatigue properties of normalized and tempered 9% Cr Mo variants.

In this chart are presented the elevated-temperature-fatigue and creep-fatigue data for the 9%Cr-1%Mo steel as a single curve in terms of total strain range against cycles to failure; also shown for direct comparison are the continuous cycling fatigue data for the low-C, 9%Cr-2%Mo variant which, although inferior at relatively high strain ranges, suggests superior endurance may be attained in the high-cycle region. From the limited evidence, it seems probable that normalized and tempered 9%Cr-1%Mo steel may be used in reactor-quality sodium at service temperatures with little effect on tensile properties and stress rupture strengths or ductility and that the short term low-cycle fatigue endurance will be increased and fatigue crack growth rate reduced. This behavior is a consequence of the structural stability of the material with respect to interstitial element transfer in liquid sodium and also the low oxygen potential of the overall system which may be expected to preclude oxide penetration and enable partial recohesion of the crack faces during fatigue.

Source: S. J. Sanderson, "Mechanical Properties and Metallurgy of 9%Cr 1%Mo Steel," in Ferritic Steels for High-Temperature Applications, Ashok K. Khare, Ed., American Society for Metals, Metals Park OH, 1983, p 95

7-13. 9.0Cr-1.0Mo Modified Steel: Stress Amplitudes Developed in Cycling

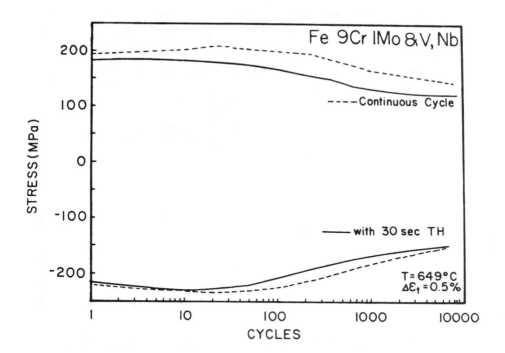

This chart shows stress amplitudes (tensile and compressive) that developed in the course of cycling the modified Fe-9.0Cr-1.0Mo steel through a total strain range of 0.5% at 649 °C (1200 °F). Fatiguing was carried out in vacuum. Dotted curve indicates continuous cycling; solid curve indicates cycling with a 30-s hold at maximum tensile strain.

Source: S. Kim, J. R. Weertman, S. Spooner, C. J. Glinka, V. Sikka and W. B. Jones, "Microstructural Evaluation of a Ferritic Stainless Steel by Small Angle Neutron Scattering," in Nondestructive Evaluation: Application to Materials Processing, Otto Buck and Stanley M. Wolf, Eds., American Society for Metals, Metals Park OH, 1984, p 175

7-14. 9.0Cr-1.0Mo Modified Steel: Effect of Deformation

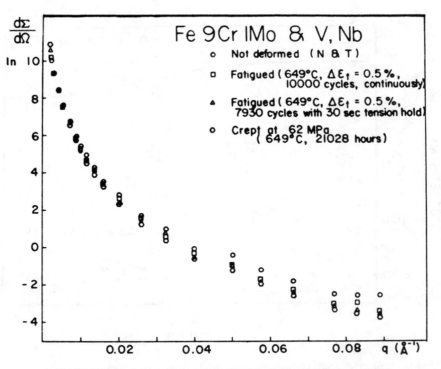

Curves of $d\Sigma/d\Omega$ vs q for specimens of modified Fe-9.0Cr-1.0Mo steel which have undergone various types of deformation. A magnetic field of ~28 kg was applied to the specimens during the SANS measurements. $\lambda = 0.48$ nm.

Source: S. Kim, J. R. Weertman, S. Spooner, C. J. Glinka, V. Sikka and W. B. Jones, "Microstructural Evaluation of a Ferritic Stainless Steel by Small Angle Neutron Scattering," in Nondestructive Evaluation: Application to Materials Processing, Otto Buck and Stanley M. Wolf, Eds., American Society for Metals, Metals Park OH, 1984, p 175

8-1. Type 301 Stainless Steel: Scatter Band for Fatigue Crack Growth Rates

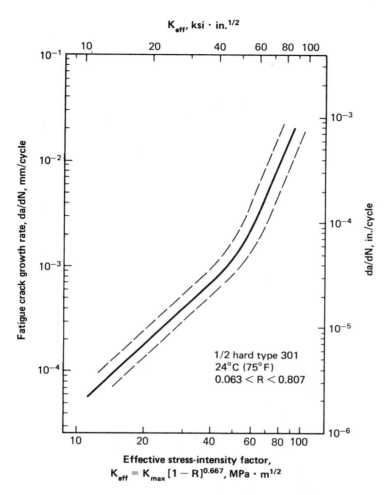

Scatter band of fatigue crack growth rates of ½-hard type 301 stainless steel, tested at 24 °C (75 °F), 10 Hz, and R ratios of 0.063 to 0.807 based on effective stress-intensity factor, K_{eff}.

Fatigue crack growth rate data reported by Walker for ½-hard type 301 stainless steel sheet are summarized in the above graph. The data were obtained in air at room temperature over a series of load ratios (R) from 0.063 to 0.807 at a frequency of 10 Hz. These data are based on the "effective stress intensity factor," K_{eff}, rather than on ΔK, to account for the effect of the range of stress ratios. K_{eff} is defined as follows:

$$K_{eff} = K_{max}(1-R)^m$$

where m is determined empirically and R is the load ratio (minimum load/maximum load) on cyclic loading. The crack growth rate law then becomes:

$$da/dN = C[K_{max}(1-R)^m]^n$$

Results of fatigue crack growth rate tests on austenitic stainless steels have shown that the crack growth rate tends to increase as the R ratio is increased, when compared at given values of ΔK. If tests are made at several load ratios to determine m, then the effects of other load ratios may be estimated.

Source: J. E. Campbell, "Fracture Properties of Wrought Stainless Steels," in Application of Fracture Mechanics for Selection of Metallic Structural Materials, James E. Campbell, William W. Gerberich and John H. Underwood, Eds., American Society for Metals, Metals Park OH, 1982, p 114

8-2. Type 301 Stainles Steel: Effects of Temperature and Environment on Fatigue Crack Growth Rate

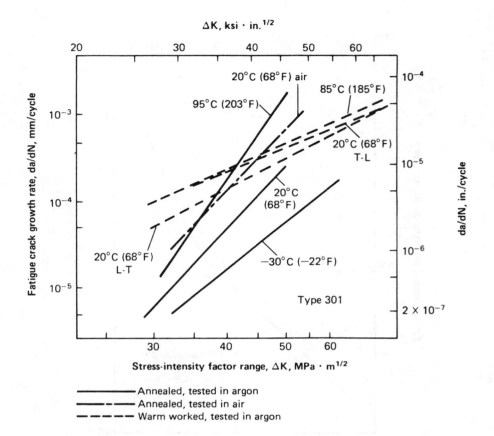

Fatigue crack growth rates for type 301 stainless steel have been reviewed by Pineau and Pelloux in the temperature range from −30 to +95 °C (−22 to +203 °F). The results, summarized in this graph, were obtained on compact specimens 7 mm (0.28 in.) thick at a cyclic frequency of 20 Hz with a sinusoidal wave form at a load ratio (R) of 0.01. All specimens were tested in dry argon except one series that was tested in laboratory air. For the annealed specimens tested in argon, fatigue crack growth rates at a given ΔK value increased as the temperature increased over the testing temperature range. Fatigue crack growth rates in laboratory air at 20 °C (68 °F) were higher than for corresponding conditions in argon, indicating that the humidity and/or oxygen in the air influenced the growth rates.

The warm worked specimens were reduced 65% at 450 to 500 °C (840 to 930 °F), resulting in a substantial increase in strength. Fatigue crack growth rates for the warm worked specimens (above) indicate that the fatigue crack propagation properties of the warm worked alloy are different from those of the annealed alloy. This effect of warm working has been observed for other austenitic stainless steels. These differences are attributed to the extent of the strain-induced transformation at the crack tip. This transformation effect would be most noticeable in type 301, because it is less stable than the other alloys in the UNS S3*xxxx* series.

Source: J. E. Campbell, "Fracture Properties of Wrought Stainless Steels," in Application of Fracture Mechanics for Selection of Metallic Structural Materials, James E. Campbell, William W. Gerberich and John H. Underwood, Eds., American Society for Metals, Metals Park OH, 1982, p 113

8-3. Type 304 Stainless Steel: Effect of Temperature on Frequency-Modified Strains

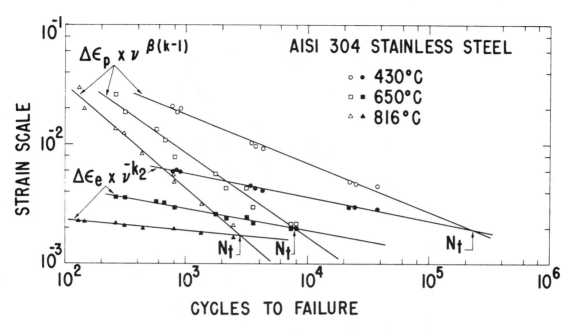

Data of Berling and Slot for AISI 304 stainless steel, showing frequency-modified elastic and plastic strains at three temperatures in air.

In contrast to most other segments of our technology, interest in the fatigue problem in the power-generation industry generally involves elevated temperature. Laboratory testing on both smooth specimens and specimens designed for crack growth is performed with temperature and frequency or strain rate as parameters. The importance of frequency or strain-rate effects is shown in this chart. These data are for solution-treated AISI 304 stainless steel subject to triangular wave shapes at equal-loading and reverse-loading strain rates. Representation of the behavior here utilizes fatigue equations known as frequency-modified fatigue equations. They describe the elastic and plastic strains versus fatigue life and include the frequency or strain rate of the test. For the present purposes they are useful in showing how increasing temperature acts to change the cyclic stress-strain response and the strain-life fatigue response of this alloy.

Source: L. F. Coffin, "Fatigue in Machines and Structures—Power Generation," in Fatigue and Microstructure, American Society for Metals, Metals Park OH, 1979, p 13

8-4. Type 304 Stainless Steel: Fatigue Crack Growth Rate— Annealed and Cold Worked

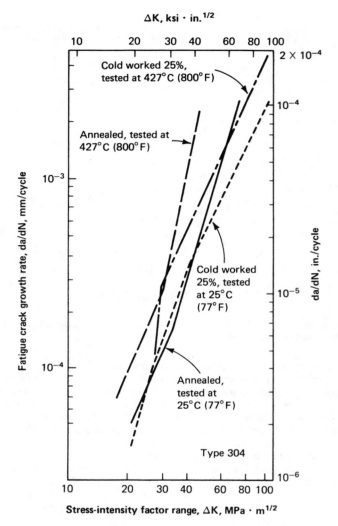

Fatigue crack growth rates for annealed and cold worked type 304 stainless steel at 25 and 427 °C (77 and 800 °F), 0.17 Hz, and an R ratio of 0.

In some applications, type 304 stainless steel components are fabricated in the cold worked condition to improve strength properties. A comparison of fatigue crack growth rate data by Shahinian, Watson, and Smith, illustrated in this graph, shows that the high-ΔK crack growth rates were lower for the cold worked specimens than for the annealed specimens. Crack growth rates were higher for the specimens tested at 427 °C (800 °F) than for corresponding specimens tested at room temperature.

Source: J. E. Campbell, "Fracture Properties of Wrought Stainless Steels," in Application of Fracture Mechanics for Selection of Metallic Structural Materials, James E. Campbell, William W. Gerberich and John H. Underwood, Eds., American Society for Metals, Metals Park OH, 1982, p 120

8-5. Type 304 Stainless Steel: Effect of Humidity on Fatigue Crack Growth Rate

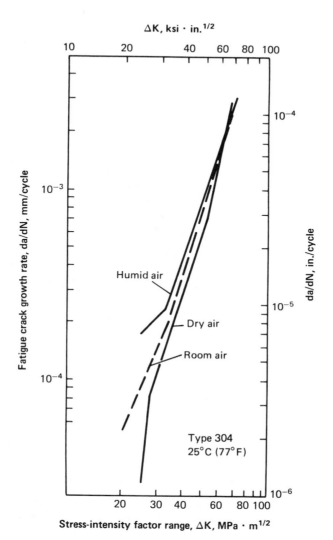

Effect of humidity on fatigue crack growth rates for type 304 stainless steel tested at room temperature, 0.17 Hz, and an R ratio of 0.

The effects of humid air environments on the room temperature fatigue crack growth rates of specimens of annealed type 304 stainless steel are shown in the above chart for specimens cycled at 0.17 Hz with an R ratio of zero (Shahinian, Watson, and Smith). At the lower end of the ΔK range, fatigue crack growth rates in humid air are substantially greater than crack growth rates in dry air. However, fatigue crack growth rates of specimens of type 304 stainless steel tested in a pressurized water reactor environment at 260 to 315 °C (500 to 600 °F) with R ratios of 0.2 and 0.7 were no greater than the fatigue crack growth rates in air at the same temperature with an R ratio less than 0.1 (Bamford). However, variations in R ratios influenced the fatigue crack growth rates in the pressurized water reactor environment.

Source: J. E. Campbell, "Fracture Properties of Wrought Stainless Steels," in Application of Fracture Mechanics for Selection of Metallic Structural Materials, James E. Campbell, William W. Gerberich and John H. Underwood, Eds., American Society for Metals, Metals Park OH, 1982, p 122

8-6. Type 304 Stainless Steel: Effect of Aging on Fatigue Crack Growth Rate

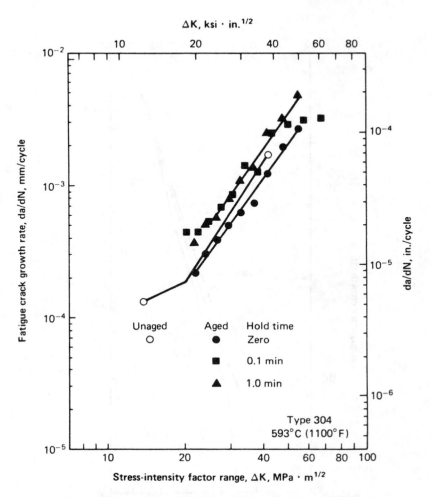

Effect of aging at 593 °C (1100 °F) for 5000 h, and hold times of 0.1 and 1.0 min for each cycle, on fatigue crack growth rates of L-T oriented specimens of type 304 stainless steel tested in air at 0.17 Hz and an R ratio of 0.

Because austenitic stainless steels are expected to give long service life, an evaluation of the effect of long-time aging at service temperatures is important. Results of fatigue crack growth rate tests on specimens that were tested in the unaged and aged conditions (5000 hours at 593 °C, or 1100 °F) are shown in this graph, as reported by Michel and Smith. After aging for 5000 hours at this temperature, precipitation of $M_{23}C_6$ carbides is essentially complete. These results indicate that at 593 °C (1100 °F) there are no deleterious effects of aging on the crack growth rates of specimens that are continuously cycled. When a holding time of 0.1 or 1.0 minute is included in each loading cycle, there tends to be a slight increase in the fatigue crack growth rate at a given ΔK level.

Source: J. E. Campbell, "Fracture Properties of Wrought Stainless Steels," in Application of Fracture Mechanics for Selection of Metallic Structural Materials, James E. Campbell, William W. Gerberich and John H. Underwood, Eds., American Society for Metals, Metals Park OH, 1982, p 121

8-7. Type 304 Stainless Steel: Effect of Temperature on Fatigue Crack Growth Rate

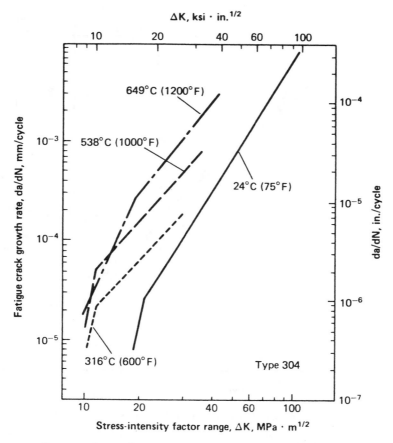

Effect of testing temperature on fatigue crack growth rates for annealed type 304 stainless steel tested in air at 0.066 Hz and an R ratio of 0 to 0.05.

Results of fatigue crack growth rate tests on types 304 and 304L stainless steel at room temperature and at elevated temperatures have been reported by James and Schwenk, and by others. As shown in this graph, increasing the exposure temperature from room temperature to 650 °C (1200 °F) increases the fatigue crack growth rates at any ΔK level within the range of the tests in an air environment. These data, reported by James and Schwenk, are for specimens of both the L-T and T-L orientations, for several different maximum alternating loads, for load ratios of 0 to 0.05, and for cyclic frequencies from 0.033 to 6.66 Hz for the room-temperature tests and 0.067 Hz for the elevated-temperature tests.

Source: J. E. Campbell, "Fracture Properties of Wrought Stainless Steels," in Application of Fracture Mechanics for Selection of Metallic Structural Materials, James E. Campbell, William W. Gerberich and John H. Underwood, Eds., American Society for Metals, Metals Park OH, 1982, p 115

8-8. Type 304 Stainless Steel: Damage Relation at 650 °C (1200 °F)

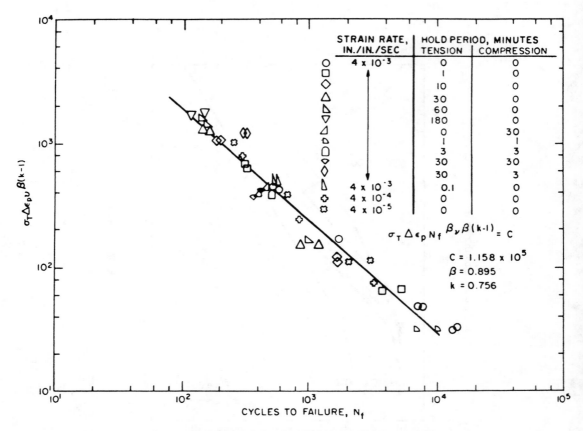

Ostergren's damage relation for AISI 304 at 650 °C (1200 °F).

The damage function was proposed by Ostergren and is based on the frequency-modified fatigue approach. A damage function is approximated by the quantity $\sigma_t \Delta\epsilon_p$, where σ_t is the maximum stress in the cycle and $\Delta\epsilon_p$ is the inelastic strain range. The tensile hysteresis energy is employed to account for the facts that low-cycle fatigue is essentially a crack-growth process and that crack growth and damage occur only during the tensile part of the cycle. The use of the tensile-stress quantity, in conjunction with the plastic-strain range, provides a means of accounting for loop unbalance, since, for the same inelastic strain, a positive mean stress provides a greater hysteresis energy than does a compressive mean stress. The method is effective in accounting for hold-time effects, as indicated in the chart above.

Source: L. F. Coffin, "Fatigue in Machines and Structures—Power Generation," in Fatigue and Microstructure, American Society for Metals, Metals Park OH, 1979, p 23

8-9. Type 304 Stainless Steel: Fatigue Crack Growth Rate at Room and Subzero Temperatures

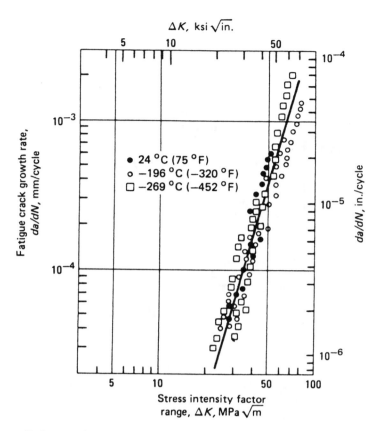

Fatigue crack growth rate data for type 304 austenitic stainless steel (annealed) at room temperature and at subzero temperatures. For this alloy, crack growth rates are nearly the same at room and cryogenic temperatures.

Source: Metals Handbook, 9th Edition, Volume 3, Properties and Selection: Stainless Steels, Tool Materials and Special-Purpose Metals, American Society for Metals, Metals Park OH, 1980, p 756

8-10. Types 304 and 304L Stainless Steel: Effect of Cryogenic Temperatures on Fatigue Crack Growth Rate

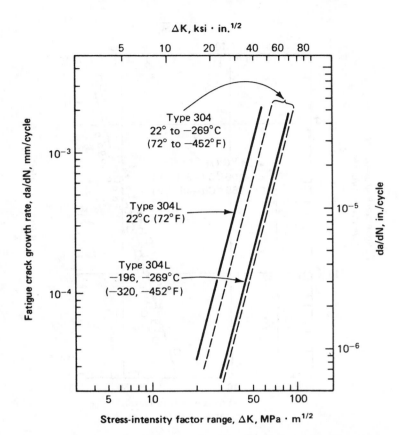

Fatigue crack growth rates for annealed types 304 and 304L stainless steel at room and cryogenic temperatures, 20 to 28 Hz, and an *R* ratio of 0.1.

Fatigue crack growth rate data obtained by Tobler and Reed on specimens of types 304 and 304L stainless steel (annealed) at temperatures in the range from room temperature to liquid helium temperature (−269 °C, or −452 °F) are shown in this graph. The data for type 304 were scattered over the range shown, while for type 304L, the data at room temperature described one curve and the data at the cryogenic temperatures described the other curve. These results indicate that cryogenic fatigue crack growth rates for type 304 do not deviate significantly from room temperature fatigue crack growth rates over the ΔK range studied. Furthermore, if design calculations for type 304L are based on room temperature fatigue crack growth rates, the calculations will be conservative for cryogenic exposure.

Source: J. E. Campbell, "Fracture Properties of Wrought Stainless Steels," in Application of Fracture Mechanics for Selection of Metallic Structural Materials, James E. Campbell, William W. Gerberich and John H. Underwood, Eds., American Society for Metals, Metals Park OH, 1982, p 123

8-11. Type 304 Stainless Steel: Fatigue Crack Growth Rate in Air With Variation in Waveforms

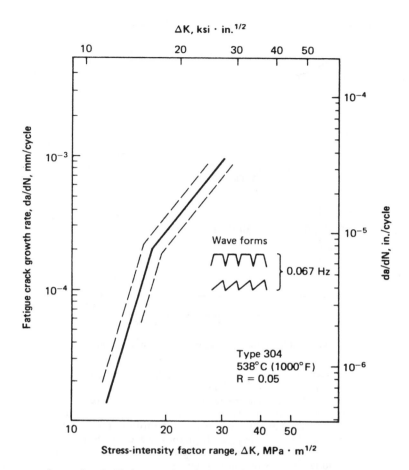

Scatter band of fatigue crack growth rates for annealed type 304 stainless steel at 538 °C (1000 °F) in air at an R ratio of 0.05 with two different waveforms at 0.067 Hz.

The data presented in this graph were obtained in tests with a sawtooth waveform. Changing from a sawtooth waveform to a waveform with a short holding period at maximum load did not influence the overall fatigue crack growth rates according to additional data reported by James and shown above.

Source: J. E. Campbell, "Fracture Properties of Wrought Stainless Steels," in Application of Fracture Mechanics for Selection of Metallic Structural Materials, James E. Campbell, William W. Gerberich and John H. Underwood, Eds., American Society for Metals, Metals Park OH, 1982, p 117

8-12. Type 304 Stainless Steel: Effect of Hold Time on Cycles to Failure

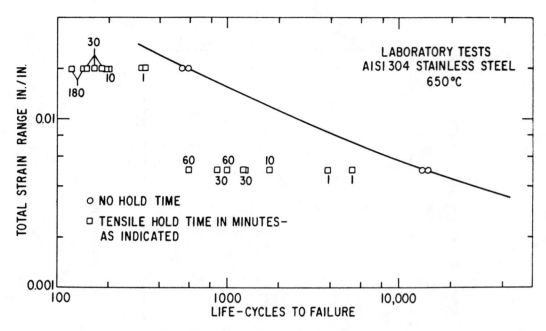

Effect of hold time on life for AISI 304 stainless steel.

Wave-shape effects are also important in fatigue crack growth, as has been studied by Barsom. He observed that the crack growth rates were greater as the loading rate increased and the unloading rate decreased, given a fixed period of cycling. Overload effects are also important in retarding crack growth. Substantial damage can result from these wave shapes, particularly when the hysteresis loop is severely unbalanced, as can occur in long tensile-strain hold-time tests.

Source: L. F. Coffin, "Fatigue in Machines and Structures—Power Generation," in Fatigue and Microstructure, American Society for Metals, Metals Park OH, 1979, p 19

8-13. Type 304 Stainless Steel: Effect of Hold Time and Continuous Cycling on Fatigue Crack Growth Rates

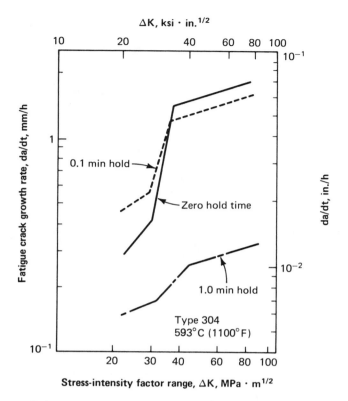

Fatigue crack growth rates per unit of time (da/dt) for annealed type 304 stainless steel for continuous cycling (0.17 Hz), for 0.1 and 1.0-min hold times at maximum load for each cycle at 593 °C (1100 °F), and for an R ratio of 0.

As shown above, the fatigue crack growth rate is greater for specimens tested with no holding time (continuous cycling) than for specimens held at maximum load for 0.1 or 1.0 minute per cycle. The lowest fatigue crack growth rates occurred for specimens with the longest holding time, based on da/dt. The same trend was observed for tests at 593 °C (1100 °F), as shown here. Therefore, cyclic loading has a more damaging effect than static loading on crack growth per unit of time.

Source: J. E. Campbell, "Fracture Properties of Wrought Stainless Steels," in Application of Fracture Mechanics for Selection of Metallic Structural Materials, James E. Campbell, William W. Gerberich and John H. Underwood, Eds., American Society for Metals, Metals Park OH, 1982, p 118

8-14. Type 304 Stainless Steel: Effect of Cyclic Frequency on Fatigue Crack Growth Rate

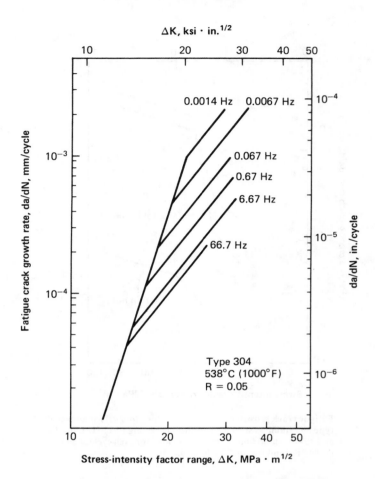

Effect of variation in cyclic frequency on fatigue crack growth rates for annealed type 304 stainless steel at 538 °C (1000 °F) for an *R* ratio of 0.05 in air with a sawtooth waveform.

For fatigue crack growth rate tests on specimens of annealed type 304 stainless steel at elevated temperatures, increasing the cyclic frequency will decrease the crack growth rate over part of the ΔK range, as shown here for tests at 538 °C (1000 °F).

Source: J. E. Campbell, "Fracture Properties of Wrought Stainless Steels," in Application of Fracture Mechanics for Selection of Metallic Structural Materials, James E. Campbell, William W. Gerberich and John H. Underwood, Eds., American Society for Metals, Metals Park OH, 1982, p 116

8-15. Type 304 Stainless Steel: Effect of Frequency on Fatigue Crack Growth Behavior

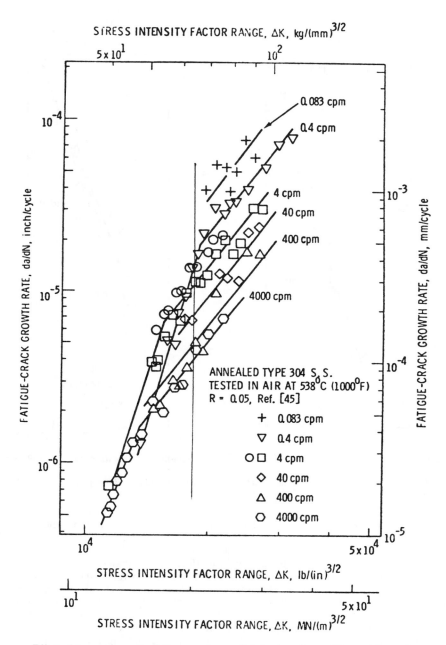

Effect of frequency on the fatigue crack growth behavior of type 304 tested in an air environment at 538 °C (1000 °F).

Source: L. F. Coffin, "Fatigue in Machines and Structures—Power Generation," in Fatigue and Microstructure, American Society for Metals, Metals Park OH, 1979, p 14

8-16. Type 304 Stainless Steel Welded With Type 308: Fatigue Crack Growth Rates

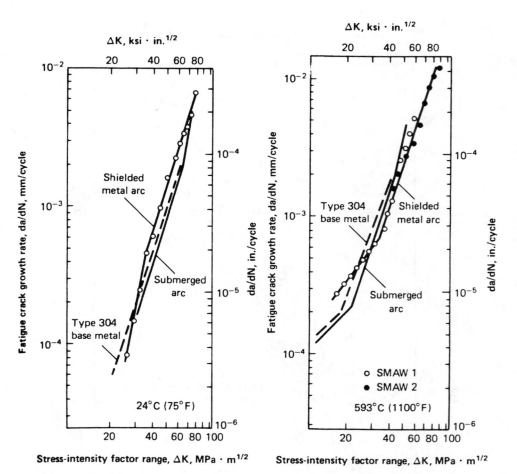

Fatigue crack growth rates for annealed type 304 base metal and type 308 weld metal at 24 and 593 °C (75 and 1100 °F), 0.17 Hz, and an R ratio of 0.

Type 308 stainless steel is the alloy that is usually used for welding rod for weldments in type 304 stainless steel when those weldments are to be exposed to room temperature or to elevated temperatures in service. Because service experience has shown that failures are more likely to originate in weld metal or in heat affected zones than in the base metal, it is important to have fracture information on weldments. In general, fatigue studies at elevated temperatures on specimens from type 304 weldments have shown that the fatigue crack growth rates in the type 308 weld metal and heat affected zones are no greater than in comparable specimens of the base metal. Fatigue crack growth rate data obtained by Shahinian for specimens of type 304 welded with type 308 rod by the submerged arc and shielded metal arc processes are shown above for tests at room temperature and at 593 °C (1100 °F).

Source: J. E. Campbell, "Fracture Properties of Wrought Stainless Steels," in Application of Fracture Mechanics for Selection of Metallic Structural Materials, James E. Campbell, William W. Gerberich and John H. Underwood, Eds., American Society for Metals, Metals Park OH, 1982, p 125

8-17. Types 304 and 310 Stainless Steel: Effect of Direction on S-N

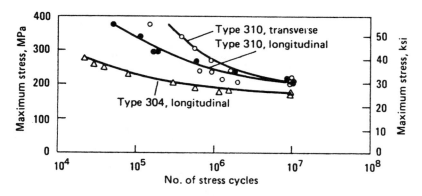

S-N curves for two grades of stainless steel.

Source: Metals Handbook, 9th Edition, Volume 3, Properties and Selection: Stainless Steels, Tool Materials and Special-Purpose Metals, American Society for Metals, Metals Park OH, 1980, p 32

8-18. Types 304, 316, 321, and 348 Stainless Steel: Effects of Temperature on Fatigue Crack Growth Rates

Fatigue crack growth rates for annealed types 304, 316, 321, and 348 stainless steel in air at room temperature and 593 °C (1100 °F), L-T orientation, 0.17 Hz, and an R ratio of 0.

As reported by Shahinian, Smith, and Watson, fatigue crack growth rate tests were made on single-edge-notch cantilever specimens of types 321 and 348 stainless steel from L-T orientation at 0.17 Hz with an R ratio of zero at room temperature and at elevated temperatures to 593 °C (1100 °F). As for types 304 and 316, fatigue crack growth rates in air increased with increasing testing temperature. The curves above show that, at room temperature, the fatigue crack growth rates for types 304, 316, 321, and 348 all fall within a narrow band. For tests at 593 °C (1100 °F), however, specimens of type 316 had the least fatigue crack propagation resistance, whereas specimens of type 348 had the highest fatigue crack propagation resistance, over the ΔK range studied. Results of tests on specimens of types 304 and 321 were nearly the same at 593 °C (1100 °F) in air.

Source: J. E. Campbell, "Fracture Properties of Wrought Stainless Steels," in Application of Fracture Mechanics for Selection of Metallic Structural Materials, James E. Campbell, William W. Gerberich and John H. Underwood, Eds., American Society for Metals, Metals Park OH, 1982, p 138

8-19. Type 309S Stainless Steel: Effect of Grain Size on Fatigue Crack Growth Rate

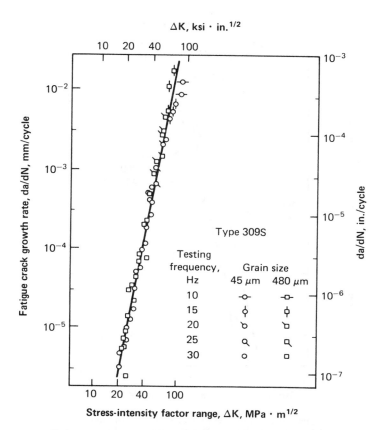

Fatigue crack growth rates for annealed type 309S stainless steel for two grain sizes, at frequencies from 10 to 30 Hz and an R ratio of 0.05 at room temperature in air.

Types 309S and 310S stainless steel are the low-carbon versions of types 309 and 310. They have higher chromium and nickel contents than type 304 and consequently have better corrosion resistance and more stable austenite than type 304. Fatigue crack growth rate data have been reported by Thompson for tests made at room temperature on compact specimens from plate of type 309S and the L-T orientation after heat treating to a grain size of 45 μm in one set and 480 μm in the second set. Specimens with the smaller grain size had substantially higher yield and ultimate tensile strengths than the specimens with the larger grain size. Fatigue crack growth rates were obtained on tension-tension loading at frequencies from 10 to 30 Hz and at an R ratio of 0.05. The results are plotted above. These data provide further evidence that a wide variation in grain size, and the associated variation in strength level, does not affect the results of fatigue crack growth rate tests.

Source: J. E. Campbell, "Fracture Properties of Wrought Stainless Steels," in Application of Fracture Mechanics for Selection of Metallic Structural Materials, James E. Campbell, William W. Gerberich and John H. Underwood, Eds., American Society for Metals, Metals Park OH, 1982, p 126

8-20. Type 310S Stainless Steel: Effect of Temperature on Fatigue Crack Growth Rate

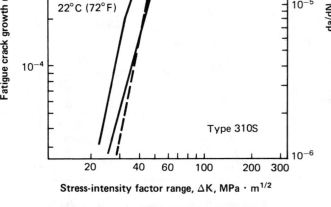

Fatigue crack growth rates for annealed type 310S stainless steel at 22, −196, and −269 °C (72, −320, and −452 °F), 10 to 28 Hz, and an R ratio of 0.1, with corresponding data for SMAW welds with type 316 filler metal.

Because of its high nickel content, type 310S stainless steel is completely stable at all cryogenic temperatures and with any amount of cold working. Therefore, it is often considered for cryogenic applications that require a high degree of austenite stability on thermal cycling and strain cycling. Fatigue crack growth rate at various temperatures is illustrated above.

Source: J. E. Campbell, "Fracture Properties of Wrought Stainless Steels," in Application of Fracture Mechanics for Selection of Metallic Structural Materials, James E. Campbell, William W. Gerberich and John H. Underwood, Eds., American Society for Metals, Metals Park OH, 1982, p 127

8-21. Type 316 Stainless Steel: Growth Rate of Fatigue Cracks in Weldments

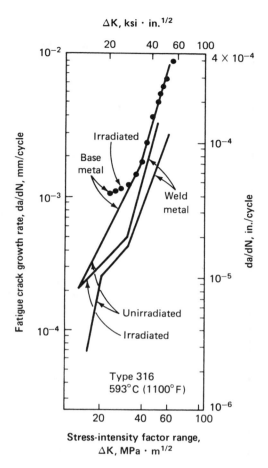

Fatigue crack growth rates in type 316 base metal and weld metal in the unirradiated and irradiated conditions at 593 °C (1100 °F) in air [fluence 1.2×10^{22} n/cm², >0.1 MeV at 410 °C (770 °F)].

Results of fatigue crack growth rate tests on weldments of type 316 stainless steel have shown that the crack growth rates in the weld metal are generally no higher than in the base metal and may be somewhat lower at elevated temperatures (Shahinian, Smith, and Hawthorne). The curve shown above for unirradiated weld metal tested at 593 °C (1100 °F) represents fatigue crack growth rates substantially lower than those for the unirradiated base metal at any given ΔK level (Shahinian). The weld was produced by the submerged arc method using type 316 welding rod. Weldments were stress-relief annealed at 482 °C (900 °F). Specimens were single-edge-notch specimens for cantilever loading and were tested at 0.17 Hz and at an R ratio of zero. Irradiation slightly reduced the fatigue crack growth resistance of the weld metal, but its fatigue crack growth resistance was better than that of the unirradiated base metal.

Source: J. E. Campbell, "Fracture Properties of Wrought Stainless Steels," in Application of Fracture Mechanics for Selection of Metallic Structural Materials, James E. Campbell, William W. Gerberich and John H. Underwood, Eds., American Society for Metals, Metals Park OH, 1982, p 134

8-22. Type 316 Stainless Steel: Fatigue Crack Growth Rates—Aged vs Unaged

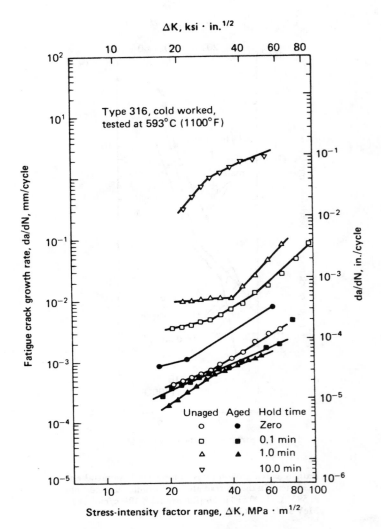

Effect of exposure at 593 °C (1100 °F) for 5000 h, and hold times during cycling, on fatigue crack growth rate of 20% cold worked type 316 stainless steel at 593 °C in air.

Results also have been reported by James for fatigue crack growth rate tests in 20% cold worked specimens of type 316 stainless steel which were cycled at frequencies of 0.0055 to 6.66 Hz, at 538 °C (1100 °F) and at an R ratio of 0.05. Over the ΔK range studied, the fatigue crack growth rates were highest for the specimens subjected to the lowest cyclic frequency.

Source: J. E. Campbell, "Fracture Properties of Wrought Stainless Steels," in Application of Fracture Mechanics for Selection of Metallic Structural Material, James E. Campbell, William W. Gerberich and John H. Underwood, Eds., American Society for Metals, Metals Park OH, 1982, p 133

8-23. Type 316 Stainless Steel: Fatigue Crack Growth Rates—Effect of Aging

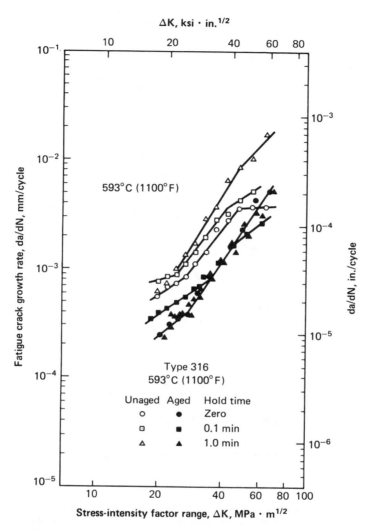

Effect of exposure in air at 593 °C (1100 °F) for 5000 h, and hold times, on fatigue crack growth rates for annealed type 316 stainless steel at 593 °C in air.

Source: J. E. Campbell, "Fracture Properties of Wrought Stainless Steels," in Application of Fracture Mechanics for Selection of Metallic Structural Materials, James E. Campbell, William W. Gerberich and John H. Underwood, Eds., American Society for Metals, Metals Park OH, 1982, p 130

8-24. Type 316 Stainless Steel: Effect of Temperature on Fatigue Crack Growth Rate

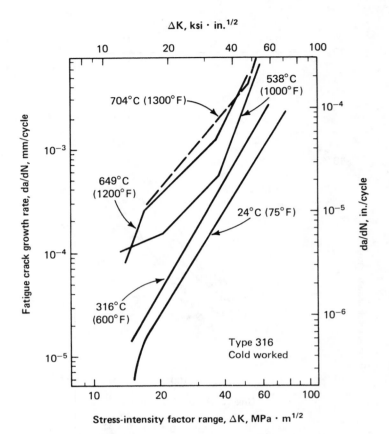

Fatigue crack growth rates of 20% cold worked type 316 stainless steel for various temperatures. Curves are averages for L-T and T-L specimens at each temperature in air; 3 Hz at 24 °C, 0.67 Hz at elevated temperatures; $R = 0.05$.

Results of tests on compact specimens of 20% cold worked type 316 stainless steel at frequencies of 0.67 and 3.0 Hz and at an R ratio of 0.05 are summarized in the above graph. Similar results have been reported by Shahinian for tests on cold worked type 316 at 427 °C (800 °F).

Source: J. E. Campbell, "Fracture Properties of Wrought Stainless Steels," in Application of Fracture Mechanics for Selection of Metallic Structural Materials, James E. Campbell, William W. Gerberich and John H. Underwood, Eds., American Society for Metals, Metals Park OH, 1982, p 132

8-25. Type 316 Stainless Steel: Effect of Cyclic Frequency on Fatigue Crack Growth Rate

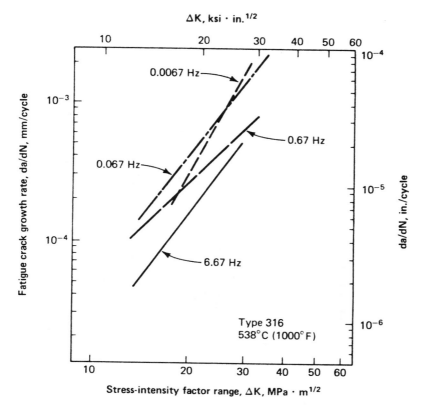

Effect of variation in cyclic frequency on fatigue crack growth rate of annealed type 316 stainless steel in air at 538 °C (1000 °F) and an R ratio of 0.05.

As may be observed above in tests at frequencies in the range from 0.0067 to 6.67 Hz at 538 °C (1000 °F), the trend is for the crack growth rate to increase as the frequency is decreased, but there is more scatter than for type 304.

In studying heat-to-heat variations in fatigue crack growth rates for specimens from three heats of type 316 stainless steel, James has shown that the spread from high to low values of fatigue crack growth rates is no greater than that represented by a factor of 2.6 over the range of ΔK values studied. One heat was produced by air melting, another by vacuum arc remelting, and a third by double vacuum melting.

Source: J. E. Campbell, "Fracture Properties of Wrought Stainless Steels," in Application of Fracture Mechanics for Selection of Metallic Structural Materials, James E. Campbell, William W. Gerberich and John H. Underwood, Eds., American Society for Metals, Metals Park OH, 1982, p 131

8-26. Type 316 Stainless Steel: Fatigue Crack Growth Rate in the Annealed Condition

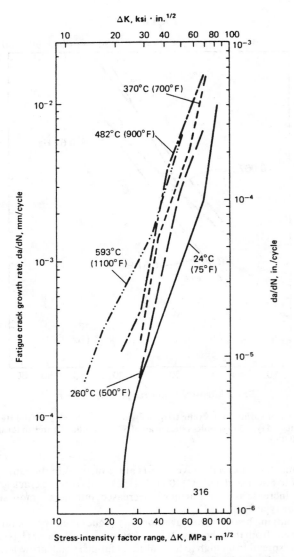

Effect of testing temperature on fatigue crack growth rates for annealed type 316 stainless steel tested in air at 0.17 Hz and an R ratio of 0.

Most of the fatigue crack growth rate testing on type 316 stainless steel has been oriented toward its use in components for nuclear reactors, but the data also are applicable to design of equipment for fossil fuel power stations, petrochemical refineries, and chemical plants. Its improved yield strength compared with that of type 304 stainless steel is an advantage for these applications. The austenite stability in type 316 is greater than that in type 304, so it is advantageous to use type 316 rather than type 304 for critical cryogenic applications. Effects of elevated temperature on crack growth rate are summarized in the graph above.

Source: J. E. Campbell, "Fracture Properties of Wrought Stainless Steels," in Application of Fracture Mechanics for Selection of Metallic Structural Materials, James E. Campbell, William W. Gerberich and John H. Underwood, Eds., American Society for Metals, Metals Park OH, 1982, p 129

8-27. Type 316 Stainless Steel: Effect of Environment (Sodium, Helium, and Air) on Cycles to Failure

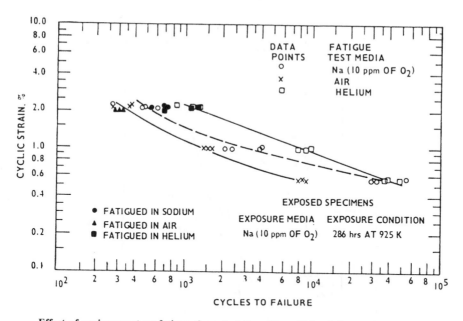

Effect of environment on fatigue characteristics of type 316 stainless steel at 925K; based on cyclic strain and cycles to failure.

Source: R. H. Cook and R. P. Skelton, "Environment-Dependence of the Mechanical Properties of Metals at High Temperature," in Source Book on Materials for Elevated-Temperature Applications, Elihu F. Bradley, Ed., American Society for Metals, Metals Park OH, 1979, p 84

8-28. Types 316 and 321 Stainless Steel: Effects of Gaseous Environments on Fatigue Crack Growth Rates

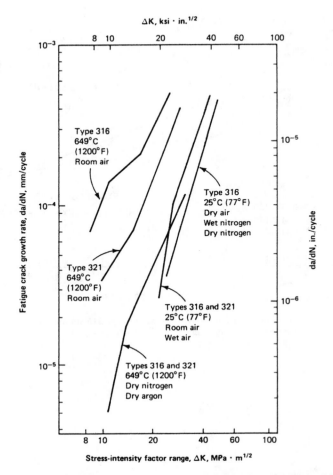

Effect of gas environments on fatigue crack growth rates for types 316 and 321 stainless steel at 25 and 649 °C (77 and 1200 °F).

Fatigue crack growth rate data at 25 °C (77 °F) show that crack growth rates increased slightly with increased humidity when oxygen was present but that high humidity in an inert gas had no significant effect. Fatigue crack growth rates in room air at room temperature were the same for types 316 and 321 stainless steel. Furthermore, in tests at 649 °C (1200 °F) in dry nitrogen, fatigue crack growth rates for types 316 and 321 also were the same. In air, however, fatigue crack growth rates in type 316 specimens increased by a factor of about 22 over rates in an inert environment at the same temperature.

Source: J. E. Campbell, "Fracture Properties of Wrought Stainless Steels," in Application of Fracture Mechanics for Selection of Metallic Structural Materials, James E. Campbell, William W. Gerberich and John H. Underwood, Eds., American Society for Metals, Metals Park OH, 1982, p 135

8-29. Type 321 Stainless Steel: Effect of Hold Time on Fatigue Crack Growth Rates

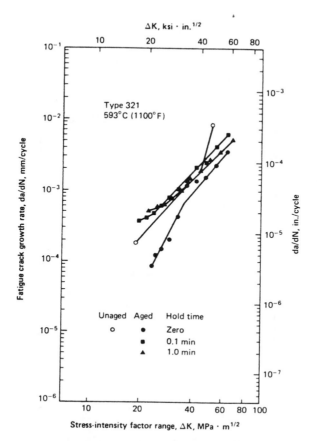

Fatigue crack growth rates for annealed type 321 stainless steel unaged and aged at 593 °C (1100 °F) for 5000 h and tested in air with continuous sawtooth waveform (0.17 Hz), with 0.1 and 1.0-min hold time at an R ratio of 0 at 593 °C (1100 °F).

Results of tests by Michel and Smith on specimens of annealed type 321 stainless steel that had been aged at 593 °C (1100 °F) for 5000 hours and then tested at 593 °C have shown that long-time exposure at the service temperature does not reduce the fatigue crack propagation resistance in air. Aged specimens tested with zero holding time had lower crack growth rates than corresponding specimens that were not aged (see above graph). Fatigue cycling with holding times of 0.1 and 1.0 minute on each cycle increased the crack growth rates slightly, as shown in the figure.

Source: J. E. Campbell, "Fracture Properties of Wrought Stainless Steels," in Application of Fracture Mechanics for Selection of Metallic Structural Materials, James E. Campbell, William W. Gerberich and John H. Underwood, Eds., American Society for Metals, Metals Park OH, 1982, p 139

8-30. Type 403 Stainless Steel: Effect of Environment on Fatigue Crack Growth Rate

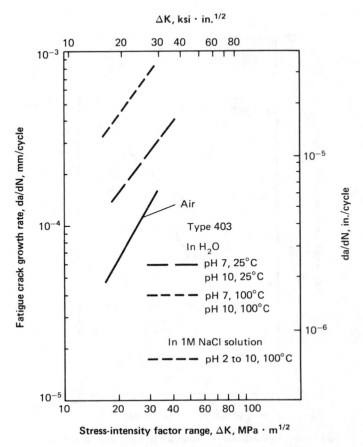

Fatigue crack growth rates in type 403 stainless steel in air, water, and a 1 M NaCl solution at 10 Hz and an R ratio of 0.5.

Exposure to water at 25 °C (77 °F) resulted in intermediate crack growth rates between those in air and those in water at 100 °C, as shown on a different scale in the above graph.

Tests in the 0.01 M (molar) and 1.0 M sodium chloride solutions were made with the solutions at pH levels of 2, 7, and 10 and with an open circuit. Fatigue crack growth rates in 0.01 M sodium chloride at pH 10 and 100 °C were the same as those in water at 100 °C. At lower cyclic frequencies, the fatigue crack growth rates were higher than at 40 Hz at ΔK values above 20 MPa·m$^{1/2}$ (18 ksi·in.$^{1/2}$). For tests in the 1.0 M sodium chloride solution at 100 °C (212 °F) (see graph), fatigue crack growth rates were the same as for water at the same temperature. At 100 °C (212 °F), fatigue crack growth rates in 1.0 M sodium phosphate solution at pH 10 and at 10 and 40 Hz and in 1.0 M sodium silicate at pH 10 and at 10 Hz were practically the same as those in air.

Source: J. E. Campbell, "Fracture Properties of Wrought Stainless Steels," in Application of Fracture Mechanics for Selection of Metallic Structural Materials, James E. Campbell, William W. Gerberich and John H. Underwood, Eds., American Society for Metals, Metals Park OH, 1982, p 147

8-31. Type 403 Modified Stainless Steel: Scatter of Fatigue Crack Growth Rates

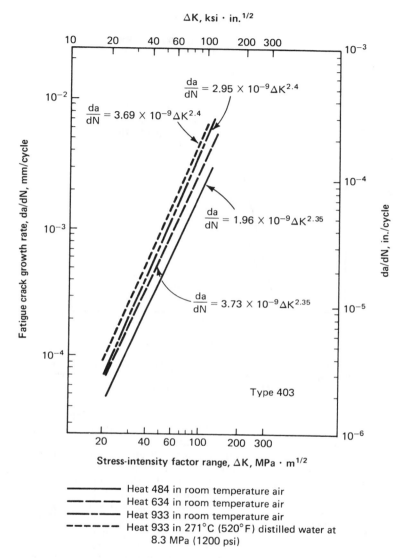

——————— Heat 484 in room temperature air
— — — — Heat 634 in room temperature air
— — — — Heat 933 in room temperature air
- - - - - - Heat 933 in 271°C (520°F) distilled water at 8.3 MPa (1200 psi)

Upper boundaries of fatigue crack growth rate scatter bands for three heats of type 403 modified stainless steel in the heat treated condition, tested at 10 Hz and an R ratio of 0.083 or 0.067.

The curves representing the upper boundaries of the scatter bands of the fatigue crack growth rate data indicate that there is some heat-to-heat variation in fatigue crack growth rate properties for these heats. Furthermore, exposure at 271 °C (520 °F) in distilled water at a pressure of 8.3 MPa (1200 psi) increased the fatigue crack growth rate.

Source: J. E. Campbell, "Fracture Properties of Wrought Stainless Steels," in Application of Fracture Mechanics for Selection of Metallic Structural Materials, James E. Campbell, William W. Gerberich and John H. Underwood, Eds., American Society for Metals, Metals Park OH, 1982, p 145

8-32. Type 422 Stainless Steel: Fatigue Crack Growth Rates in Precracked Specimens

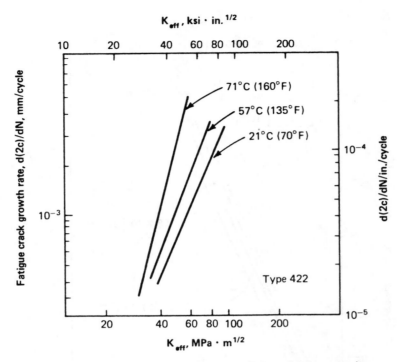

Fatigue crack growth rates in precracked round rotating beam specimens of type 422 stainless steel in 4.5% NaCl solution at room and elevated temperatues, 10 Hz, and an R ratio of -1.

Type 422 stainless steel contains nickel, molybdenum, and tungsten, as well as 12% chromium to improve properties. The effects of sodium chloride solutions and elevated-temperature exposure on fatigue crack growth rates were determined by Eisenstadt and Rajan in tests of notched round rotating beam specimens in which the numbers of test cycles were marked by minor stress interruptions that produced marking rings. Calculations for maximum stress-intensity factors were based on equations for solid round bars subjected to bending loads. The material for these tests apparently had been heat treated to a yield strength of approximately 827 MPa (120 ksi). The specimens were one inch in diameter in the test sections. Each specimen was rotated at 600 cycles per minute (10 Hz) while at constant load with the salt water solution flowing over the notched section. Tests with several concentrations of salt solution indicated that the maximum corrosive effect was obtained with the 4.5% solution. Results of tests with specimens in the 4.5% sodium chloride solution at room temperature, 57°C (135°F), and 71°C (160°F) are shown above. Increasing the temperature of the solution substantially increased the fatigue crack growth rates.

Source: J. E. Campbell, "Fracture Properties of Wrought Stainless Steels," in Application of Fracture Mechanics for Selection of Metallic Structural Materials, James E. Campbell, William W. Gerberich and John H. Underwood, Eds., American Society for Metals, Metals Park OH, 1982, p 150

8-33. Type 422 Stainless Steel: Fatigue Strength—Longitudinal vs Transverse

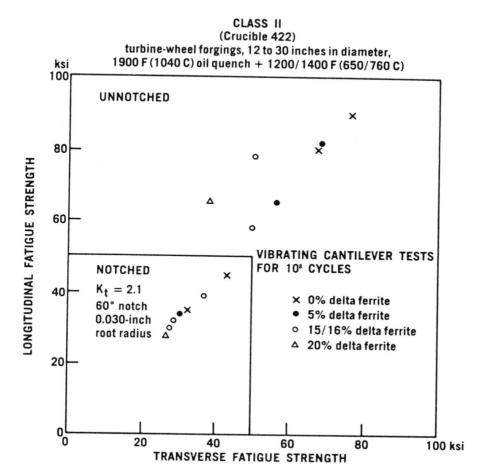

Transverse fatigue strength as related to longitudinal fatigue strength for type 422 stainless steel, including effects of varying amounts of delta ferrite.

Source: J. Z. Briggs and T. D. Parker, "The Super 12% Cr Steels," in Source Book on Materials for Elevated-Temperature Applications, Elihu F. Bradley, Ed., American Society for Metals, Metals Park OH, 1979, p 121

8-34. Type 422 Stainless Steel: Effect of Temperature on Fatigue Strength

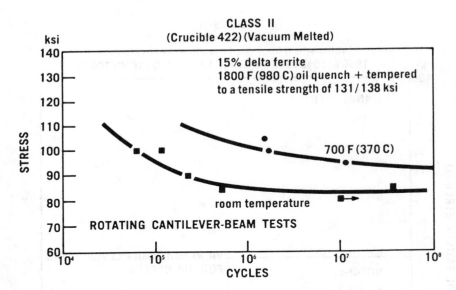

S-N curves for vacuum-melted type 422 stainless steel with 15% delta ferrite, showing effect of temperature on fatigue strength.

Source: J. Z. Briggs and T. D. Parker, "The Super 12% Cr Steels," in Source Book on Materials for Elevated-Temperature Applications, Elihu F. Bradley, Ed., American Society for Metals, Metals Park OH, 1979, p 121

8-35. Type 422 Stainless Steel: Effects of Delta Ferrite on Fatigue Strength

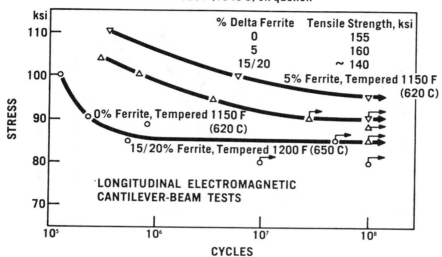

S-N curves for type 422 stainless steel, which demonstrate the adverse effects of delta ferrite on fatigue strength.

Source: J. Z. Briggs and T. D. Parker, "The Super 12% Cr Steels," in Source Book on Materials for Elevated-Temperature Applications, Elihu F. Bradley, Ed., American Society for Metals, Metals Park OH, 1979, p 121

8-36. 17-4 PH Stainless Steel: Fatigue Crack Growth Rates in Air vs Salt Solution

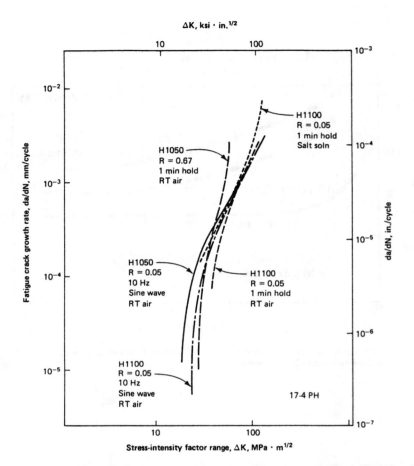

Fatigue crack growth rates in WOL specimens of 17-4 PH stainless steel in the H1050 and H1100 conditions in room temperature air and in a 3.5% NaCl solution.

Results of fatigue crack growth rate tests on specimens of 17-4 PH stainless steel under comparable conditions are presented here. Those specimens that were tested in the H1050 condition at a stress ratio of 0.67 with a one-minute holding period at maximum load in each cycle had the highest fatigue crack growth rates (as for 15-5 PH) in the upper levels of ΔK values. Specimens in the H1100 condition tested in a salt solution with a one-minute holding period, however, had fatigue crack growth rates only slightly higher than those of comparable specimens tested in air with continuous cycling.

Source: J. E. Campbell, "Fracture Properties of Wrought Stainless Steels," in Application of Fracture Mechanics for Selection of Metallic Structural Materials, James E. Campbell, William W. Gerberich and John H. Underwood, Eds., American Society for Metals, Metals Park OH, 1982, p 156

8-37. 15-5 PH Stainless Steel: Fatigue Crack Growth Rates in Air vs Salt Solution

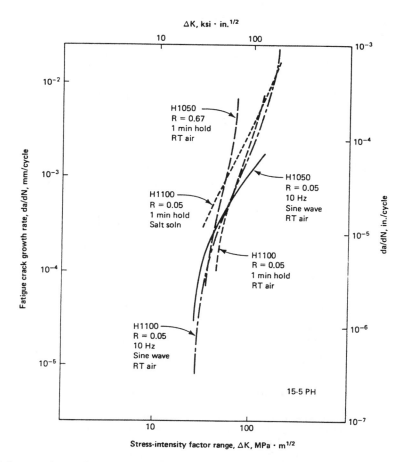

Fatigue crack growth rates in WOL specimens of 15-5 PH stainless steel in the H1050 and H1100 conditions in room temperature air and in a 3.5% NaCl solution.

For specimens in the H1050 condition, increasing the R ratio from 0.05 to 0.67 and incorporating a one-minute holding period at maximum load in each cycle substantially increased the crack growth rates at ΔK values over 40 MPa·m$^{1/2}$ (36 ksi·in.$^{1/2}$). For specimens in the H1100 condition, exposure to a salt solution environment during tests with a one-minute holding period at maximum load increased the fatigue crack growth rates over those of specimens tested in air with one-minute holding time or with continuous cycling (see graph).

Source: J. E. Campbell, "Fracture Properties of Wrought Stainless Steels," in Application of Fracture Mechanics for Selection of Metallic Structural Materials, James E. Campbell, William W. Gerberich and John H. Underwood, Eds., American Society for Metals, Metals Park OH, 1982, p 155

8-38. PH 13-8 Mo Stainless Steel: Fatigue Crack Growth Rates at Room Temperature

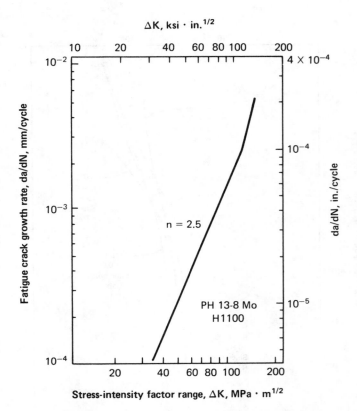

Fatigue crack growth rates in cantilever beam specimens of PH 13-8 Mo (H1100) stainless steel, at L-T orientation, 0.17 Hz, and an R ratio of 0, in room temperature air. Data are based on the stress-intensity-factor range as shown.

Source: J. E. Campbell, "Fracture Properties of Wrought Stainless Steels," in Application of Fracture Mechanics for Selection of Metallic Structural Materials, James E. Campbell, William W. Gerberich and John H. Underwood, Eds., American Society for Metals, Metals Park OH, 1982, p 159

8-39. PH 13-8 Mo Stainless Steel: Fatigue Crack Growth Rates in Air and Sump Tank Water

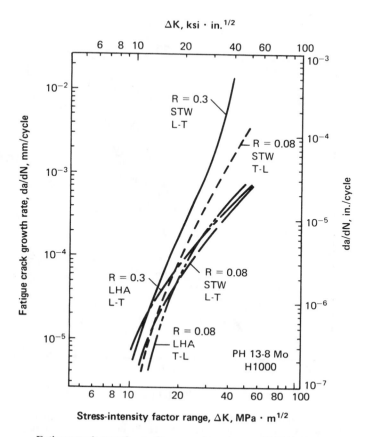

Fatigue crack growth rates in compact specimens of PH 13-8 Mo stainless steel in the H1000 condition for room temperature tests at 1 Hz, R ratios of 0.08 and 0.3, L-T and T-L orientations, in low-humidity air (LHA) or sump tank water (STW).

Effects of increasing the load ratio, R, on fatigue crack growth rates in low humidity air (LHA) in sump tank residue water (STW) for specimens of PH 13-8 Mo (H1000) are shown above. The highest fatigue crack growth rates in this series were obtained on specimens tested at an R ratio of 0.3 in STW. Increasing the load ratio from 0.08 to 0.3 had a marked effect on the growth rates.

Source: J. E. Campbell, "Fracture Properties of Wrought Stainless Steels," in Application of Fracture Mechanics for Selection of Metallic Structural Materials, James E. Campbell, William W. Gerberich and John H. Underwood, Eds., American Society for Metals, Metals Park OH, 1982, p 158

8-40. PH 13-8 Mo Stainless Steel: Fatigue Crack Growth Rates at Subzero Temperatures

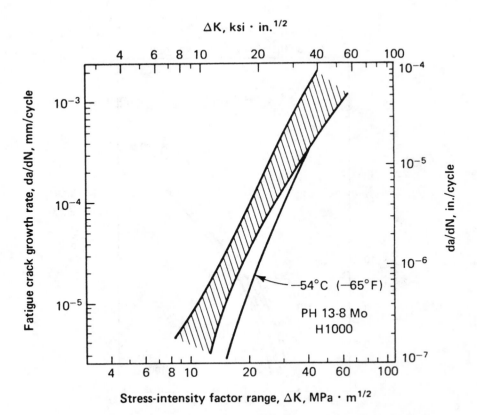

Fatigue crack growth rate scatter band for compact specimens from rolled bar and extrusions of PH 13-8 Mo stainless steel in the H1000 condition for room temperature tests in low-humidity air and in sump tank water at frequencies of 1 and 6 Hz and an R ratio of 0.08 for L-T and T-L orientations.

Fatigue crack growth rate data for room temperature tests on specimens from rolled bar and extrusions of PH 13-8 Mo (H1000) stainless steel make up the scatter band in the above graph. Specimens of L-T and T-L orientations were tested in low-humidity air and in sump tank residue water at frequencies of 1 and 6 Hz and at an R ratio of 0.08. Under these conditions, variations in frequency and environment had little effect on fatigue crack growth rates. For tests at $-54\,°C\,(-65\,°F)$, the rates of fatigue crack growth were lower than those at room temperature over most of the ΔK range.

Source: J. E. Campbell, "Fracture Properties of Wrought Stainless Steels," in Application of Fracture Mechanics for Selection of Metallic Structural Materials, James E. Campbell, William W. Gerberich and John H. Underwood, Eds., American Society for Metals, Metals Park OH, 1982, p 157

8-41. PH 13-8 Mo Stainless Steel: Constant-Life Fatigue Diagram

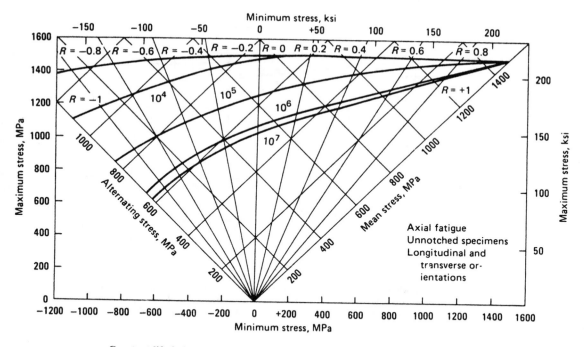

Constant-life fatigue diagram for PH 13-8 Mo stainless steel, condition H1000.

Source: Metals Handbook, 9th Edition, Volume 3, Properties and Selection: Stainless Steels, Tool Materials and Special-Purpose Metals, American Society for Metals, Metals Park OH, 1980, p 32

8-42. Types 600 and 329 Stainless Steel: S-N Curves for Two Processing Methods

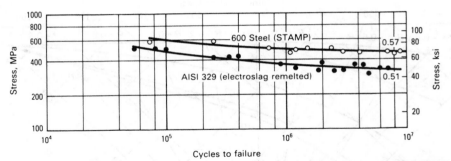

Steel	Tensile strength, MPa (ksi)	Yield strength (0.2% offset), MPa (ksi)	Mechanical properties Elongation in 50 mm (2 in.), %	Reduction in area, %	Impact energy, J (ft·lb)	Fatigue strength, MPa (ksi)
STAMP 600	760 (110)	600 (87)	26	54	25 (18)	430 (62)
Electroslag-remelted 329	630 (91)	500 (73)	29	65	35 (25)	320 (46)

S-N curves showing test results and mechanical properties of STAMP-processed 600 stainless steel and electroslag-remelted AISI 329 stainless steel. Fatigue ratio ($\sigma\,10^7/Rm$) for 600 steel: 0.57. Fatigue ratio for electroslag-remelted 329 steel: 0.51.

Source: Metals Handbook, 9th Edition, Volume 7, Powder Metallurgy, American Society for Metals, Metals Park OH, 1984, p 549

8-43. Grade 21-6-9 Stainless Steel: Effect of Temperature on Fatigue Crack Growth Rates

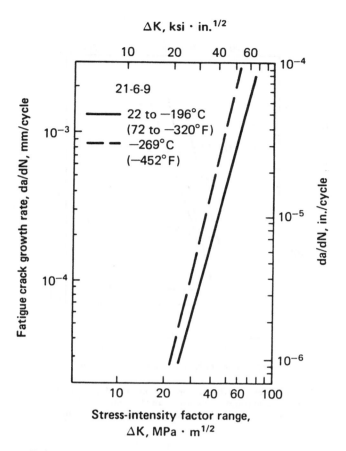

Fatigue crack growth rates in specimens of annealed 21-6-9 stainless steel at 22, −196 and −269 °C (72, −320 and 452 °F), 20 and 28 Hz, and an R ratio of 0.1.

Similar tests made with specimens of 22-13-5 stainless steel showed fatigue crack growth rates that were nearly the same as shown here for 21-6-9.

Source: J. E. Campbell, "Fracture Properties of Wrought Stainless Steels," in Application of Fracture Mechanics for Selection of Metallic Structural Materials, James E. Campbell, William W. Gerberich and John H. Underwood, Eds., American Society for Metals, Metals Park OH, 1982, p 140

8-44. Kromarc 58 Stainless Steel: Effect of Cryogenic Temperatures on Weldments

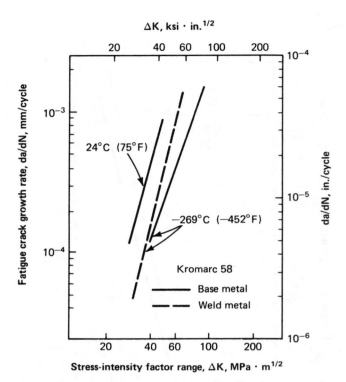

Fatigue crack growth rates for solution treated Kromarc 58 base metal in air at room temperature, and base metal and weld metal at −269 °C (−452 °F) in liquid helium, at 10 Hz and an R ratio of 0.1.

For the fusion zone of a gas tungsten arc weld made with Kromarc 58 filler metals, the $K_{Ic}(J)$ value was 156 MPa · m$^{1/2}$ (141 ksi · in.$^{1/2}$) at −269 °C (−452 °F). Fatigue crack growth rate data for the base metal at room temperature and at −269 °C and for the weld metal at −269 °C are shown above. The data were obtained on compact specimens at 10 Hz and at an R ratio of 0.1. Fatigue crack growth rates for tests in liquid helium were lower than at room temperature at the same ΔK values. Therefore, if room temperature crack growth rate data are used to estimate crack growth at cryogenic temperatures, the estimated values will be conservative.

Source: J. E. Campbell, "Fracture Properties of Wrought Stainless Steels," in Application of Fracture Mechanics for Selection of Metallic Structural Materials, James E. Campbell, William W. Gerberich and John H. Underwood, Eds., American Society for Metals, Metals Park OH, 1982, p 142

8-45. Pyromet 538 Stainless Steel: Effects of Welding Methods on Fatigue Crack Growth Rates

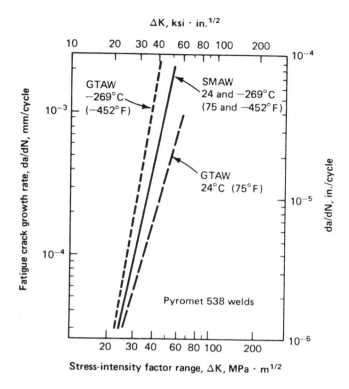

Fatigue crack growth rates in weld metal in Pyromet 538 stainless steel at room temperature and −269 °C (−452 °F) and at 10 Hz.

The base metal was solution annealed prior to welding. One set of welds was made by the gas tungsten arc welding (GTAW) process with 21-6-9 filler wire, and the other was made by the shielded metal arc welding (SMAW) process with IN 182 covered electrodes. Results of these tests are summarized in the graph above. Specimens with SMAW welds had the same fatigue crack growth rates at room temperature and at −269 °C (−452 °F). Specimens welded by the GTAW process had higher crack growth rates at −269 °C than at room temperature. Examination of the microstructures near the fracture surfaces for the specimens tested at −269 °C showed that there was 6 to 7% delta ferrite (produced by welding) in the weld metals along with induced martensite. The SMAW weld metal was fully austenitic.

Source: J. E. Campbell, "Fracture Properties of Wrought Stainless Steels," in Application of Fracture Mechanics for Selection of Metallic Structural Materials, James E. Campbell, William W. Gerberich and John H. Underwood, Eds., American Society for Metals, Metals Park OH, 1982, p 141

8-46. Duplex Stainless Steel KCR 171: Corrosion Fatigue

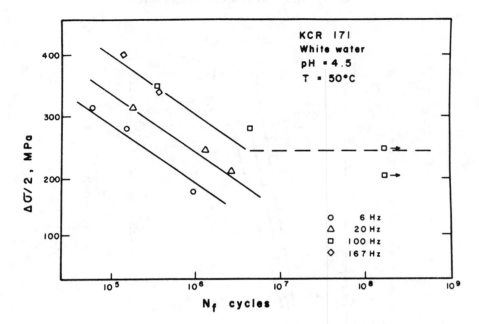

Rotating bending S-N tests were carried out in 50 °C (122 °F) white water at different frequencies (6, 20, 100, and 167 Hz) for samples polished with 240 grit emery paper and the results obtained are presented in the above S-N diagram. The results thus far obtained for the two highest frequencies appear to fall on the same S-N curve, and the indication is that this curve would present a quite horizontal fatigue limit. In the short life regime ($N_f < 10^6$ cycles), the results suggest that decreasing the frequency below 100 Hz displaces this portion of the S-N curve to shorter lives without significantly changing its slope.

Source: M. Ait Bassidi, J. Masounave and J. I. Dickson, "The Corrosion Fatigue Behaviour in White Water of KCR 171," in Duplex Stainless Steels, R. A. Lula, Ed., American Society for Metals, Metals Park OH, 1983, p 455

9-1. Grades 200, 250, and 300 Maraging Steel: S-N Curves for Smooth and Notched Specimens

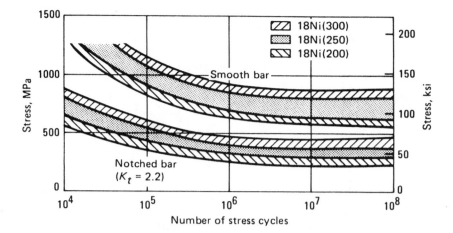

Fatigue properties of maraging steels are comparable to those of other high-strength steels. Smooth-bar and notched-bar fatigue properties for 18Ni(200), 18Ni(250), and 18Ni(300) grades are summarized in the S-N curves shown above. Fatigue crack growth rates in maraging steels obey the $da/dN = (\Delta K)^m$ relationship commonly observed in steels and are similar to those of conventional steels. Improved fatigue properties can be obtained by shot peening and by nitriding.

Source: Metals Handbook, 9th Edition, Volume 1, Properties and Selection: Irons and Steels, American Society for Metals, Metals Park OH, 1978, p 451

9-2. Grade 300 Maraging Steel: Fatigue Life in Terms of Total Strain

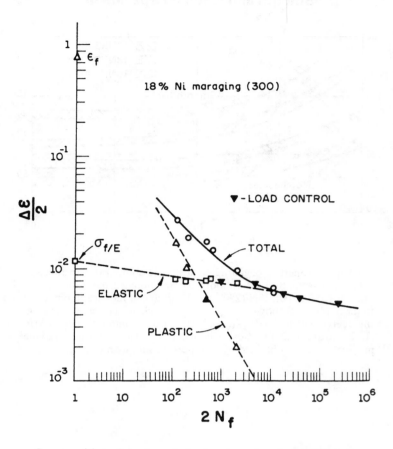

Superposition of elastic and plastic curves gives fatigue life in terms of total strain. An actual example for this method of determining fatigue life is presented above for a maraging steel.

Source: Marc André Meyers and Krishan Kumar Chawla, "Mechanical Metallurgy: Principles and Applications," Prentice-Hall, Inc., Englewood Cliffs NJ, 1984, p 700

10-1. Fatigue of Cast Irons as a Function of Structure-Sensitive Parameters

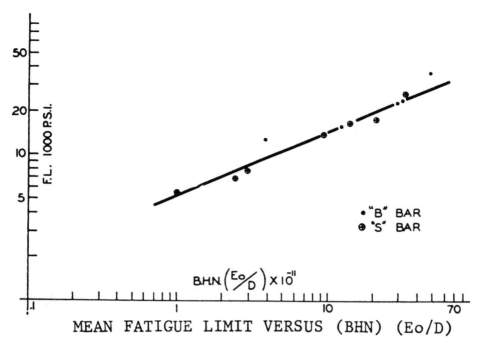

Fatigue of cast irons as a function of structure-sensitive parameters: Bhn, elastic modulus (Eo) and damping capacity (D).

Walter has shown that the fatigue properties of irons are highly dependent on volume of graphite and its morphology and distribution, as well as the matrix structure. He was able to reduce these factors to some easily measurable parameters, Eo, D, and Bhn, which gave good correlation with fatigue properties over a rather wide range of irons (see graph). It is reasonable that these parameters relate to fatigue performance, since they are measures of fatigue-related properties. Eo, the modulus at very small strains, is controlled mostly by the volume of free graphite and to some degree by the graphite shape. Since the graphite present detracts from the matrix load-carrying area, the more graphite, the higher the stress on the remaining matrix—thus lower fatigue performance. D, the damping capacity, is controlled mostly by the graphite morphology and to some degree by the graphite volume. Sharp-edged flakes are greater stress raisers than rounded-edge flakes and spheroids; thus the higher the D, the poorer the fatigue performance. Bhn is largely a measure of the matrix hardness and, to some degree, of the graphite volume; thus the higher the Bhn, the better the fatigue performance. These easily measured properties are put to good use in industry as specification means and process-control criteria.

Source: D. H. Breen and E. M. Wene, "Fatigue in Machines and Structures—Ground Vehicles," in Fatigue and Microstructure, American Society for Metals, Metals Park OH, 1979, p 86.

10-2. Gray Iron: Fatigue Life, and Fatigue Limit as a Function of Temperature

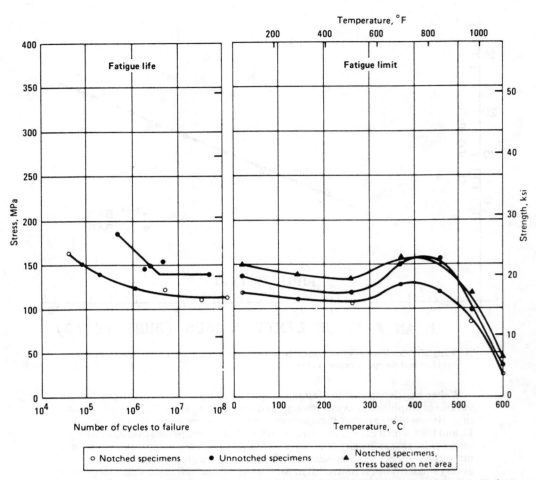

Composition: 2.84 C, 1.52 Si, 1.05 Mn, 0.07 P, 0.12 S, 0.31 Cr, 0.20 Ni, 0.37 Cu. (Ref 5)

Typical fatigue life for as-cast gray iron of the above composition (left). Effect of temperature on fatigue limit for the same gray iron (right).

Source: Metals Handbook, 9th Edition, Volume 1, Properties and Selection: Irons and Steels, American Society for Metals, Metals Park OH, 1978, p 21

10-3. Gray Iron: *S-N* Curves for Unalloyed vs Alloyed

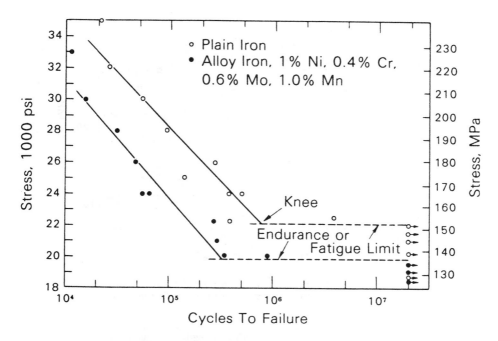

A fatigue crack starts in an area of high stress concentration after a large number of loading cycles. It is always a brittle type of fracture even when occurring in ductile metals. As the crack progresses it increases the stress concentration, and the rate of propagation under the cyclic loading increases. When the cross section of the remaining metal becomes insufficient to support the maximum load, complete failure occurs as it would under an excessive steady stress.

The number of stress applications that will induce a fatigue failure is less at higher maximum stress values, and conversely a larger number of stress cycles can occur at a lower maximum stress level before a fatigue crack is initiated. A plot of this relation for a metal is called an *S-N* curve and relates the maximum applied stress to the logarithm of the number of cycles for failure. When the number of cycles without failure exceeds ten million, the endurance life is considered infinite for body-centered-cubic ferrous metals. The maximum stress that will allow this number of cycles is established as the endurance limit, or the fatigue strength or fatigue limit. Two typical *S-N* curves for a plain and alloy high-strength gray iron are presented above.

Source: Iron Castings Handbook, Charles F. Walton, Ed., Iron Castings Society, Inc., 1981, p 246

10-4. Gray Iron: Effect of Environment

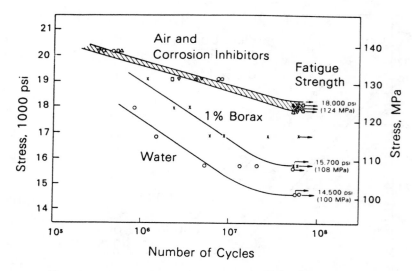

The effect of various environments and corrosion inhibitors listed in the table below on the corrosion fatigue properties of gray iron.

Environment	Fatigue strength psi	Fatigue strength MPa	Fatigue strength reduction factor
Air	17,920	124	—
Water	14,560	100	1.23
3% sodium chloride	5,600	39	3.20
1% borax	15,680	108	1.14
3% "Sobenite"*	17,920	124	1.00
3% sodium carbonate	17,920	124	1.00
3% soluble oil	17,920	124	1.00
0.25% potassium chromate	17,920	124	1.00

* "Sobenite" is a mixture of 10 parts sodium benzoate to 1 part sodium nitrite.

The corrosion fatigue program involved testing in air, a spray of demineralized water, and a spray of three-percent sodium chloride solution; additional tests were made with a demineralized water spray and various known corrosion inhibitors. The *S-N* curves and table above indicate that both the demineralized water and three-percent sodium chloride sprays reduced the fatigue strength of a pearlitic gray iron. Of the various alkaline inhibitors and soluble oils investigated, only borax was not completely effective for the pearlitic irons. Annealed ferritic gray irons were similarly affected by the sodium chloride solution.

Source: Iron Castings Handbook, Charles F. Walton, Ed., Iron Castings Society, Inc., 1981, p 255

10-5. Class 30 Gray Iron: Modified Goodman Diagram

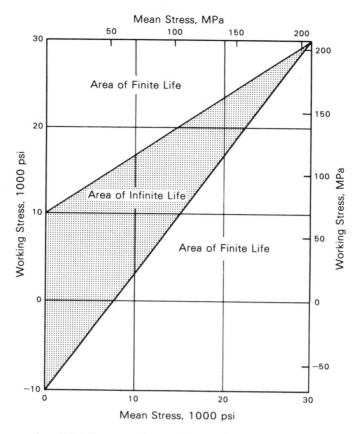

A modified Goodman diagram relates the endurance limit to an allowable working stress when it is superimposed on a steady stress.

In many engineering applications, alternating stress is superimposed on a steady stress and requires special consideration. A method of relating the effect of the combined static and alternating stresses on the endurance limit has been developed into the Goodman diagram, of which a modified form is shown here.

Source: Iron Castings Handbook, Charles F. Walton, Ed., Iron Castings Society, Inc., 1981, p 251

10-6. Class 30 Gray Iron: Fatigue Crack Growth Rates

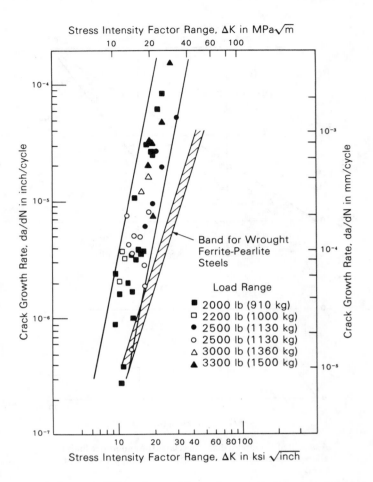

Fatigue crack growth rate. The endurance limit approach to design utilizes fatigue data taken on smooth, defect-free test specimens. For such specimens, fatigue crack initiation may take 80 to 90 percent of the total lifetime while crack growth is only 10 to 20 percent of the lifetime. Such flaws allow fatigue cracks to initiate in a relatively small number of cycles so that the lifetime of the component depends principally on the crack growth rate. If the initial flaw size can be determined from experience or by utilizing nondestructive inspection and the critical flaw size calculated using the fracture toughness value K_{Ic}, then crack growth rate data may be used to calculate the number of cycles required to grow a crack from an initial size to a critical size where final fracture occurs.

Only limited fatigue crack growth rate data are available on cast irons. These results are presented in the above chart for a class 30 gray iron, where da/dN is the crack growth per cycle and ΔK is the stress intensity range.

Source: Iron Castings Handbook, Charles F. Walton, Ed., Iron Castings Society, Inc., 1981, p 250

10-7. Gray Irons: Torsional Fatigue for Various Tensile Strength Values

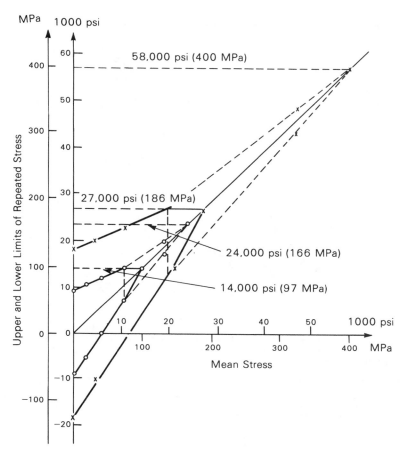

Torsional fatigue strength for three levels of tensile strength with various mean stresses.

Source: Iron Castings Handbook, Charles F. Walton, Ed., Iron Castings Society, Inc., 1981, p 253

10-8. Gray Irons: Torsional Fatigue Data for Five Different Compositions

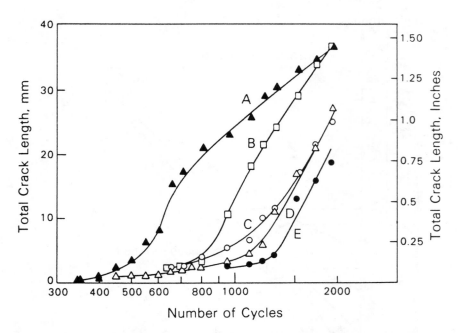

Total length of six cracks (the first three cracks in each of two specimens of each iron) as a function of the number of thermal cycles between 1100 and 400 °C (590 and 200 °C). Iron compositions are as follows:

Iron	Composition, %							
	C	Si	Mn	Cr	Mo	Ni	Cu	Sn
A	3.43	1.65	0.57	—	—	—	—	—
B	3.45	1.74	0.59	0.49	—	0.60	0.59	—
C	3.45	1.68	0.63	0.30	0.30	0.97	0.87	—
D	3.44	1.69	0.58	0.21	0.38	—	0.30	0.077
E	3.43	1.66	0.58	0.50	0.39	—	—	—

Source: Iron Castings Handbook, Charles F. Walton, Ed., Iron Castings Society, Inc., 1981, pp 288, 289

10-9. Gray Irons: Thermal Fatigue—Effect of Aluminum Additions

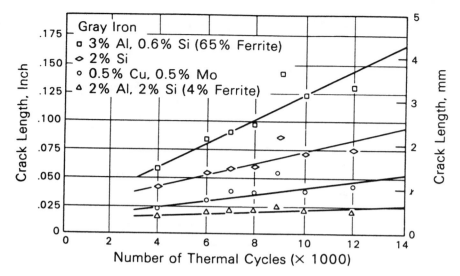

Thermal fatigue resistance of different alloyed gray irons.

This graph shows results of a thermal fatigue test in which notched disc specimens were alternately heated to 800 °F (425 °C) and cooled to 200 °F (95 °C) in two fluid beds, demonstrating that a pearlitic gray iron containing 3.4% carbon and 2% aluminum was highly resistant to thermal crack propagation.

Source: Iron Castings Handbook, Charles F. Walton, Ed., Iron Castings Society, Inc., 1981, p 434

10-10. Gray Irons: Thermal Fatigue—Effect of Chromium and Molybdenum Additions

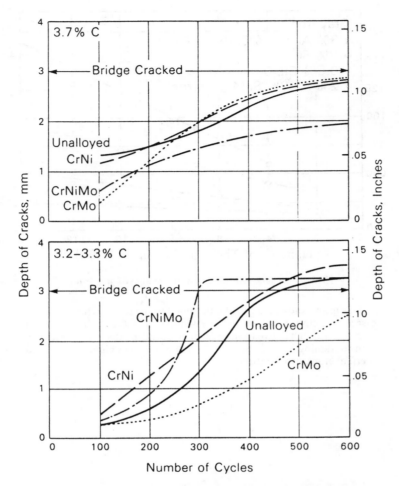

The depth of cracks resulting from the thermal cycling of gray irons between 860 °F (460 °C) and room temperature.

Alloying with molybdenum and chromium provided superior thermal fatigue resistance compared with irons that contained other alloying additions or no alloying at all. In this case, the improved thermal fatigue resistance is believed to be directly related to the higher elevated-temperature tensile strength and better stability of the chromium-molybdenum irons. However, it must be remembered that this improvement is related to and dependent on the temperature cycle and base iron composition, as shown above. It has also been indicated that the development of an acicular matrix structure, by adding relatively large quantities of molybdenum and copper, supplies a less than desirable influence on thermal fatigue cracking.

Source: Iron Castings Handbook, Charles F. Walton, Ed., Iron Castngs Society, Inc., 1981, p 288

10-11. Gray Irons: Thermal Fatigue—Room Temperature and 540 °C (1000 °F)

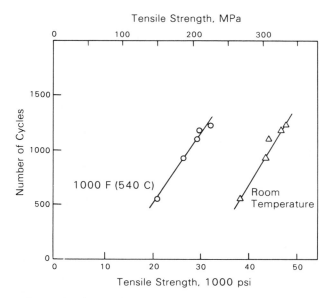

Curves showing relation between the number of thermal cycles for cracking and tensile strength at room temperature and 1000 °F (540 °C).

For good resistance to thermal fatigue, gray irons should have high thermal conductivity, a low modulus of elasticity, high strength at both room and elevated temperatures, and for temperatures above 900 °F (500 °C), resistance to oxidation and structural change. Because some of these properties are in opposition, a compromise must be made in selecting the most appropriate metal for each type of service. As the maximum temperature to which the gray iron is cycled and number of cycles increase, the number and size of thermal fatigue cracks become larger. The above curves illustrate the influence of room- and elevated-temperature strength on the thermal fatigue resistance of irons having similar carbon equivalents, thermal conductivities, matrix structures, and elastic moduli. Those irons with higher room and elevated-temperature tensile strengths (achieved by alloying) generally display higher thermal fatigue strength.

Source: Iron Castings Handbook, Charles F. Walton, Ed., Iron Castings Society, Inc., 1981, p 285

10-12. Gray Irons: Thermal Fatigue Properties—Comparisons With Ductile Cast Iron and Carbon Steel

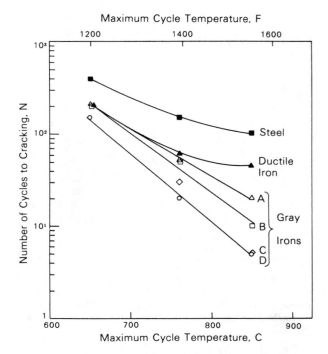

The above curves show the variation of the number of cycles to cracking with the maximum temperature of the cycles for gray iron, ductile iron, and carbon steel. Compositions of the four gray irons are as follows:

Iron	Analysis, %					
	C	Si	Mn	Cr	Mo	Other
A	3.43	2.37	0.78	0.22	0.32	0.21 Sn
B	3.49	2.37	0.84	0.24	0.22	—
C	3.48	0.60	0.88	0.23	0.20	2.37 Al
D	3.50	2.38	0.83	0.30	0.77	1.51 Cu

Source: Iron Castings Handbook, Charles F. Walton, Ed., Iron Castings Society, Inc., 1981, pp 286, 287

10-13. Cast Irons: Thermal Fatigue Properties for Six Grades

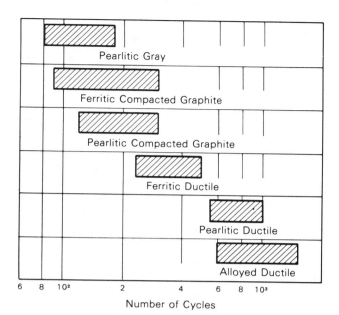

The number of thermal cycles required to produce thermal fatigue cracking in cast irons. Compositions are tabulated below.

Analysis	%C	%S	%Mn	%P	%Mg	Alloys
Class 35 Gray Iron	2.96	2.90	0.78	0.07	—	0.12Cr
Ferritic Compacted Graphite	3.52	2.61	0.25	0.05	0.015	—
Pearlitic Compacted Graphite	3.52	2.25	0.40	0.05	0.015	1.47Cu
Ferritic Ductile	3.67	2.55	0.13	0.06	0.030	—
Pearlitic Ductile	3.60	2.34	0.50	0.05	0.030	0.54 Cu
Alloyed Ferritic Ductile	3.48	4.84	0.31	0.07	0.030	1.02Mo

Source: Iron Castings Handbook, Charles F. Walton, Ed., Iron Castings Society, Inc., 1981, pp 393, 396

10-14. Ductile Iron: Effect of Microstructure on Endurance Ratio–Tensile Strength Relationship

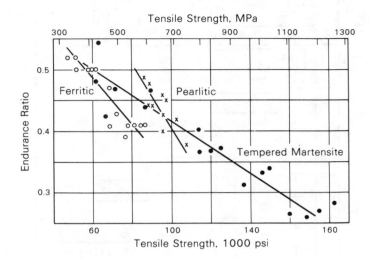

In general the fatigue limit for ductile iron increases with tensile strength, but as with other ferrous metals, the increase is less than proportional. The relation between the tensile strength and the endurance ratio for the annealed, ferritic irons is different from that of the irons with a matrix of pearlite or tempered martensite, as illustrated above.

Source: Iron Castings Handbook, Charles F. Walton, Ed., Iron Castings Society, Inc., 1981, p 341

10-15. Ductile Iron: Effect of Microstructure on Endurance Ratio-Tensile Strength Relationship

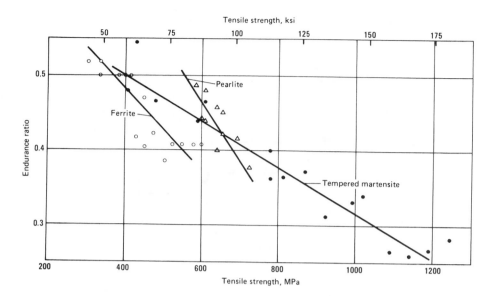

The influence of tensile strength and structure on the endurance ratio of ductile iron is indicated in this graph. Endurance ratio is defined as endurance limit divided by tensile strength. Because the endurance ratio of ductile iron decreases as tensile strength increases, regardless of structure, there may be little value in specifying a higher-strength ductile iron for a structure that is prone to fatigue failure. For tempered martensite ductile iron, the improvement in fatigue strength due to an increase in tensile strength is greater than for pearlitic or ferritic structures. This is indicated in the graph above by the shallower slope for martensite.

Source: Metals Handbook, 9th Edition, Volume 1, Properties and Selection: Irons and Steels, American Society for Metals, Metals Park OH, 1978, p 45

10-16. Ductile Iron: S-N Curves for Ferritic and Pearlitic Grades, Using V-Notched Specimens

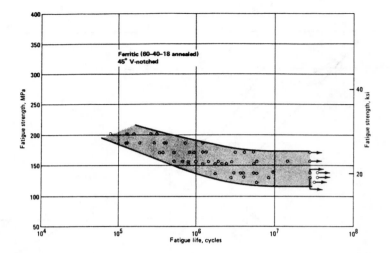

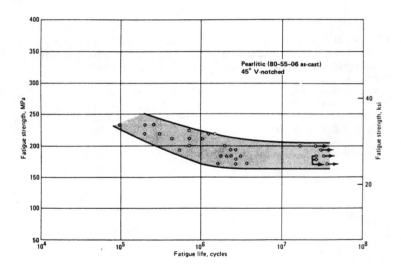

Top: S-N curves, including scatter bands, for annealed ductile iron. Bottom: Similar to above except for as-cast pearlitic ductile iron. All test specimens were V-notched (45°).

Source: Metals Handbook, 9th Edition, Volume 1, Properties and Selection: Irons and Steels, American Society for Metals, Metals Park OH, 1978, p 43

10-17. Ductile Iron: S-N Curves for Ferritic and Pearlitic Grades, Using Unnotched Specimens

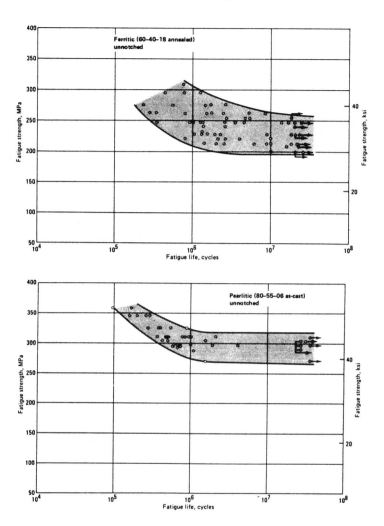

Grade	Tensile strength MPa	ksi	Unnotched Endurance limit MPa	ksi	Endurance ratio	Notched Endurance limit MPa	ksi	Endurance ratio	Stress concentration factor
60-40-18	480	70	205	30	0.43	125	18	0.26	1.67
80-55-06	680	99	275	40	0.40	165	24	0.24	1.67

Top: Similar to upper graph on the opposite page, but here the specimens were unnotched. **Bottom:** Similar to lower graph on the opposite page, but here the specimens were unnotched. Data in table pertain to graphs on this and the opposite page.

Source: Metals Handbook, 9th Edition, Volume 1, Properties and Selection: Irons and Steels, American Society for Metals, Metals Park OH, 1978, p 44

10-18. Ductile Iron: Fatigue Diagrams for Bending Stresses and Tension-Compression Stresses

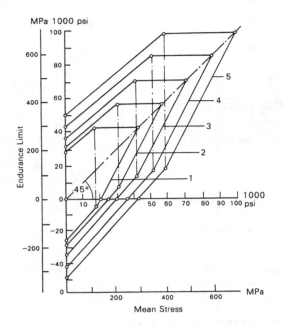
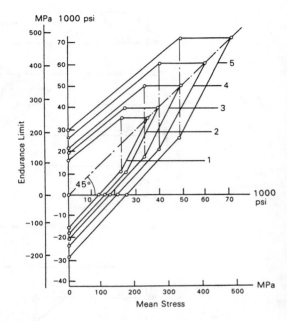

Fatigue diagrams indicating endurance limits for five grades of ductile iron under bending stresses (left) and tension-compression stresses (right). Minimum properties of the irons are given in the table below.

Iron No.	Min. Tensile Strength		Min. Yield Strength		Min. Elongation Percent
	1000 psi	MPa	1000 psi	MPa	
1	55	38	36	25	17
2	61	42	41	28	12
3	72	50	51	35	7
4	87	60	61	42	2
5	102	70	72	50	2

Source: Iron Castings Handbook, Charles F. Walton, Ed., Iron Castings Society, Inc., 1981, pp 344, 345 and 346

10-19. Ductile Iron: Effect of Surface Conditions— As-Cast vs Polished Surface

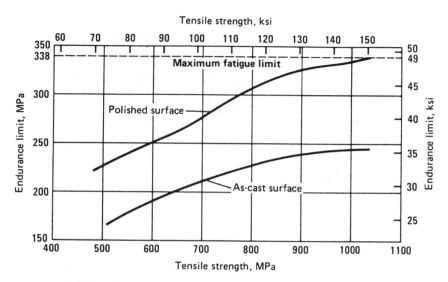

Tests made on 10.6-mm (0.417-in.) diameter specimens. Fully reversed stress ($R = -1$).

Data given in the above graph show that the endurance limit for any given strength level of ductile iron is significantly affected by surface conditions of unnotched specimens. The endurance limit is much higher for the polished specimens than it is for the as-cast specimens, which have relatively rough surfaces.

Source: Metals Handbook, 9th Edition, Volume 1, Properties and Selection: Irons and Steels, American Society for Metals, Metals Park OH, 1978, p 45

10-20. Ductile Iron: Fatigue Limit in Rotary Bending as Related to Hardness

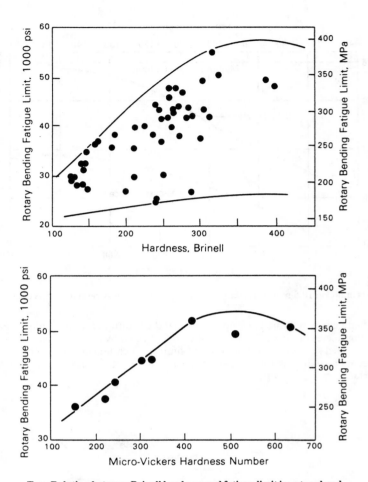

Top: Relation between Brinell hardness and fatigue limit in rotary bending for ductile iron. Bottom: Relation between rotary bending fatigue limit and matrix hardness for ductile iron.

Source: Iron Castings Handbook, Charles F. Walton, Ed., Iron Castings Society, Inc., 1981, p 347

10-21. Ductile Iron: Effect of Rolling on Fatigue Characteristics

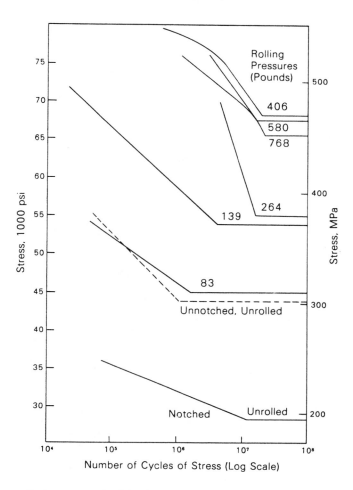

Fatigue strength of ductile iron can be increased substantially by cold working, especially when this method is applied to stressed radii or notches. More than a 60% increase in the endurance limit was obtained with a rolling pressure that was insufficient to depress the surface a measurable amount. The improvement in fatigue properties obtained by various rolling pressures on ductile iron is indicated above.

Source: Iron Castings Handbook, Charles F. Walton, Ed., Iron Castings Society, Inc., 1981, p 348

10-22. Ductile Iron: Effect of Notches on a 65,800-psi-Tensile-Strength Grade

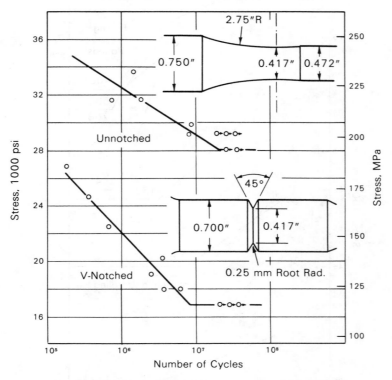

The unnotched and notched fatigue properties of an annealed ductile iron with a tensile strength of 65,800 psi (454 MPa). The endurance ratio is 0.41 and the notch sensitivity ratio is 1.67.

Source: Iron Castings Handbook, Charles F. Walton, Ed., Iron Castings Society, Inc., 1981, p 341

10-23. Ductile Iron: Fatigue Crack Growth Rate Compared With That of Steel

267

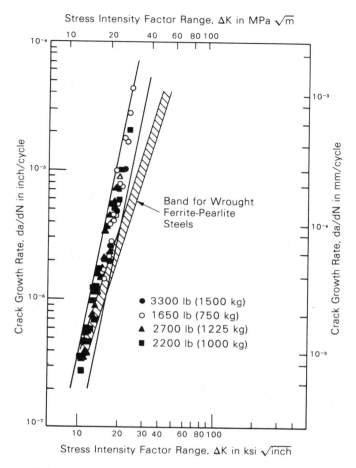

Fatigue crack growth rate of annealed ferritic ductile iron, compared with that of ferritic-pearlitic steels.

Source: Iron Castings Handbook, Charles F. Walton, Ed., Iron Castings Society, Inc., 1981, p 349

10-24. Malleable Iron: S-N Curve Comparisons of Four Grades

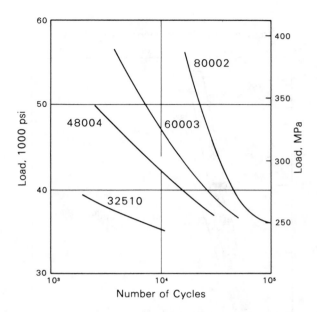

The effect of cast surfaces on four grades of malleable iron was also studied in high-stress, low-cycle fatigue. The results with a 95% confidence limit are presented in this S-N diagram. Unmachined and notched surfaces do reduce the fatigue strength. The reduction factor is as low as 1.2 for the lower-strength irons to over 2.0 for the higher-strength irons. Inducing compressive stresses into the surface by rolling, coining, or shot peening can increase the fatigue life of a component significantly. Design with adequate sections that are well blended to reduce stress concentrations is most effective in reducing the possibility of a fatigue failure.

Source: Iron Castings Handbook, Charles F. Walton, Ed., Iron Castings Society, Inc., 1981, p 311

10-25. Pearlitic Malleable Iron: Effect of Surface Conditions on S-N Curves

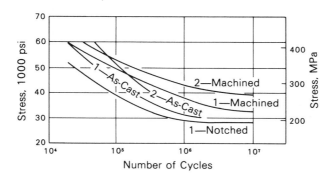

The influence of as-cast surfaces, smooth machined surfaces, and machined notches on the fatigue behavior of pearlitic malleable irons. Iron 1 is grade 60003 and Iron 2 is grade 80002.

Surface finish has an important influence on fatigue properties, as shown above. Samples of malleable grades 60003 and 80002 were tested in fatigue with "as-cast" and machined surfaces. Samples of the 60003 grade were also included with a machined surface containing a sixty-degree notch that was 0.050 in. (1.25 mm) deep. The resulting data are shown in this diagram.

Source: Iron Castings Handbook, Charles F. Walton, Ed., Iron Castings Society, Inc., 1981, p 310

10-26. Pearlitic Malleable Iron: Effect of Nitriding

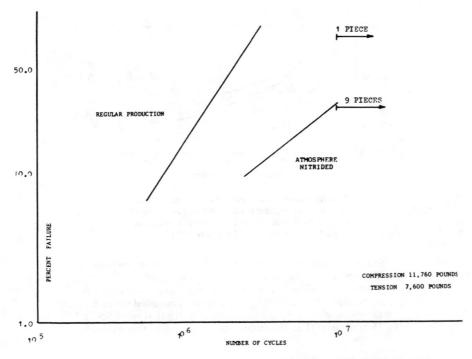

Effect of gaseous atmosphere nitriding on fatigue characteristics of pearlitic malleable iron, tested by tension-compression.

Samples were austenitized, oil quenched, and tempered to 241–269 HB prior to nitriding or testing without nitriding. This chart indicates an increase in fatigue life of 750,000 to 2,700,000 cycles attained by nitriding.

Source: J. A. Riopelle, "Short Cycle Atmosphere Nitriding," in Source Book on Nitriding, American Society for Metals, Metals Park OH, 1977, p 287

10-27. Ferritic Malleable Iron: Effect of Notch Radius and Depth

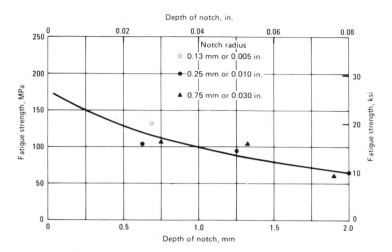

Effect of notch radius and notch depth on fatigue strength of ferritic malleable iron.

Fatigue strength of unnotched ferritic malleable iron is approximately 50% of the tensile strength, or from 170 to 205 MPa (25 to 30 ksi). The graph above summarizes the effects of notches on fatigue strength. As a rule, notch radius has little effect on fatigue strength, but fatigue strength decreases as notch depth increases.

Source: Metals Handbook, 9th Edition, Volume 1, Properties and Selection: Irons and Steels, American Society for Metals, Metals Park OH, 1978, p 65

11-1. A286: Effect of Environment

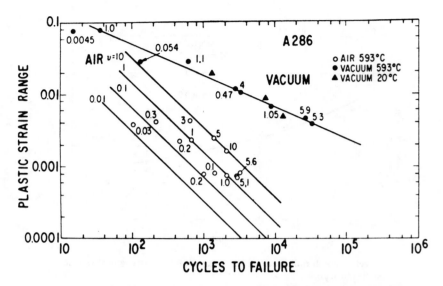

Plastic-strain range versus fatigue life for A286 ferrous alloy in air and in vacuum at 593 °C (1095 °F). Numbers adjacent to test points indicate frequency in cycles per minute. Note absence of frequency effects in vacuum.

Coffin has suggested that for a number of materials, virtually all of the degradation in fatigue life at elevated temperatures can be attributed to environmental interactions. He noted that frequency effects in the low-cycle-fatigue law could be eliminated for a large number of metals and alloys by testing in vacuum (note above). Additionally, it was noted that tests performed in vacuum showed transgranular crack nucleation and propagation versus intergranular nucleation and propagation in air at elevated temperatures. These results are not unambiguous, since Koburger has shown a frequency effect in high-cycle fatigue for directionally solidified eutectic alloys when tested in air and in vacuum, particularly at elevated temperatures. The primary difference in these results may be related to the lack of intergranular cracking in eutectic alloys.

Source: D. J. Duquette, "Environmental Effects I: General Fatigue Resistance and Crack Nucleation in Metals and Alloys," in Fatigue and Microstructure, American Society for Metals, Metals Park OH, 1979, p 343

11-2. A286: Effect of Frequency on Life at 593 °C (1095 °F)

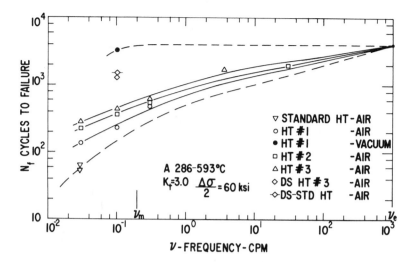

Effect of frequency on life of notched fatigue bars of A286 at 593 °C (1095 °F) in air and vacuum. As indicated, decreasing frequency has a degrading effect on fatigue life of samples tested in air, with little or no effect on samples tested in a vacuum.

Source: L. F. Coffin, "Fatigue in Machines and Structures—Power Generation," in Fatigue and Microstructure, American Society for Metals, Metals Park OH, 1979, p 13

11-3. A286: Fatigue Crack Growth Rates at Room and Elevated Temperatures

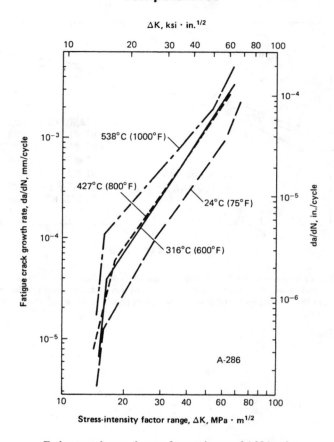

Fatigue crack growth rates for specimens of A286 stainless steel at room temperature and elevated temperatures for tests in air at 3 Hz (RT) and 0.67 Hz (elevated temperatures), an R ratio of 0.05, and at L-T, T-L, R-L, and R-C orientations.

The austenitic precipitation-hardening stainless steel A286 (heat-resistant alloy) is the main representative in this category. It contains titanium and small amounts of vanadium and aluminum, which precipitate as intermetallic compounds such as Ni_3(Al, Ti) and Ni_4Mo(Fe, Cr) Ti on aging. Various mill forms of the alloy are usually supplied in the annealed condition—Condition A (980 °C, or 1800 °F, for one hour followed by quenching in oil or water). Precipitation hardening occurs on aging in the range from 700 to 760 °C (1300 to 1400 °F) for 16 hours. Other combinations of heat treatments may be used depending on the application. One variation is to re-solution treat at 900 °C (1650 °F) for two hours, quench in oil or water, and age at 700 °C (1300 °F) for 16 hours. This variation results in improved room temperature properties but less desirable stress-rupture properties.

Source: J. E. Campbell, "Fracture Properties of Wrought Stainless Steels," in Application of Fracture Mechanics for Selection of Metallic Structural Materials, James E. Campbell, William W. Gerberich and John H. Underwood, Eds., American Society for Metals, Metals Park OH, 1982, p 161

11-4. Astroloy: S-N Curves for Powder vs Conventional Forgings

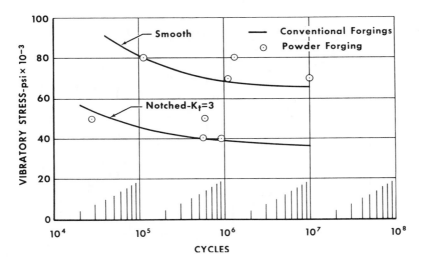

S-N curves for conventional and powder forgings of Astroloy (notched versus smooth).

Testing was performed using standard methods at 705 °C (1300 °F) and a combination of steady and vibratory stresses for which comparative data were available. Cycles to first indication (crack) were comparable to conventional material. Crack propagation as judged by the number of additional cycles from first indication to failure was slower than conventional material, as shown above.

Source: M. M. Allen, R. L. Athey and J. B. Moore, "Application of Powder Metallurgy to Superalloy Forgings," in Source Book on Powder Metallurgy, Samuel Bradbury, Ed., American Society for Metals, Metals Park OH, 1979, p 97

11-5. Astroloy: Powder vs Conventional Forgings Tested at 705 °C (1300 °F)

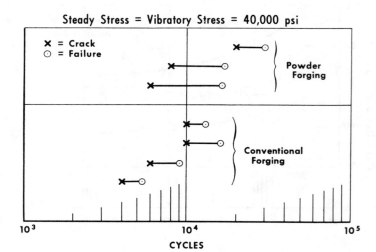

Astroloy tested in high-cycle fatigue at 705 °C (1300 °F). Vibratory stress levels were selected to facilitate a direct comparison between conventional and powder forgings.

Source: M. M. Allen, R. L. Athey and J. B. Moore, "Application of Powder Metallurgy to Superalloy Forgings," in Source Book on Powder Metallurgy, Samuel Bradbury, Ed., American Society for Metals, Metals Park OH, 1979, p 97

11-6. FSX-430: Effect of Grain Size on Cycles to Cracking

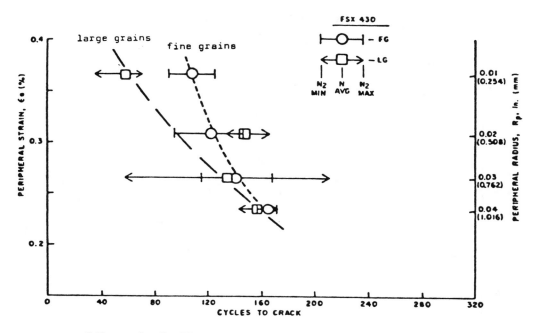

S-N curves for alloy FSX-430, showing effect of grain size on cycles to cracking.

Source: Eric Bachelet and Gerard Lesoult, "Quality of Castings of Superalloys," in Superalloys: Source Book, Matthew J. Donachie, Jr., Ed., American Society for Metals, Metals Park OH, 1984, p 336

11-7. FSX-430: Effect of Grain Size on Fatigue Crack Propagation Rate

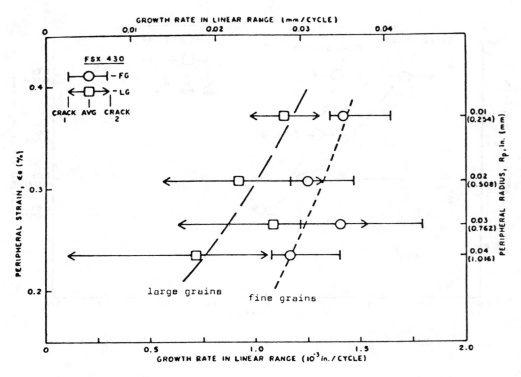

Fatigue crack propagation rate—effect of grain size on fatigue characteristics of FSX-430.

Source: Eric Bachelet and Gerard Lesoult, "Quality of Castings of Superalloys," in Superalloys: Source Book, Matthew J. Donachie, Jr., Ed., American Society for Metals, Metals Park OH, 1984, p 337

11-8. HS-31: Effect of Testing Temperature

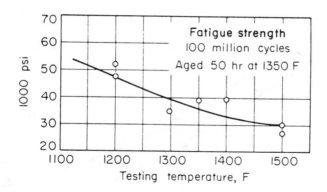

Effect of testing temperature on fatigue strength of HS-31 casting alloy, after aging at 730 °C (1350 °F), for 100 million cycles.

11-9. IN 738 LC Casting Alloy: Standard vs HIP'd Material

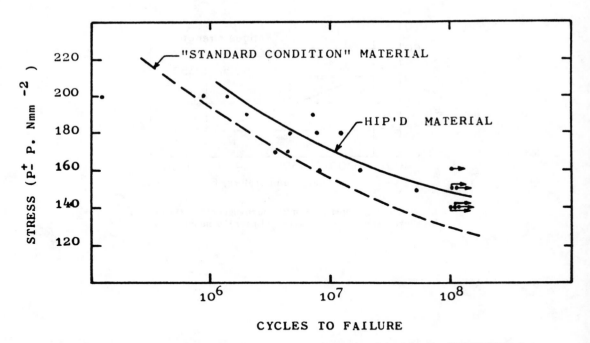

S-N curves for casting alloy IN 738 LC. High-cycle fatigue properties of nimocast alloy IN 738 LC tested at 850 °C (1560 °F).

Source: Eric Bachelet and Gerard Lesoult, "Quality of Castings of Superalloys," in Superalloys: Source Book, Matthew J. Donachie, Jr., Ed., American Society for Metals, Metals Park OH, 1984, p 340

11-10. IN 738 LC: Effect of Grain Size on Cycles to Failure

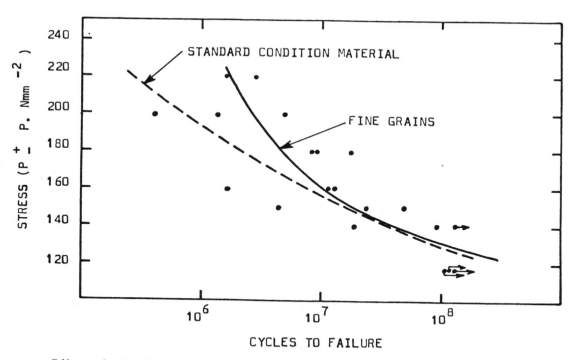

S-N curves for alloy IN 738 LC. High-cycle fatigue properties of extra-fine-grain and conventional material tested at 850 °C (1560 °F).

Source: Eric Bachelet and Gerard Lesoult, "Quality of Castings of Superalloys," in Superalloys: Source Book, Matthew J. Donachie, Jr., Ed., American Society for Metals, Metals Park OH, 1984, p 340

11-11. IN 738 LC: Effect of Grain Size on Cycles to Cracking

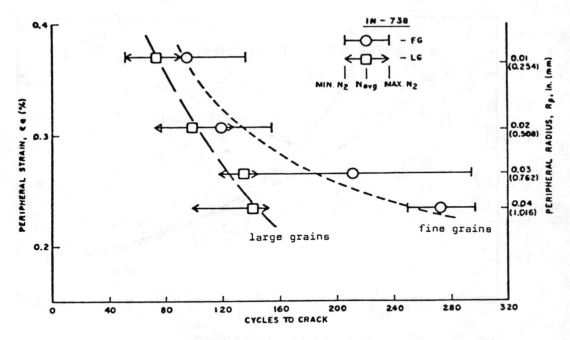

S-N curves for alloy IN 738 LC, showing the effect of grain size on number of cycles to cracking.

Source: Eric Bachelet and Gerard Lesoult, "Quality of Castings of Superalloys," in Superalloys: Source Book, Matthew J. Donachie, Jr., Ed., American Society for Metals, Metals Park OH, 1984, p 335

11-12. IN 738 LC: Effect of Grain Size on Crack Propagation Rate

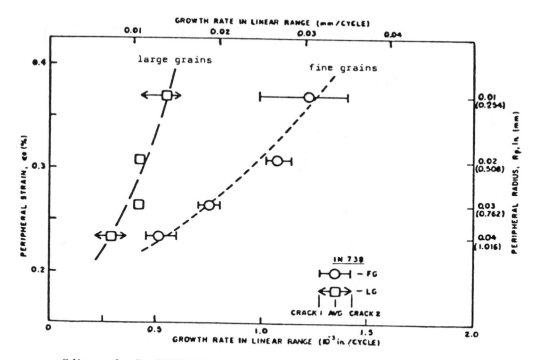

S-N curves for alloy IN 738 LC, showing the effect of grain size on crack propagation rate.

Source: Eric Bachelet and Gerard Lesoult, "Quality of Castings of Superalloys," in Superalloys: Source Book, Matthew J. Donachie, Jr., Ed., American Society for Metals, Metals Park OH, 1984, p 336

11-13. IN 738 LC: Fatigue Crack Growth Rate at 850 °C (1560 °F)

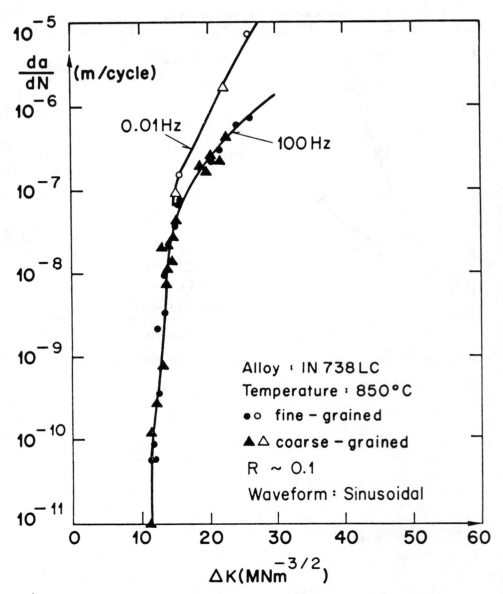

Fatigue crack growth rate at 850 °C (1560 °F) in various grain sizes of alloy IN 738 LC.

Source: Eric Bachelet and Gerard Lesoult, "Quality of Castings of Superalloys," in Superalloys: Source Book, Matthew J. Donachie, Jr., Ed., American Society for Metals, Metals Park OH, 1984, p 341

11-14. Inconel 550: Axial Tensile Fatigue Properties in Air and Vacuum at 1090 K

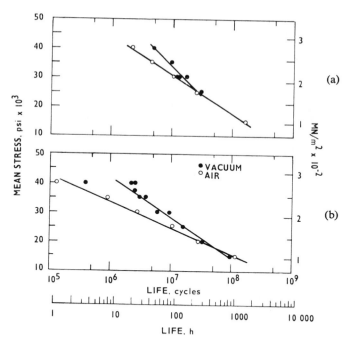

Axial tensile fatigue properties of Inconel 550 at 1090 K in air and vacuum. (a) Ratio of cyclic to mean stress = 0.125. (b) Ratio of cyclic to mean stress = 0.667. Testing frequency = 33 Hz.

In reversed bend tests on lead at 500 cycles/min, Snowden demonstrated a difference of two orders of magnitude in fatigue life between vacuum, air, and pure oxygen. At all strain levels vacuum endurances exceeded those in air, which exceeded those in oxygen. Intermittent stress-free exposure to air had no effect on the lifetime in vacuum. At high temperature (1090 K) vacuum also improved endurance, relative to air, of the Co-base alloy S-816 and the Ni-base alloy Inconel 550, although the effect was much smaller than that seen in lead. Endurances for the nickel-base alloy converged at low stresses, indicating a possible strengthening effect of air, as shown above.

Source: R. H. Cook and R. P. Skelton, "Environment-Dependence of the Mechanical Properties of Metals at High Temperature," in Source Book on Materials for Elevated-Temperature Applications, Elihu F. Bradley, Ed., American Society for Metals, Metals Park OH, 1979, p 81

11-15. Inconel 625: Effect of Temperature on Cycles to Failure

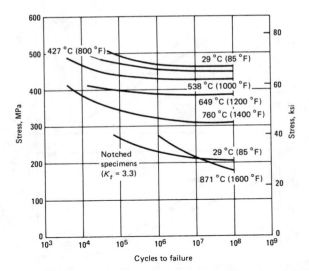

S-N curves for hot rolled solution treated Inconel 625 bar 15.9 mm (0.625 in.) in diameter at various temperatures. Average grain size was 0.10 mm (0.004 in.).

Source: Metals Handbook, 9th Edition, Volume 3, Properties and Selection: Stainless Steels, Tool Materials and Special-Purpose Metals, American Society for Metals, Metals Park OH, 1980, p 143

11-16. Inconel 706: Effect of Temperature on Fatigue Crack Growth Rate

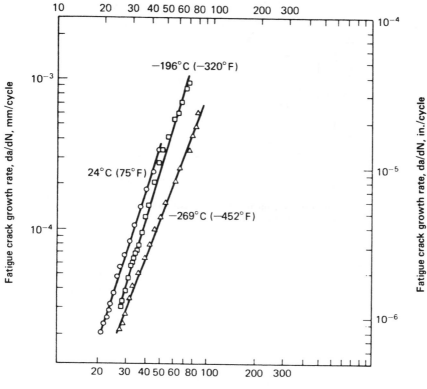

Fatigue crack growth rates of Inconel 706 forged billet (vacuum induction melted/vacuum are remelted) at an R ratio of 0.1 and a frequency of 10 Hz. Heat treatment: 980 °C (1800 °F) 1 h, AC; double aged 730 °C (1350 °F) 8 h, FC to 620 °C (1150 °F), hold 8 h, AC.

Results of FCP tests at room temperature and at temperatures as low as −269 °C (−452 °F) for Inconel 706 are shown above. At equivalent ΔK values, the fatigue crack growth rates for this alloy are slightly lower at subzero temperatures than at room temperature.

Source: Stephen D. Antolovich and J. E. Campbell, "Fracture Properties of Superalloys," in Application of Fracture Mechanics for Selection of Metallic Structural Materials, James E. Campbell, William W. Gerberich and John H. Underwood, Eds., American Society for Metals, Metals Park OH, 1982, p 297

11-17. Inconel "713C": Effect of Elevated Temperatures on Fatigue Characteristics

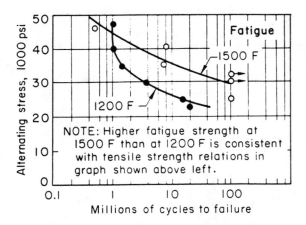

S-N curves for Inconel "713C." Tests were performed at two different elevated temperatures as shown.

Source: ASM Committee on Heat-Resistant Castings, "Heat-Resistant Alloy Castings," in Source Book on Materials for Elevated-Temperature Applications, Elihu F. Bradley, Ed., American Society for Metals, Metals Park OH, 1979, p 235

11-18. Inconel "713C" and As-Cast HS-31: Comparison of Two Alloys for Number of Cycles in Thermal Fatigue to Initiate Cracks

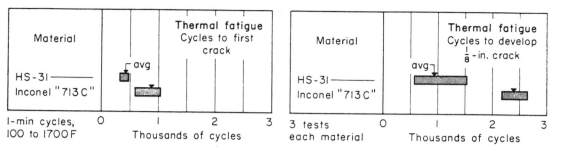

Thermal fatigue properties of HS-31 compared with those of Inconel "713C." Left: Number of cycles required to initiate cracks. Right: Number of cycles required to develop ⅛-in. crack.

Source: ASM Committee on Heat-Resistant Castings, "Heat-Resistant Alloy Castings," in Source Book on Materials for Elevated-Temperature Applications, Elihu F. Bradley, Ed., American Society for Metals, Metals Park OH, 1979, p 235

11-19. Inconel 718: Effect of Frequency on Fatigue Crack Propagation Rate

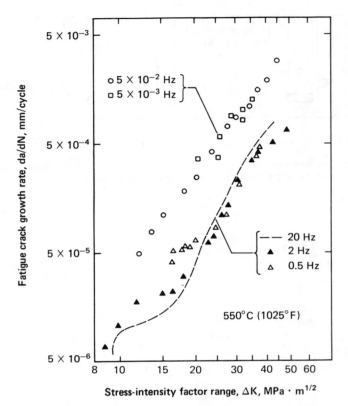

Variation of FCP rate (da/dN) with stress-intensity factor range (ΔK) and frequency at 550 °C (1025 °F) (sinusoidal load) for specimens of Inconel 718.

The effect of frequency at 550 °C (1025 °F) was studied using a sine wave; the results are shown above. Below 0.5 Hz, the FCP rate was more rapid, and the crack surfaces showed an increased amount of intergranular fracture with decreasing frequency, with the crack path following the boundaries of the largest grains. One may be inclined to attribute the increase in FCP to either creep or environment, but this may not be the case, because different modes of deformation may have occurred at different strain rates.

Source: Stephen D. Antolovich and J. E. Campbell, "Fracture Properties of Superalloys," in Application of Fracture Mechanics for Selection of Metallic Structural Materials, James E. Campbell, William W. Gerberich and John H. Underwood, Eds., American Society for Metals, Metals Park OH, 1982, p 294

11-20. Inconel 718: Relationship of Fatigue Crack Propagation Rate With Stress Intensity

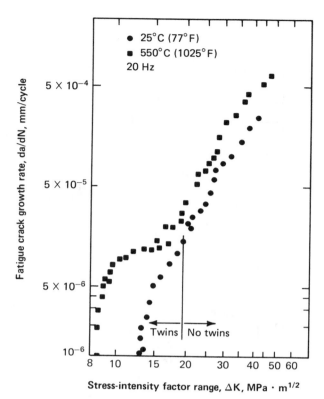

Dependence of FCP rate (da/dN) on stress-intensity factor range (ΔK) and temperature at 20 Hz (sinusoidal wave shape signal) for specimens of Inconel 718.

An important effect is the hydrostatic state of stress in the tip region. This idea has been considered for Inconel 718. The FCP response is shown above for 20 Hz, where the effect of temperature is to increase the FCP rate, especially at ΔK levels.

Source: Stephen D. Antolovich and J. E. Campbell, "Fracture Properties of Superalloys," in Application of Fracture Mechanics for Selection of Metallic Structural Materials, James E. Campbell, William W. Gerberich and John H. Underwood, Eds., American Society for Metals, Metals Park OH, 1982, p 290

11-21. Inconel 718: Relationship of Fatigue Crack Growth Rate With Load/Time Waveforms

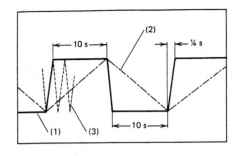

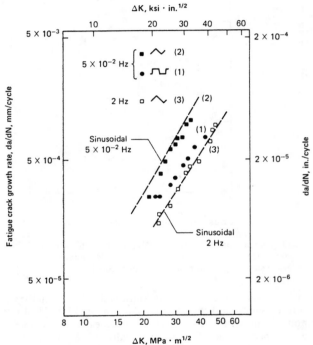

Load/time waveforms and FCP rates for specimens of Inconel 718. Top: Various forms of cyclic stress fluctuations used at 550 °C (1025 °F) at a frequency of 5×10^{-2} Hz. Bottom: FCP rates at 550 °C under sinusoidal, triangular and square loads.

To separate out the possible effects of creep or environment from deformation mode, the authors used triangular and square wave shapes, as shown in the top graph. The data obtained using the triangular wave at 2 Hz were the same as the data obtained in other tests using the sine wave at the same frequency which resulted in the lowest FCP rate. The effect of loading at the same rate but imposing a 10-second hold time at maximum load was to increase the FCP rate only slightly, as shown in the lower graph.

Source: Stephen D. Antolovich and J. E. Campbell, "Fracture Properties of Superalloys," in Application of Fracture Mechanics for Selection of Metallic Structural Materials, James E. Campbell, William W. Gerberich and John H. Underwood, Eds., American Society for Metals, Metals Park OH, 1982, p 295

11-22. Inconel 718: Fatigue Crack Growth Rate in Air vs Helium

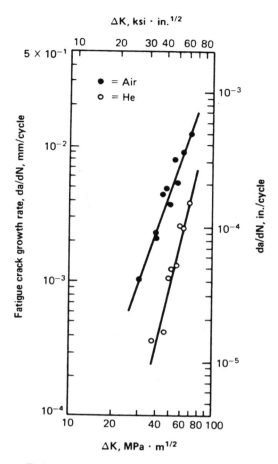

Fatigue crack growth rate data for Inconel 718 in air and in helium. Frequency, 0.1 Hz. Temperature, 650 °C (1200 °F). Here it is evident that crack growth rate at constant temperature is lower in inert gas.

Source: Stephen D. Antolovich and J. E. Campbell, "Fracture Properties of Superalloys," in Application of Fracture Mechanics for Selection of Metallic Structural Materials, James E. Campbell, William W. Gerberich and John H. Underwood, Eds., American Society for Metals, Metals Park OH, 1982, p 287

11-23. Inconel 718: Effect of Environment on Fatigue Crack Growth Rate

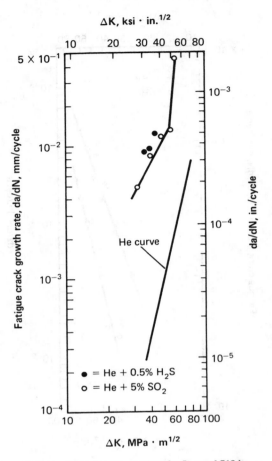

Fatigue crack growth rate data for Inconel 718 in helium + 0.5% hydrogen sulfide and helium + 5% sulfur dioxide. Frequency, 0.1 Hz. Temperature, 650 °C (1200 °F).

From the data above it becomes obvious that fatigue crack growth rate increases greatly in aggressive environments compared with exposure to helium alone.

Source: Stephen D. Antolovich and J. E. Campbell, "Fracture Properties of Superalloys," in Application of Fracture Mechanics for Selection of Metallic Structural Materials, James E. Campbell, William W. Gerberich and John H. Underwood, Eds., American Society for Metals, Metals Park OH, 1982, p 288

11-24. Inconel 718: Fatigue Crack Growth Rate in Air Plus 5% Sulfur Dioxide

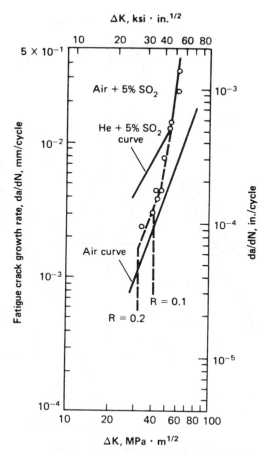

Fatigue crack growth rate data for Inconel 718 in air + 5% sulfur dioxide.

(The effect of air plus 5% SO_2 was similar to the effect of air alone.) It was observed that in the helium atmosphere, which was used to establish a baseline, cracking was generally transgranular with well-defined striations. In the air, oxygen-bearing and sulfur-bearing environments, the crack path changed from transgranular to intergranular, indicating that an important effect of the environment was to degrade the boundary strength by mechanisms that were not clearly defined. It was suggested that oxygen diffusion along grain boundaries and localized oxidation may have occurred. Another very important observation was that the effect of a given environment on FCP could not be predicted on the basis of unstressed exposure tests. The attack on the surfaces of unstressed specimens in aggressive SO_2 environments was minimal, but the SO_2 environments caused substantial increases in FCP.

Source: Stephen D. Antolovich and J. E. Campbell, "Fracture Properties of Superalloys," in Application of Fracture Mechanics for Selection of Metallic Structural Materials, James E. Campbell, William W. Gerberich and John H. Underwood, Eds., American Society for Metals, Metals Park OH, 1982, p 289

11-25. Inconel 718: Fatigue Crack Growth Rate in Air at Room Temperature

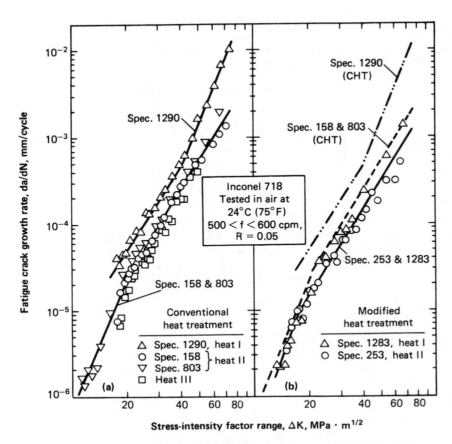

Fatigue crack growth rate behavior of Inconel 718 tested in air at 24 °C (75 °F). CHT = conventional heat treatment. All testing was done at $R = 0.05$ and at a frequency of 0.67 Hz.

Source: Stephen D. Antolovich and J. E. Campbell, "Fracture Properties of Superalloys," in Application of Fracture Mechanics for Selection of Metallic Structural Materials, James E. Campbell, William W. Gerberich and John H. Underwood, Eds., American Society for Metals, Metals Park OH, 1982, p 276

11-26. Inconel 718: Fatigue Crack Growth Rate in Air at 316 °C (600 °F)

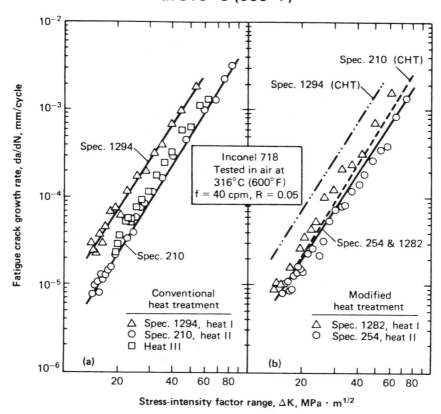

Fatigue crack growth rate behavior of Inconel 718 tested in air at 316 °C (600 °F). CHT = conventional heat treatment. All testing was done at $R = 0.05$, and at a frequency of 0.67 Hz.

Source: Stephen D. Antolovich and J. E. Campbell, "Fracture Properties of Superalloys," in Application of Fracture Mechanics for Selection of Metallic Structural Materials, James E. Campbell, William W. Gerberich and John H. Underwood, Eds., American Society for Metals, Metals Park OH, 1982, p 277

11-27. Inconel 718: Fatigue Crack Growth Rate in Air at 427 °C (800 °F)

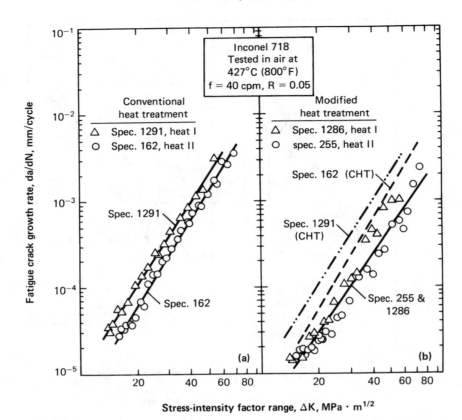

Fatigue crack growth rate behavior of Inconel 718 tested in air at 427 °C (800 °F). CHT = conventional heat treatment. All testing was done at $R = 0.05$ and at a frequency of 0.67 Hz.

Source: Stephen D. Antolovich and J. E. Campbell, "Fracture Properties of Superalloys," in Application of Fracture Mechanics for Selection of Metallic Structural Materials, James E. Campbell, William W. Gerberich and John H. Underwood, Eds., American Society for Metals, Metals Park OH, 1982, p 278

11-28. Inconel 718: Fatigue Crack Growth Rate in Air at 538 °C (1000 °F)

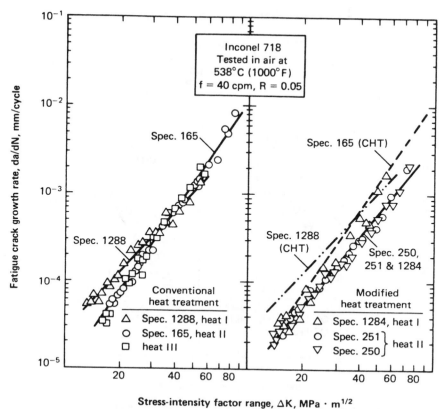

Fatigue crack growth rate behavior of Inconel 718 tested in air at 538 °C (1000 °F). CHT = conventional heat treatment. All testing was done at $R = 0.05$ and at a frequency of 0.67 Hz.

Source: Stephen D. Antolovich and J. E. Campbell, "Fracture Properties of Superalloys," in Application of Fracture Mechanics for Selection of Metallic Structural Materials, James E. Campbell, William W. Gerberich and John H. Underwood, Eds., American Society for Metals, Metals Park OH, 1982, p 279

11-29. Inconel 718: Fatigue Crack Growth Rate in Air at 649 °C (1200 °F)

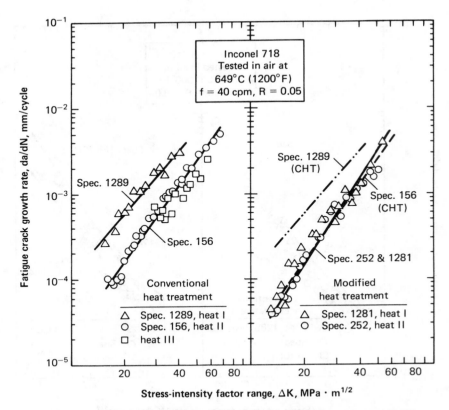

Fatigue crack growth rate behavior of Inconel 718 tested in air at 649 °C (1200 °F). CHT = conventional heat treatment. All testing was done at $R = 0.05$ and at a frequency of 0.67 Hz.

Source: Stephen D. Antolovich and J. E. Campbell, "Fracture Properties of Superalloys," in Application of Fracture Mechanics for Selection of Metallic Structural Materials, James E. Campbell, William W. Gerberich and John H. Underwood, Eds., American Society for Metals, Metals Park OH, 1982, p 280

11-30. Inconel 718: Fatigue Crack Growth Rates at Cryogenic Temperatures

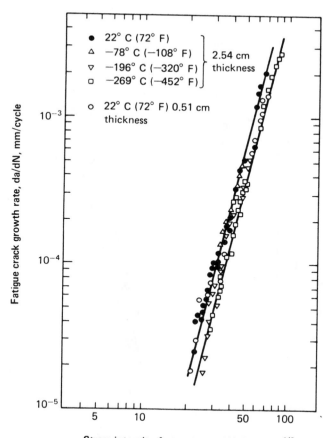

Fatigue crack growth rates of Inconel 718 forged bar at an R ratio of 0.1 and a frequency of 20 Hz. Heat treatment: 980 °C (1800 °F) ¾ h, AC; double aged 720 °C (1325 °F) 8 h, FC to 620 °C (1150 °F), hold 10 h, AC. At the constant frequency the effect of higher temperature is to increase the FCP rate.

Source: Stephen D. Antolovich and J. E. Campbell, "Fracture Properties of Superalloys," in Application of Fracture Mechanics for Selection of Metallic Structural Materials, James E. Campbell, William W. Gerberich and John H. Underwood, Eds., American Society for Metals, Metals Park OH, 1982, p 298

11-31. Inconel 718 and X-750: Fatigue Crack Growth Rates at Cryogenic Temperatures

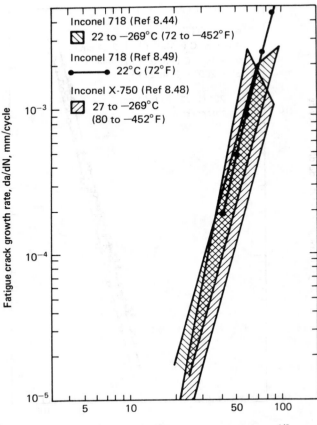

Fatigue crack growth rates for Inconel 718 and Inconel X-750 in the subzero temperature range.

A comparison of FCP values from room temperature to −269 °C (−452 °F) for Inconel 718 and Inconel X-750 is shown in the above chart, along with room temperature FCP data for Inconel 718 from Shahinian et al. The FCP data for these two alloys overlap in the ΔK range shown. Under some conditions, the FCP rate for Inconel 706 is slightly less than those for Inconel 718 and Inconel X-750 at corresponding temperatures and ΔK levels. However, results of FCP tests depend on both melting practice and thermomechanical processing.

Source: Stephen D. Antolovich and J. E. Campbell, "Fracture Properties of Superalloys," in Application of Fracture Mechanics for Selection of Metallic Structural Materials, James E. Campbell, William W. Gerberich and John H. Underwood, Eds., American Society for Metals, Metals Park OH, 1982, p 300

11-32. Inconel X-750: Effect of Temperature on Fatigue Crack Growth Rates

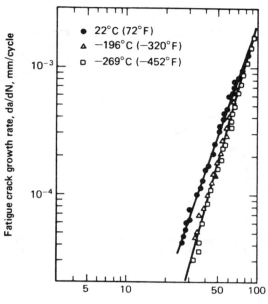

Fatigue crack growth rates of Inconel X-750 at an R ratio of 0.1 and at frequencies of 20 to 28 Hz. Heat treatment: solution treated and double aged. Within this frequency range, the effect of higher temperature is to increase the FCP rate.

Source: Stephen D. Antolovich and J. E. Campbell, "Fracture Properties of Superalloys," in Application of Fracture Mechanics for Selection of Metallic Structural Materials, James E. Campbell, William W. Gerberich and John H. Underwood, Eds., American Society for Metals, Metals Park OH, 1982, p 299

11-33. Jethete M152: Interrelationship of Tempering Treatment, Alloy Class, and Testing Temperature With Fatigue Characteristics

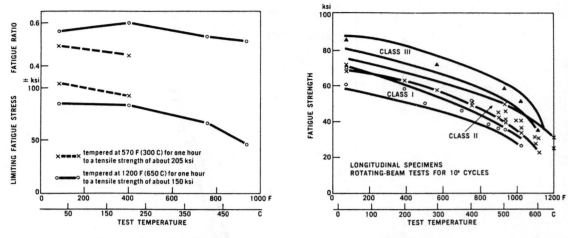

Left: Interrelationship of prior tempering treatment and testing temperature with limiting fatigue stress, and with fatigue ratio for Jethete M152. Right: Influence of alloy class and testing temperature on fatigue strength for the same alloy.

Source: J. Z. Briggs and T. D. Parker, "The Super 12% Cr Steels," in Source Book on Materials for Elevated-Temperature Applications, Elihu F. Bradley, Ed., American Society for Metals, Metals Park OH, 1979, p 123

11-34. Lapelloy: Interrelationship of Hardness and Strength With Fatigue Characteristics

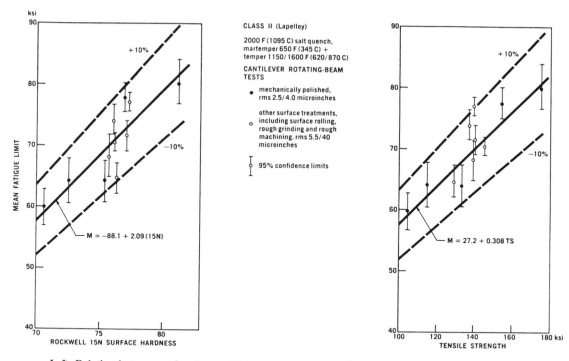

Left: Relation between surface hardness and mean fatigue limit for Lapelloy. Right: Relation between tensile strength and mean fatigue limit for the same alloy.

Source: J. Z. Briggs and T. D. Parker, "The Super 12% Cr Steels," in Source Book on Materials for Elevated-Temperature Applications, Elihu F. Bradley, Ed., American Society for Metals, Metals Park OH, 1979, p 123

11-35. Mar-M200: Effect of Atmosphere on Cycles to Failure

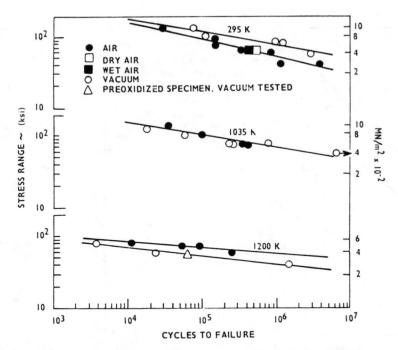

S-N curves showing fatigue life at 10 Hz of single-crystal low-carbon alloy Mar-M200 at 295-1200 K.

Convergence of air and vacuum data was noted for AISI 316 steel at 1090 K, and a crossover of the air and vacuum curves occurred for nickel, where it was suggested that oxide in cracks could prolong life in air at low stresses. Crossovers have also been seen in a ferritic stainless steel and a Ni/Cr alloy in the range 875-1025 K, where tests in purified argon gave shorter endurances than those in air, impure argon, or sulfur dioxide. Also, in single crystals of the alloy Mar-M200, air endurances were less than those in vacuum at room temperature whilst the reverse was true at high temperature (above). A thin oxide film, formed during testing, suppressed surface crack initiation, but oxide formed during pre-exposure did not.

Source: R. H. Cook and R. P. Skelton, "Environment-Dependence of the Mechanical Properties of Metals at High Temperature," in Source Book on Materials for Elevated-Temperature Applications, Elihu F. Bradley, Ed., American Society for Metals, Metals Park OH, 1979, p 81

11-36. Mar-M509: Correlation of Initial Crack Propagation and Dendrite Arm Spacing

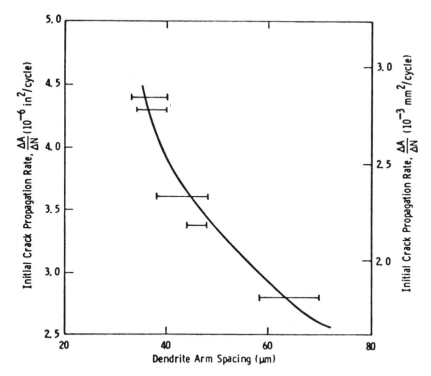

Correlation between the initial crack propagation rate and the dendrite arm spacing for Mar-M509.

Source: Eric Bachelet and Gerard Lesoult, "Quality of Castings of Superalloys," in Superalloys: Source Book, Matthew J. Donachie, Jr., Ed., American Society for Metals, Metals Park OH, 1984, p 338

11-37. Mar-M509: Correlation Between Number of Cycles Required to Initiate a Crack and Dendrite Arm Spacing

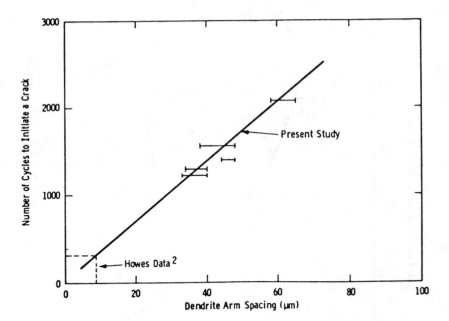

Correlation between the number of cycles required to initiate a crack and the dendritic arm spacing for cast alloy Mar-M509.

Source: Eric Bachelet and Gerard Lesoult, "Quality of Castings of Superalloys," in Superalloys: Source Book, Matthew J. Donachie, Jr., Ed., American Society for Metals, Metals Park OH, 1984, p 337

11-38. MERL 76, P/M: Axial Low-Cycle Fatigue Life of As-HIP'd Alloy at 540 °C (1000 °F)

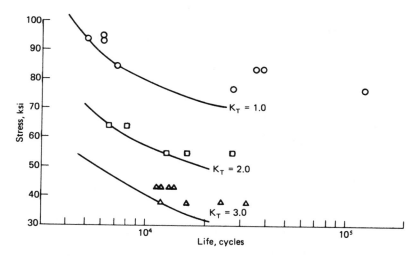

Axial low-cycle fatigue life of as-HIP'd P/M alloy MERL 76 at 540 °C (1000 °F) at notch severities as indicated.

Source: J. H. Moll, V. C. Petersen and E. J. Dulis, "Powder Metallurgy Parts for Aerospace Applications," in Powder Metallurgy—Applications, Advantages and Limitations, Erhard Klar, Ed., American Society for Metals, Metals Park OH, 1983, p 275

11-39. Nickel-Base Alloys: Effect of Solidification Conditions on Cycles to Onset of Cracking

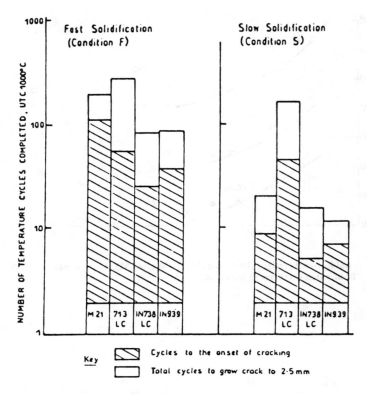

Bar chart showing effect of solidification conditions on cycles to onset of cracking and total number of cycles required to grow cracks to 2.5 mm in several nickel-base casting alloys.

Source: Eric Bachelet and Gerard Lesoult, "Quality of Castings of Superalloys," in Superalloys: Source Book, Matthew J. Donachie, Jr., Ed., American Society for Metals, Metals Park OH, 1984, p 339

11-40. René 95 (As-HIP): Cyclic Crack Growth Behavior Under Continuous and Hold-Time Conditions

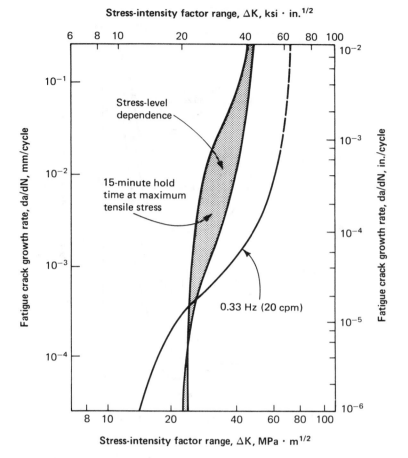

Cyclic crack growth behavior for as-HIP René 95 under both continuous and hold-time conditions at 650 °C (1200 °F).

The effect of environment need not always lead to more rapid crack growth. It has been proposed that oxidation products could form in the crack tip region and prevent crack resharpening during the unloading portion of the cycle. If the stresses are sufficiently low, the oxidation products in the crack tip region will not be cracked and, in some systems, an elevation of the threshold might occur. Such effects would be pronounced at high temperatures and long hold times and have actually been observed in René 95, as shown in the above chart. Once the stress intensity is high enough to crack the oxides, the rate of crack growth would be expected to increase due to the severely degraded region in the crack tip zone.

Source: Stephen D. Antolovich and J. E. Campbell, "Fracture Properties of Superalloys," in Application of Fracture Mechanics for Selection of Metallic Structural Materials, James E. Campbell, William W. Gerberich and John H. Underwood, Eds., American Society for Metals, Metals Park OH, 1982, p 284

11-41. René 95: Effect of Temperature on Fatigue Crack Growth Rate

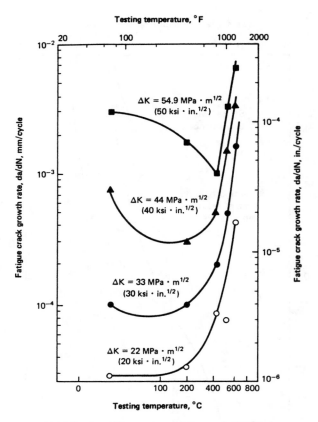

Effect of temperature on fatigue crack growth rate at constant ΔK for René 95.

That the effect of environment can be large may be inferred from some low-cycle fatigue studies of René 95 in which surface and subsurface cracking was observed at comparable strain ranges and defect sizes. As expected, the life of the subsurface crack was much greater than that of the surface crack, leading to the hypothesis of a strong environmental effect. This possibility is considered in more detail in an analysis of FCP properties of René 95. The FCP rate was plotted as a function of temperature for a given ΔK range, as shown in the above chart. It is noteworthy that there is a minimum in the FCP rate at all ΔK levels except 22 MPa·M$^{1/2}$ (20 ksi·in.$^{1/2}$), where the data are at least suggestive of a minimum. Because any environmental interaction is thermally activated, the crack growth rate at a given ΔK level and frequency may be written as:

$$\frac{da}{dN} = A \exp - Q(\Delta K)/RT$$

where A is a constant and $Q(\Delta K)$ is the apparent activation energy.

Source: Stephen D. Antolovich and J. E. Campbell, "Fracture Properties of Superalloys," in Application of Fracture Mechanics for Selection of Metallic Structural Materials, James E. Campbell, William W. Gerberich and John H. Underwood, Eds., American Society for Metals, Metals Park OH, 1982, p 282

11-42. S-816: Effect of Notches on Cycles to Failure at 900 °C (1650 °F)

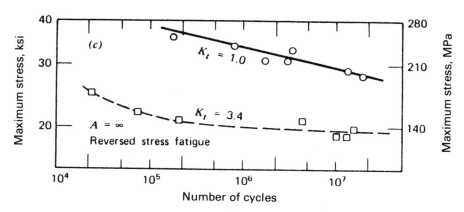

S-N diagram for S-816 heat-resisting alloy tested at 900 °C (1650 °F), notched (broken curve) versus unnotched (solid curve).

11-43. Udimet 700: Fatigue Crack Growth Rates at 850 °C (1560 °F)

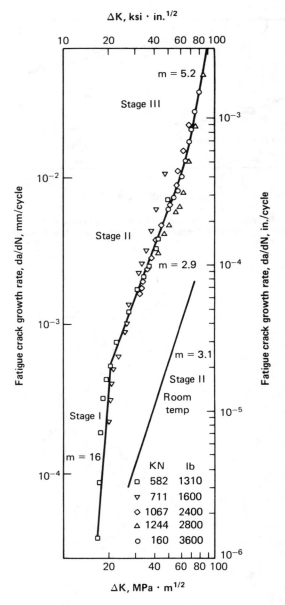

Crack growth rates in terms of stress-intensity factor range for Udimet 700 at 850 °C (1560 °F). Crack growth rates for this alloy are greatly accelerated by increases in temperature.

Source: Stephen D. Antolovich and J. E. Campbell, "Fracture Properties of Superalloys," in Application of Fracture Mechanics for Selection of Metallic Structural Materials, James E. Campbell, William W. Gerberich and John H. Underwood, Eds., American Society for Metals, Metals Park OH, 1982, p 285

11-44. U-700 and Mar-M200: Comparison of Fatigue Properties

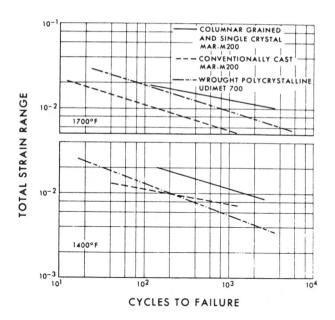

Comparison of fatigue properties at 760 °C (1400 °F) and 925 °C (1700 °F) for a typical wrought nickel-base alloy (U-700) with conventionally cast, directionally solidified and monocrystal Mar-M200.

Source: Francis L. Versnyder and M. E. Shank, "The Development of Columnar Grain and Single Crystal High Temperature Materials Through Directional Solidification," in Source Book on Materials for Elevated-Temperature Applications, Elihu F. Bradley, Ed., American Society for Metals, Metals Park OH, 1979, p 358

11-45. Waspaloy: Stress-Response Curves

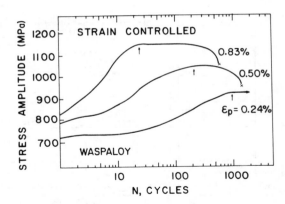

Stress-response curves for Waspaloy having non-shearable precipitates.

During aging of precipitation-hardenable alloys, the coherent precipitates grow, and accommodation strains build up. At some point the energy associated with the accommodation strains exceeds that necessary to create a precipitate-matrix interface, and the precipitates become partly incoherent. This is accompanied by a change in precipitate-dislocation interaction from one of shearing to that of dislocation looping or bypassing the precipitates. Since the reasons for strain localization have been removed, deformation becomes more homogeneous. Local softening is thus prevented, and the cyclic-response curve shows hardening to saturation, or to failure, as illustrated above.

Source: Edgar A. Starke, Jr., and Gerd Lütjering, "Cyclic Plastic Deformation and Microstructure," in Fatigue and Microstructure, American Society for Metals, Metals Park OH, 1979, p 217

11-46. X-40: Effect of Grain Size and Temperature on Fatigue Characteristics

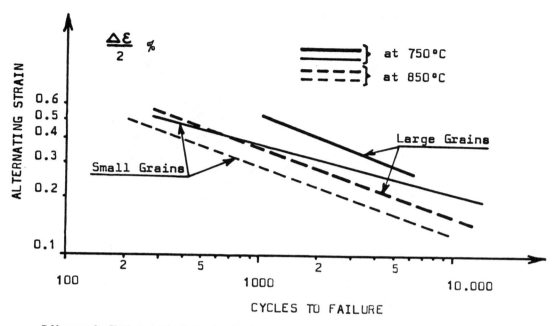

S-N curves for X-40 showing effects of grain size and temperature on fatigue characteristics of this alloy.

Source: Eric Bachelet and Gerard Lesoult, "Quality of Castings of Superalloys," in Superalloys: Source Book, Matthew J. Donachie, Jr., Ed., American Society for Metals, Metals Park OH, 1984, p 335

11-47. Cast Heat-Resisting Alloys: Ranking for Resistance to Thermal Fatigue

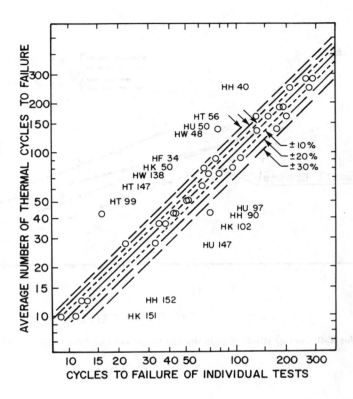

The design of components that are subject to considerable temperature cycling must also include consideration of thermal fatigue. This is particularly true if the temperature changes are frequent or rapid, and nonuniform within or between casting sections. Fatigue is a condition in which failure results from alternating load applications in shorter times, or at lower stresses, than expected from constant-load properties. "Thermal fatigue" denotes the condition when the stresses are primarily due to hindered expansion or contraction. Good design helps minimize the external restraint to expansion and contraction. Rapid heating and cooling may, however, impose temperature gradients within the part causing the cooler elements of the component to restrain the hotter elements. Finite-element computer analysis has shown that, for some industrial applications, these thermally induced stresses may exceed those resulting from the mechanical loads.

An example of results from thermal fatigue data is presented above. This graph offers a ranking of many cast heat-resistant high-alloy grades relative to their resistance to thermal fatigue. Such rankings are indicative of general alloy properties only because most thermal fatigue tests are based on an arbitrary set of experimental conditions rather than on their fundamental material behavior. Nevertheless, such test results have been useful in considering alloy selection questions, and in identifying the superior thermal fatigue resistance of nickel predominating grades and the good performance of some HH type compositions.

Source: Steel Castings Handbook, 5th Edition, Peter F. Weiser, Ed., Steel Founders' Society of America, Rocky River OH, 1980, p 19-7

12-1. Corrosion-Fatigue Properties of Aluminum Alloys Compared With Those of Other Alloys

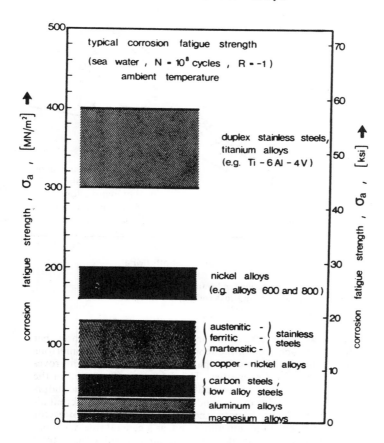

As shown above the corrosion-fatigue strength of bare aluminum alloys is superior only to that of magnesium alloys. Careful surface protection may bring the corrosion-fatigue strength up into the range of bare stainless steels or copper-nickel alloys.

Source: Markus O. Speidel, "Aluminum as a Corrosion Resistant Material," in Aluminum Transformation Technology and Applications (Proceedings of the International Symposium at Puerto Madryn, Chubut, Argentina), C. A. Pampillo, H. Biloni and D. E. Embury, Eds., American Society for Metals, Metals Park OH, 1980, p 617

12-2. Comparisons of Aluminum Alloys With Magnesium and Steel: Tensile Strength vs Endurance Limit

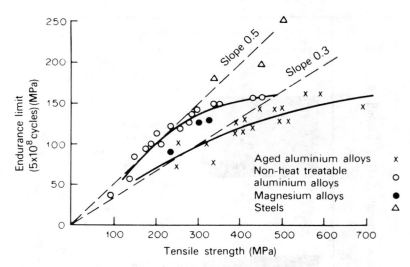

Fatigue ratios (endurance limit/tensile strength) for aluminum alloys compared with those of magnesium and steel.

It is well known that, in contrast to steels, the increases that have been achieved in the tensile strength of most nonferrous alloys have not been accompanied by proportionate improvements in fatigue properties. This feature is illustrated in the graph above, which shows relationships between fatigue endurance limit (5×10^8 cycles) and tensile strength for different alloys. It should also be noted that the fatigue ratios are lowest for age-hardened aluminum alloys and, as a general rule, the more an alloy is dependent upon precipitation hardening for its total strength, the lower this ratio becomes.

Source: I. J. Polmear, Light Alloys, Edward Arnold Ltd, London, England, and American Society for Metals, Metals Park OH, 1981, p 39

12-3. Aluminum Alloys (General): Yield Strength vs Fatigue Strength

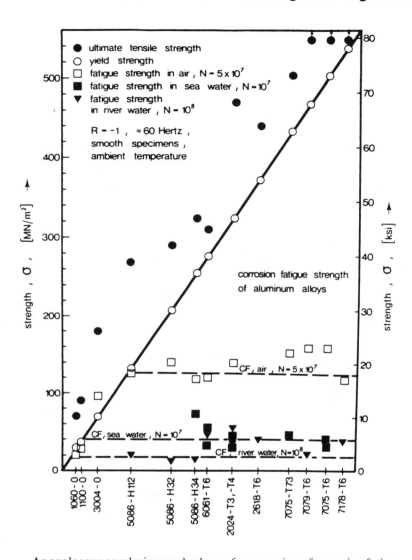

An analogous conclusion can be drawn from a review of corrosion fatigue tests with smooth aluminum alloy specimens as shown in the above graph. Here aluminum alloys are listed in order of increasing yield strength. As the yield strength goes up, so does the ultimate tensile strength, but the fatigue strength in air soon reaches a limit which is roughly the same for alloys of greatly different yield strength. In other words, medium- and high-strength aluminum alloys all have about the same fatigue strength. The above graph shows that the same is true for the corrosion-fatigue strength: there is as yet not a single commercial aluminum alloy available with a high-cycle corrosion-fatigue strength significantly higher than all the other aluminum alloys. Thus, corrosion fatigue is still a limiting factor for the application of aluminum alloys.

Source: Markus O. Speidel, "Aluminum as a Corrosion Resistant Material," in Aluminum Transformation Technology and Applications (Proceedings of the International Symposium at Puerto Madryn, Chubut, Argentina), C. A. Pampillo, H. Biloni and D. E. Embury, Eds., American Society for Metals, Metals Park OH, 1980, p 616

12-4. Comparison of Aluminum Alloy Grades for Crack Propagation Rate

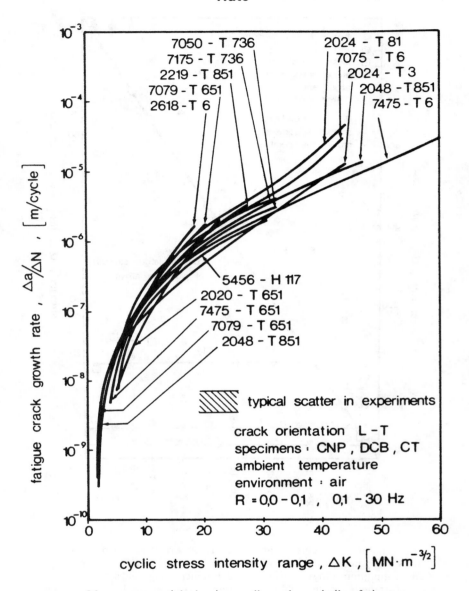

Many commercial aluminum alloys show similar fatigue crack propagation rates in air, as indicated in the above comparison.

Source: Markus O. Speidel, "Aluminum as a Corrosion Resistant Material," in Aluminum Transformation Technology and Applications (Proceedings of the International Symposium at Puerto Madryn, Chubut, Argentina), C. A. Pampillo, H. Biloni and D. E. Embury, Eds., American Society for Metals, Metals Park OH, 1980, p 613

12-5. Alloy 1100: Relationship of Fatigue Cycles and Hardness for H0 and H14 Tempers

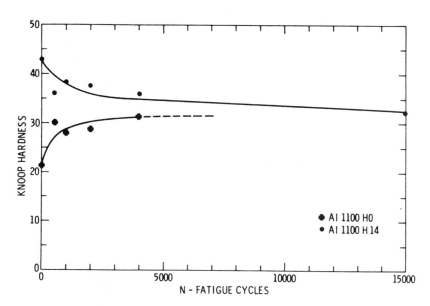

Comparison of the Knoop hardness for well-annealed (H0) and cold-rolled (H14) aluminum as a function of fatigue.

Microcrack initiation is often triggered by a dislocation rearrangement. For instance, in the case of well-annealed Al 1100 (H0), the material will harden in the early stages of fatigue (see S-N curves above) as the dislocation density in the bulk of the material increases, accompanied by pronounced slip-step formation on the surface. On the other hand, in the case of the cold-worked material Al 1100 (H14), the material will soften in the early stages of fatigue (above curves) as the dislocation density, introduced by the cold work, decreases. Slip-step formation in this situation is much less pronounced than it is during hardening, because the initial dislocation-loop length is much shorter. In either case, during this initial rearrangement, the dislocations form a cell structure with individual dislocations of long loop length shuttling to and fro between the cell walls. This latter part of the fatigue life is called the saturation stage of fatigue, during which the dislocation shuttling leads to local instabilities, or "extrusions-intrusions," and finally to microcracks, which can be observed after about 25% of the fatigue life has been expended. The microcrack density is about the same for both materials.

Source: O. Buck and G. A. Alers, "New Techniques for Detection and Monitoring of Fatigue Damage," in Fatigue and Microstructure, American Society for Metals, Metals Park OH, 1979, p 128

12-6. Alloy 1100: Interrelationship of Fatigue Cycles, Acoustic Harmonic Generation and Hardness

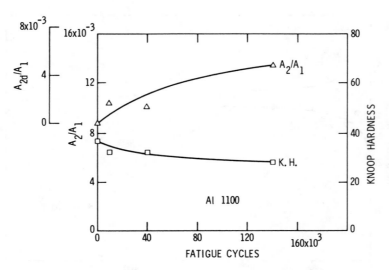

Normalized second harmonic displacement and Knoop hardness as a function of fatigue.

The effects of dislocation rearrangements on harmonic generation within the bulk of the material during fatigue are shown in the above chart. Using 30-MHz longitudinal waves, the normalized second harmonic amplitude of an initially compression-deformed Al 1100 single crystal was monitored and found to increase as a function of compression-compression fatigue. At the same time, the surface hardness (Knoop) decreased. Apparently, the dislocation-loop length prior to fatigue was quite short, since the initial amplitude of harmonic generation was small. During fatigue softening, the cell structure that developed (with its individual dislocations within the cells) became quite large, so that a change of the dislocation-induced harmonic generation, A_{2d}, increased. Application of this technique to high-strength aluminum alloys failed, however, apparently because of an immediate repinning of the long loops by interstitials in this alloy.

Source: O. Buck and G. A. Alers, "New Techniques for Detection and Monitoring of Fatigue Damage," in Fatigue and Microstructure, American Society for Metals, Metals Park OH, 1979, p 131

12-7. Alloy 2014-T6: Notched vs Unnotched Specimens; Effect on Cycles to Failure

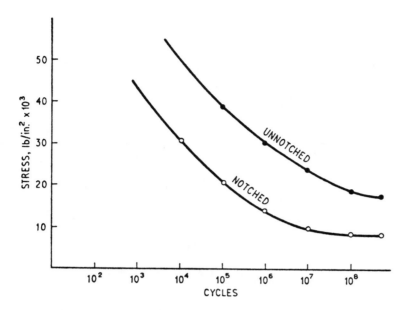

Effect of notch on fatigue of 2014-T6. As is true for most alloys, notches greatly reduce the fatigue properties of aluminum alloys.

Source: P. C. Varley, The Technology of Aluminium and Its Alloys, Butterworth & Co. Ltd., London, England, 1970, p 43

12-8. Alloy 2024-T3: Effect of Air vs Vacuum Environments on Cycles to Failure

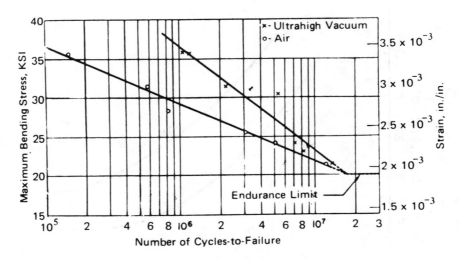

The effects of air versus vacuum on the fatigue life of a 2024-T3 aluminum alloy.

For most materials, environment appears to be most effective early in the crack-growth process, with little or no effect at high crack-growth rates. Additionally, the majority of S-N curves diverge at decreasing stresses, the increase in fatigue life caused by vacuum becoming greater at lower stresses. In contrast to this behavior, however, aluminum and aluminum alloys have been shown to exhibit conflicting results. For example, a 2017-T4 alloy tested in air and at 2×10^{-6} torr and a 2024-T3 alloy tested in air and at 10^{-10} torr in rotating bending exhibit convergence of S-N curves at low stresses, the effect of environment apparently becoming less important at decreasing stresses, as shown in the above chart.

Source: D. J. Duquette, "Environmental Effects I: General Fatigue Resistance and Crack Nucleation in Metals and Alloys," in *Fatigue and Microstructure*, American Society for Metals, Metals Park OH, 1979, p 337

12-9. Alloy 2024-T4 Alclad Sheet: Effect of Bending on Cycles to Failure

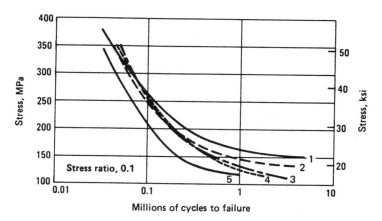

Effects of bending on fatigue characteristics of aluminum alloy sheet.

For the data here, sheet 1.02 mm (0.040 in.) thick was annealed, solution heat treated and quenched, and then fatigue tested. The sheet represented by curve 1 was not bent. All other sheet was bent 90° in the annealed condition. Flattening (unbending) was done in either the annealed condition (curve 2) or the solution heat treated and quenched condition (curves 3, 4 and 5). Details of bending and flattening were as follows: (1) Not bent. (2) Bend radius, 3.18 mm (⅛ in.); flattened in annealed condition. (3) Bend radius, 3.18 mm (⅛ in.); flattened in quenched condition after 3 days of storage at −18 to −12 °C (0 to 10 °F). (4) Bend radius, 3.18 mm (⅛ in.); flattened in quenched condition after 14 days of storage at −18 to −12 °C (0 to 10 °F). (5) Bend radius, 1.59 mm (1/16 in.); flattened in quenched condition after 3 days of storage at −18 to −12 °C (0 to 10 °F).

Source: Metals Handbook, 9th Edition, Volume 2, Properties and Selection: Nonferrous Alloys and Pure Metals, American Society for Metals, Metals Park OH, 1979, p 35

12-10. Alloy 2024-T4: High-Cycle vs Low-Cycle Fatigue

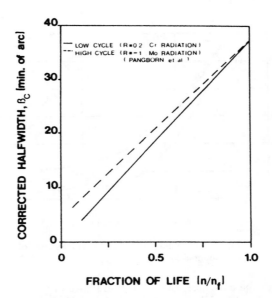

Dependence of $\bar{\beta}$ on n/n_f for low-cycle fatigue and bulk properties of high-cycle fatigue of Al 2024.

After correcting for the difference in initial β_0 values, it can be seen in the above diagram that the two fatigue processes, although radically different in strain history, exhibit similar behavior throughout most of the fatigue life.

Source: Sigmund Weissmann and William E. Mayo, "Determination of Strain Distributions and Failure Prediction by Novel X-ray Methods," in Nondestructive Evaluation: Application to Materials Processing, Otto Buck and Stanley M. Wolf, Eds., American Society for Metals, Metals Park OH, 1984, p 195

12-11. Alloy 2024-T4: Relationship of Stress and Fatigue Cycles

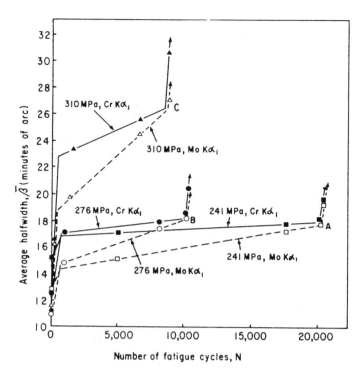

Dependence of $\bar{\beta}$ on number of cycles N at various stress levels of Al 2024-T4.

Here is shown that for the maximum stress of 241 MPa with $R = 0.1$, the $\bar{\beta}$ value increased during the first several hundred cycles. This was more pronounced for the surface grains (Cr Kα_1 radiation).

Source: Sigmund Weissmann and William E. Mayo, "Determination of Strain Distributions and Failure Prediction by Novel X-ray Methods," in Nondestructive Evaluation: Application to Materials Processing, Otto Buck and Stanley M. Wolf, Eds., American Society for Metals, Metals Park OH, 1984, p 194

12-12. Alloy 2024-T4: Dependence of the Average Rocking Curve Halfwidth $\bar{\beta}$ on Distance From the Surface

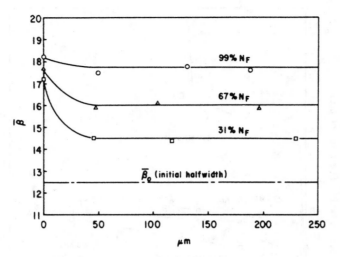

Dependence of the average rocking curve halfwidth $\bar{\beta}$ on depth distance from surface for different fractions of corrosion fatigue lives, N_F, of Al 2024-T4.

X-ray rocking curve measurements were carried out as a function of depth distance from the surface, and typical results of the dependence of $\bar{\beta}$ on depth distance for an alloy cycled with $\sigma = 276$ MPa, corresponding to the static yield stress, are shown above. It may be seen that the minimum $\bar{\beta}$ values at the surface layers were larger than those in the interior. The $\bar{\beta}$ values declined up to a depth of about 50 μm from the surface and subsequently retained a plateau value throughout the interior of the specimen for each fraction of the life.

Source: Sigmund Weissmann and William E. Mayo, "Determination of Strain Distributions and Failure Prediction by Novel X-ray Methods," in Nondestructive Evaluation: Application to Materials Processing, Otto Buck and Stanley M. Wolf, Eds., American Society for Metals, Metals Park OH, 1984, p 193

12-13. Alloys 2024 and X2024: Effect of Alloy Purity on Cycles to Failure

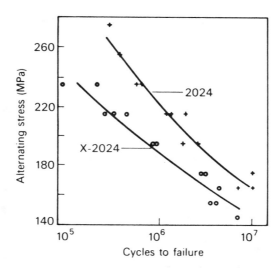

Effect of reducing the concentration of submicron particles in an Al-Cu-Mg alloy. X2024 is a high-purity version of the commercial alloy 2024.

The disappointing fatigue properties of age-hardened aluminum alloys are also attributed to an additional factor, which is the metastable nature of the metallurgical structure under conditions of cyclic stressing. Localization of strain is particularly harmful because the precipitate may be removed from certain slip bands, which causes softening there and leads to a further concentration of stress, so that the whole process of cracking is accelerated.

The fatigue behavior of age-hardened aluminum alloys would therefore be improved if fatigue deformation could be dispersed more uniformly. Factors which prevent the formation of coarse slip bands should assist in this regard. Thus it is to be expected that commercial-purity alloys should perform better than equivalent high-purity compositions because the presence of inclusions and intermetallic compounds would tend to disperse slip. This effect has been confirmed for an Al-Cu-Mg alloy, and fatigue curves for commercial-purity and high-purity compositions are shown in the above *S-N* diagram. Here the superior fatigue behavior of the former alloy arises because slip is more uniformly dispersed by submicron dispersoids such as $MnAl_6$.

Source: I. J. Polmear, Light Alloys, Edward Arnold Ltd, London, England, and American Society for Metals, Metals Park OH, 1981, p 40

12-14. Alloys 2024 and 2124: Relationship of Particle Size and Fatigue Characteristics

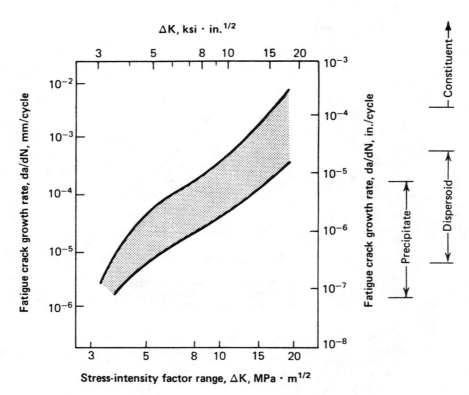

Comparison of typical particle sizes in aluminum alloys with crack advance per cycle on fatigue loading.

The above graph represents Staley's work in summarizing the role of particle size on fatigue crack growth in aluminum alloys.

Source: J. G. Kaufman and J. S. Santner, "Fracture Properties of Aluminum Alloys," in Application of Fracture Mechanics for Selection of Metallic Structural Materials, James E. Campbell, William W. Gerberich and John H. Underwood, Eds., American Society for Metals, Metals Park OH, 1982, p 191

12-15. Alloys 2024-T4 and 2124-T4: Comparison of Resistance to Fatigue Crack Initiation

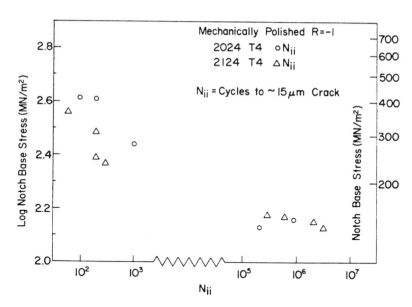

Cycles to fatigue crack initiation for specimens of aluminum alloys 2024-T4 and 2124-T4 versus stress at notch base (computed using Neuber stress-concentration factor).

The 2124 alloy studied had 1/10 the inclusions of the 2024 alloy studied (0.2 vol% compared to 2 vol%) but a larger grain size (45 μm compared to 20 μm in the transverse direction normal to the loading direction). With 2124-T4, slip-band cracks not associated with inclusions formed at the lowest stress studied. They also formed more easily in 2124-T4 than in 2024-T4 at high stresses, in keeping with the larger grain size. Thus, as shown in the above chart, at high stresses 2124-T4 is less resistant to fatigue crack initiation than is 2024-T4, but it is more resistant at low stresses.

Source: M. E. Fine and R. O. Ritchie, "Fatigue-Crack Initiation and Near-Threshold Crack Growth," in *Fatigue and Microstructure*, American Society for Metals, Metals Park OH, 1979, p 251

12-16. Alloys 2024-T3 and 7075-T6: Summary of Fatigue Crack Growth Rates

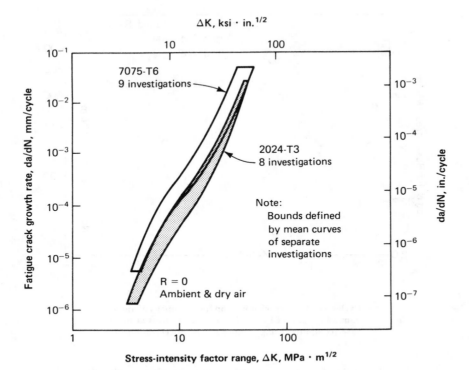

Summary of fatigue crack growth rate data for aluminum alloys 7075-T6 and 2024-T3.

Considerable use has been made of the fracture mechanics approach in measurement of fatigue crack growth rates in aluminum alloys. These data have been generated by methods comparable to those of ASTM Method E647 for measuring fatigue crack growth rates. In general, fatigue crack growth rates are found to fall within a relatively narrow scatter band, with only small systematic effects of composition, fabricating practice or strength level, as illustrated by the data in the above chart.

Source: J. G. Kaufman and J. S. Santner, "Fracture Properties of Aluminum Alloys," in Application of Fracture Mechanics for Selection of Metallic Structural Materials, James E. Campbell, William W. Gerberich and John H. Underwood, Eds., American Society for Metals, Metals Park OH, 1982, p 189

12-17. Alloys 2024-T4 and 7075-T6: Effect of Product Form and Notches

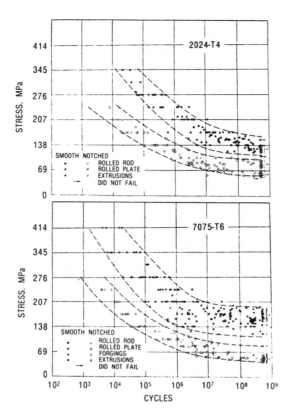

Fatigue performance of smooth and notched ($K_t \gtrsim 17$) rotating-beam specimens from various product forms of 2024-T4 and 7075-T6 aluminum alloys.

Numerous methods have been developed to evaluate response of materials to cyclic deformation. The earliest method was by use of S-N plots. Typical examples are depicted above. Basic specimens include rotating beam, axially stressed and sheet flexure. Notches have been employed to provide stress concentration, and special specimens have been used to simulate a variety of other conditions. The S-N response is strongly influenced by a number of conditions, including surface condition, stress ratio, and environment. The various alloys differ widely in their response to fatigue testing—specifically, in the number of cycles where a "level out" condition is attained. As shown in the above S-N diagrams, the S-N response for aluminum alloys tends to level out as the number of applied cycles approaches 500 million.

Based on S-N data of smooth and sharply notched specimens and of similar tests of specimens designed to simulate joints in structures, the following conclusions have been drawn. From fatigue results for aluminum alloys obtained with smooth specimens . . . rather wide variations can exist without causing appreciable differences in fatigue strengths. . . . When severely notched specimens are used, the effects of composition and temper are even less pronounced and generally are of no practical significance. . . . As in the case of simple notch fatigue tests, there is a lack of significant differences in the fatigue strength of the joints of the various alloys.

Despite these laboratory data, users discovered that certain aluminum alloys performed decidedly better than others in service when fluctuating loads were encountered. For example, airframe manufacturers determined that fatigue performance of alloy 7075-T6 was unquestionably inferior to that of alloy 2024-T3.

Source: T. H. Sanders, Jr., and J. T. Staley, "Review of Fatigue and Fracture Research on High-Strength Aluminum Alloys," in Fatigue and Microstructure, American Society for Metals, Metals Park OH, 1979, p 470

12-18. Alloys 2024-T351 and 7075-T73XXX: Comparison of P/M Extrusions and Rod

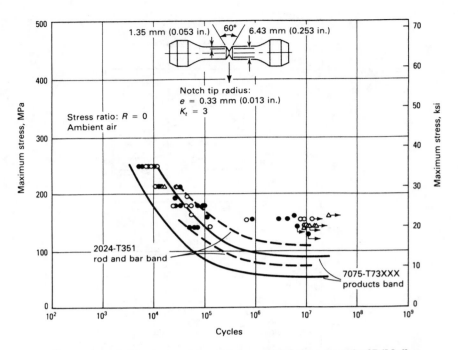

Comparison of room-temperature axial stress notch fatigue strength of P/M alloy extrusions and ingot metallurgy alloy rod, bar and products, ○, X7090-T7E71 in the longitudinal direction; ●, X7091-T7E69 in the longitudinal direction; △, X7091-T7E69 in the long transverse direction; → denotes test specimen did not fail.

Source: Metals Handbook, 9th Edition, Volume 7, Powder Metallurgy, American Society for Metals, Metals Park OH, 1984, p 468

12-19. Alloy 2048-T851: Longitudinal vs Transverse for Axial Fatigue

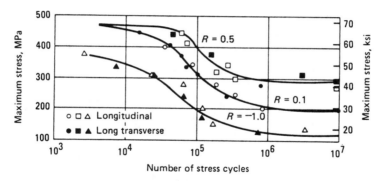

S-N axial fatigue curves for unnotched specimens of aluminum alloy 2048-T851 plate, showing effects of *R* value and direction upon fatigue properties.

Source: Metals Handbook, 9th Edition, Volume 2, Properties and Selection: Nonferrous Alloys and Pure Metals, American Society for Metals, Metals Park OH, 1979, p 80

12-20. Alloy 2048-T851: Notched vs Unnotched Specimens at Room and Elevated Temperatures

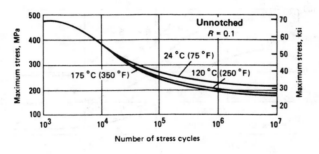

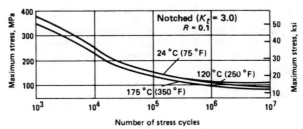

S-N curves for unnotched (upper graph) versus notched (lower graph) specimens of aluminum alloy 2048-T851 plate.

Source: Metals Handbook, 9th Edition, Volume 2, Properties and Selection: Nonferrous Alloys and Pure Metals, American Society for Metals, Metals Park OH, 1979, p 82

12-21. Alloy 2048-T851: Fatigue Crack Propagation Rates in LT and TL Orientations

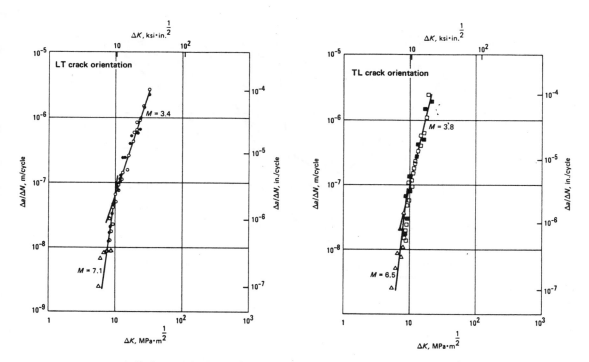

Fatigue crack propagation in aluminum alloy 2048-T851 plate, showing propagation data for both LT and TL (longitudinal and transverse) crack orientations.

Source: Metals Handbook, 9th Edition, Volume 2, Properties and Selection: Nonferrous Alloys and Pure Metals, American Society for Metals, Metals Park OH, 1979, p 81

12-22. Alloy 2048-T851: Modified Goodman Diagram for Axial Fatigue

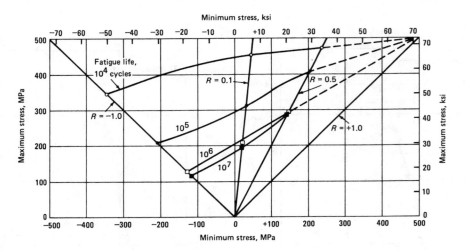

Modified Goodman diagram for axial fatigue of unnotched specimens of aluminum alloy 2048-T851 plate.

Source: Metals Handbook, 9th Edition, Volume 2, Properties and Selection: Nonferrous Alloys and Pure Metals, American Society for Metals, Metals Park OH, 1979, p 81

12-23. Alloy 2219-T851: Dependence of Relaxation Behavior on the Cyclic Hardening Parameter

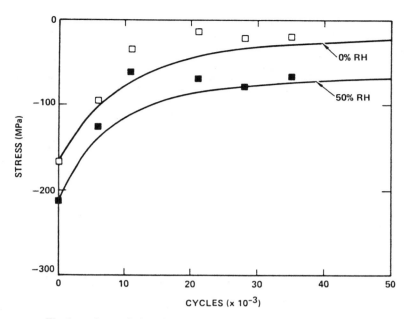

The dependence of relaxation behavior on the cyclic hardening parameter, θ. θ was varied by changing the relative humidity (RH), which affects the near surface ductility in this alloy. Values used were: $\theta = 6 \times 10^{-5}$ for 50% RH and $\theta = 2 \times 10^{-5}$ for 0% RH. The cyclic stress amplitude was 0.88 σ_{yield} for both samples.

Source: M. R. James and W. L. Morris, "The Relaxation of Machining Stresses in Aluminum Alloys During Fatigue," in Residual Stress for Designers and Metallurgists, Larry J. Vander Walle, Ed., American Society for Metals, Metals Park OH, 1981, p 184

12-24. Alloy 2219-T851: Effect of Strain Amplitude on the Relaxation of Residual Surface Stress With Fatigue

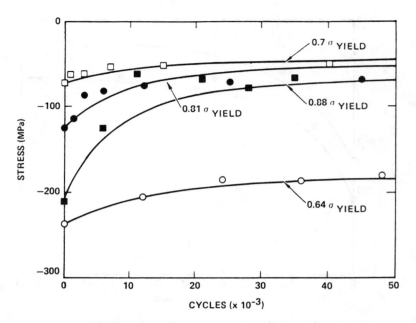

The effect of strain amplitude on the relaxation of surface residual stress with fatigue. The symbols are the residual stress value measured by the x-ray diffraction peak shift technique. The solid curves are the predicted mean residual stress values during fatigue.

Surface milling produced the shallowest stress gradient and resulted in the slowest rate of relaxation of the surface stresses. A comparison of measured to predicted values of residual stress during fatigue is made for four "as machined" specimens in the above chart. The residual stress values were measured parallel to the external stress axis. A value of $\beta = 0.0004$ was used to fit the data for all specimens. Residual stress measurements were also made in a direction transverse to the applied stress axis. Within experimental error, the cyclic relaxation rate was the same as in the longitudinal direction.

Source: M. R. James and W. L. Morris, "The Relaxation of Machining Stresses in Aluminum Alloys During Fatigue," in Residual Stress for Designers and Metallurgists, Larry J. Vander Walle, Ed., American Society for Metals, Metals Park OH, 1981, p 182

12-25. Alloy 2219-T851: Relationship of Fatigue Cycles to Different Depth Distributions of Surface Stress

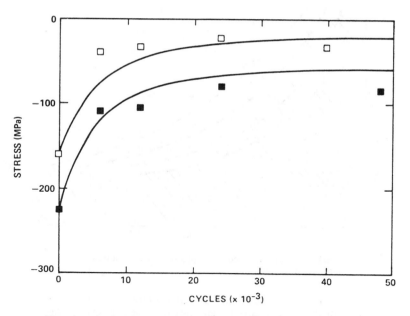

The relaxation behavior of two samples having different depth distributions of residual stress. Note the difference in the peak cyclic stress, σ. □ = rolling (10% reduction); $\beta = 0.012$; $\sigma = 0.91\ \sigma_{yield}$. ■ = sand blasting; $\beta = 0.003$; $\sigma = 0.71\ \sigma_{yield}$.

Relaxation of a compressive surface stress requires an expansion of the material normal to the surface. Of necessity, this involves slip at an acute angle to the surface. If the slip does not penetrate the surface, the residual stress cannot relax. Supporting this picture are our observations that the relaxation rate in Al 2219-T851 is more rapid in dry air. It is known that humidity increases the rate of cyclic hardening of a thin (less than 1 μm) layer at the surface. The effect of humidity on relaxation is therefore simply to make it more difficult for dislocations to penetrate to the surface.

Source: M. R. James and W. L. Morris, "The Relaxation of Machining Stresses in Aluminum Alloys During Fatigue," in Residual Stress for Designers and Metallurgists, Larry J. Vander Walle, Ed., American Society for Metals, Metals Park OH, 1981, p 183

12-26. Alloy 2219-T851: Probability of Fatigue Failure

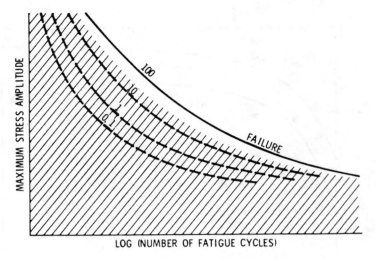

Schematic curves of constant probability for failure (actual failure = 100%).

The solid line in the graph represents failure; the dashed lines indicate the percentage of fatigue life expended. The exact location of these lines is highly sensitive to the material and its microstructure as well as the influences of environment.

Source: O. Buck and G. A. Alers, "New Techniques for Detection and Monitoring of Fatigue Damage," in Fatigue and Microstructure, American Society for Metals, Metals Park OH, 1979, p 104

12-27. Alloys 3003-O, 5154-H34 and 6061-T6: Effect of Alloy on Fatigue Characteristics of Weldments

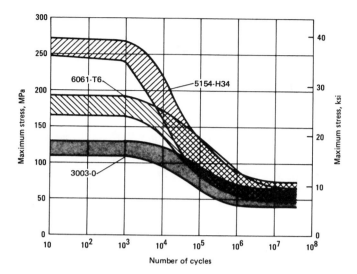

The fatigue life of welded joints at high loads varies with the alloy. As the load is decreased, differences disappear until, at about one to ten million cycles of axial loading ($R = 0$), the fatigue strength of an arc-welded joint is approximately the same regardless of alloy and is 50 to 70% that of the unwelded alloy. Typical data are given in the above graph for three aluminum alloys. Specimens were from 9.5-mm (⅜-in.) plate; weld reinforcement removed; axial loading; $R = 0$.

Source: Metals Handbook, 9th Edition, Volume 2, Properties and Selection: Nonferrous Alloys and Pure Metals, American Society for Metals, Metals Park OH, 1979, p 195

12-28. Alloy 5083-O Plate: Effect of Orientation on Fatigue Crack Growth Rates

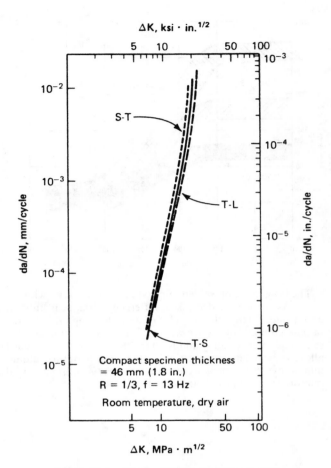

Effect of orientation on fatigue crack growth rates in 180- and 196-mm (7.0- and 7.7-in.) 5083-O plate.

From the data shown above there is obviously no great effect of specimen orientation on fatigue crack growth rates.

Source: J. G. Kaufman and J. S. Santner, "Fracture Properties of Aluminum Alloys," in Application of Fracture Mechanics for Selection of Metallic Structural Materials, James E. Campbell, William W. Gerberich and John H. Underwood, Eds., American Society for Metals, Metals Park OH, 1982, p 193

12-29. Alloy 5083-O Plate: Effect of Temperature and Humidity on Fatigue Crack Growth Rates

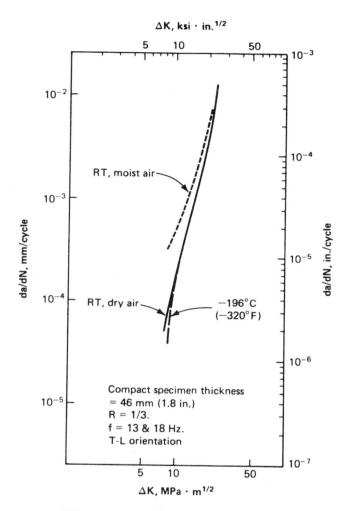

Effect of temperature and humidity on fatigue crack growth in 180-mm (7.0-in.) 5083-O plate.

As shown in the above graph, growth rates for alloy 5083-O are appreciably higher in moist air than in dry air. Growth rates in water solutions of sodium chloride are similar to those in moist air.

Source: J. G. Kaufman and J. S. Santner, "Fracture Properties of Aluminum Alloys," in Application of Fracture Mechanics for Selection of Metallic Structural Materials, James E. Campbell, William W. Gerberich and John H. Underwood, Eds., American Society for Metals, Metals Park OH, 1982, p 195

12-30. Alloys 5086-H34, 5086-H36, 6061-T6, 7075-T73 and 2024-T3: Comparative Resistance to Axial-Stress Fatigue

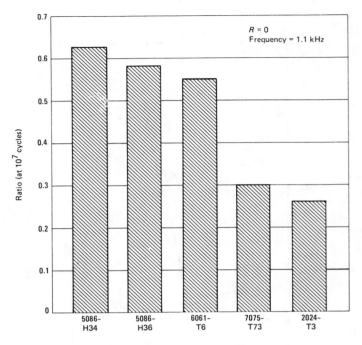

Ratio of axial-stress fatigue strength of aluminum alloy sheet in 3% NaCl solution to that in air.

Fatigue strengths of aluminum alloys are lower in corrosive environments such as seawater and other salt waters than they are in air, especially when evaluated by low-stress, long-period tests. As shown in the above bar chart, such corrosive environments produce smaller reductions in fatigue strength in alloys of the more corrosion-resistant types, such as 5xxx and 6xxx alloys, than in less resistant alloys, such as those of the 2xxx and 7xxx series. Like stress-corrosion cracking of aluminum alloys, corrosion fatigue requires the presence of water. In contrast to stress-corrosion cracking, corrosion fatigue is not appreciably affected by test direction, because fracture resulting from this type of attack is predominantly transgranular.

Source: Metals Handbook, 9th Edition, Volume 2, Properties and Selection: Nonferrous Alloys and Pure Metals, American Society for Metals, Metals Park OH, 1979, p 220

12-31. Alloys 5083-O/5183: Fatigue Life Predictions and Experimental Data Results for Double V-Butt Welds

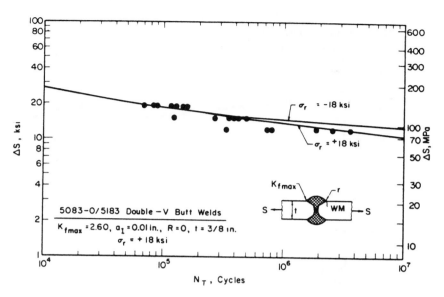

Total fatigue life predictions and experimental results for 5083-O/5183 3/8-in. (10-mm) butt welds.

Source: F. V. Lawrence, "The Predicted Influence of Weld Residual Stresses on Fatigue Crack Initiation," in *Residual Stress for Designers and Metallurgists*, Larry J. Vander Walle, Ed., American Society for Metals, Metals Park OH, 1981, p 114

12-32. Alloys 5083-O/5183: Predicted Effect of Stress Relief and Stress Ratio on Fatigue Life of Butt Welds

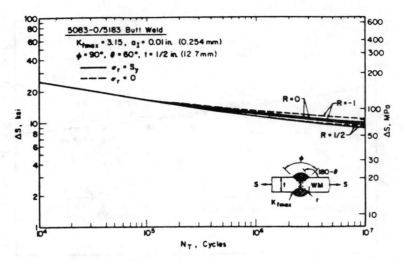

Predicted effect of stress relief and stress ratio on 5083-O/5183 butt weld fatigue life.

Because of the high notch-root plasticity during the first few cycles, before the material cyclically hardens, the aluminum weld considered here (5083/5183) exhibits little dependence upon either residual stress or stress ratio, even though the relaxation of the stabilized mean stress (σ_{os}) is very slow—as indicated in the above chart.

Source: F. V. Lawrence, "The Predicted Influence of Weld Residual Stresses on Fatigue Crack Initiation," in Residual Stress for Designers and Metallurgists, Larry J. Vander Walle, Ed., American Society for Metals, Metals Park OH, 1981, p 113

12-33. 7XXX Alloys: Cyclic Strain vs Crack Initiation Life

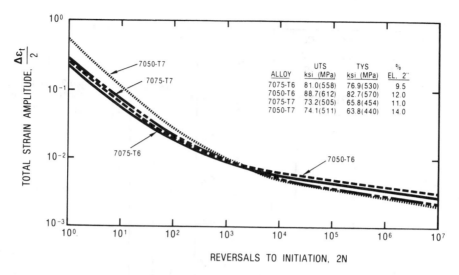

Cyclic strain versus initiation life for laboratory-fabricated high-strength 7XXX aluminum alloys.

Plots of elastic-strain amplitude versus life have seen relatively little use for commercial alloys, but plots of log total strain amplitude versus life have been used more frequently to compare materials. This approach offers the advantage that both high- and low-strain fatigue may be characterized with one plot. As illustrated above, fatigue resistance at low total strain amplitude is governed by the elastic-strain amplitude. Fatigue lives for total strain amplitudes less than about 5×10^{-3} generally increase with increasing strength. On the other hand, fatigue lives for total strain amplitudes greater than about 10^{-2} generally increase with increasing ductility.

Source: T. H. Sanders, Jr., and J. T. Staley, "Review of Fatigue and Fracture Research on High-Strength Aluminum Alloys," in Fatigue and Microstructure, American Society for Metals, Metals Park OH, 1979, p 472

12-34. Alloy 7050: Influence of Alloy Composition and Dispersoid Effect on Mean Calculated Fatigue Life

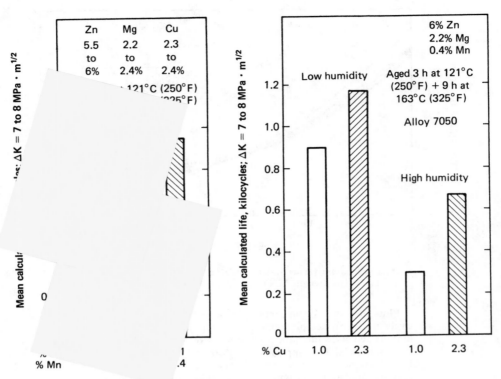

Effect of dispersoid type (based on composition) on fatigue crack propagation life of 7050 alloy sheet.

The influence of alloy composition on dispersoid effect is shown in the above bar chart. The general trend in this chart is that for more finely dispersed particles the fatigue crack propagation life is increased. Whereas dispersoid type appears to have a relatively small effect on mean calculated life, the smaller precipitates provided by aging produce a much larger effect. There is some evidence that new processing practices may provide the fine microstructures needed to enhance fatigue resistance. The potential of intermediate working (commonly referred to as ITMT treatments) remains attractive but has not been proven for notched specimens.

Source: J. G. Kaufman and J. S. Santner, "Fracture Properties of Aluminum Alloys," in Application of Fracture Mechanics for Selection of Metallic Structural Materials, James E. Campbell, William W. Gerberich and John H. Underwood, Eds., American Society for Metals, Metals Park OH, 1982, p 192

12-35. Alloy 7050: Effect of Grain Shape on Cycles to Failure

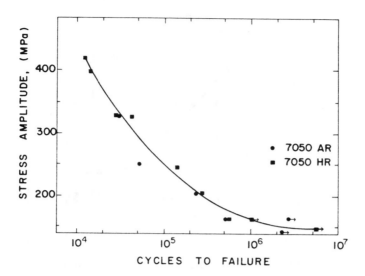

Stress-life curves for two 7050 alloys having fine, equiaxed grains (AR) and pancake-shaped grains (HR).

As indicated in the above graph, grain shape showed no perceptible difference in life over a range of stress amplitudes.

Source: Edgar A. Starke, Jr., and Gerd Lütjering, "Cyclic Plastic Deformation and Microstructure," in Fatigue and Microstructure, American Society for Metals, Metals Park OH, 1979, p 238

12-36. Alloy 7075 (TMP, T6 and T651): Effect of Thermomechanical Processing on Cycles to Failure

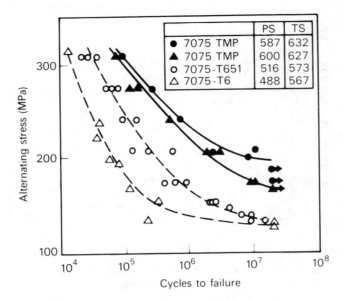

Effect of thermomechanical processing (TMP) on the unnotched fatigue properties of the commerical Al-Zn-Mg-Cu alloy 7075. PS = proof stress (MPa); TS = tensile strength.

Detailed studies of the processes of fatigue in metals and alloys have shown that the initiation of cracks normally occurs at the surface. It is here that strain becomes localized due to the presence of pre-existing stress concentrations such as mechanical notches or corrosion pits, coarse (persistent) slip bands in which minute extrusions and intrusions may form, or at relatively soft zones such as the precipitate-free regions adjacent to grain boundaries. Density has also been found to improve the fatigue performance of certain alloys, although this effect arises in part from an increase in tensile properties caused by such a treatment (see above diagram). It should be noted, however, that the promising results mentioned above were obtained for smooth specimens. The improved fatigue behavior has not been sustained for severely notched conditions, and it seems that the resultant stress concentrations override the more subtle microstructural effects that have been described.

Source: I. J. Polmear, Light Alloys, Edward Arnold Ltd, London, England, and American Society for Metals, Metals Park OH, 1981, p 41

12-37. Alloys 7075 and 7475: Effect of Inclusion Density on Cycles to Failure

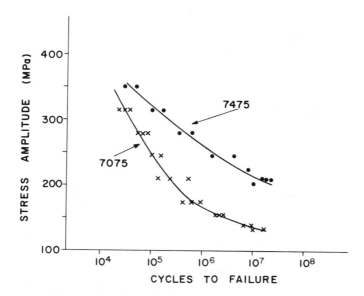

Effect of inclusion density on the stress-life behavior of two 7XXX alloys: high-inclusion density, alloy 7075; low-inclusion density, alloy 7475.

Source: Edgar A. Starke, Jr., and Gerd Lütjering, "Cyclic Plastic Deformation and Microstructure," in Fatigue and Microstructure, American Society for Metals, Metals Park OH, 1979, p 233

12-38. Alloy 7075: Effect of TMT on Cycles to Failure

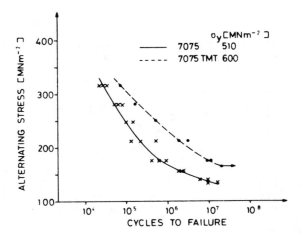

Influence of TMT on *S-N* curves ($R = -1$).

There is evidence in the literature that a uniform dislocation density introduced by cold working improves the fatigue life also in connection with FTMT. The above graph shows an example taken from the work of Ostermann. Most of these improvements are due to an increased yield stress.

Source: G. Lütjering and A. Gysler, "Fatigue and Fracture of Aluminum Alloys," in Aluminum Transformation Technology and Applications (Proceedings of the International Symposium at Puerto Madryn, Chubut, Argentina), C. A. Pampillo, H. Biloni and D. E. Embury, Eds., American Society for Metals, Metals Park OH, 1980, p 195

12-39. Alloys 7075 and 7050: Relative Ranking for Constant Amplitude and Periodic Overload

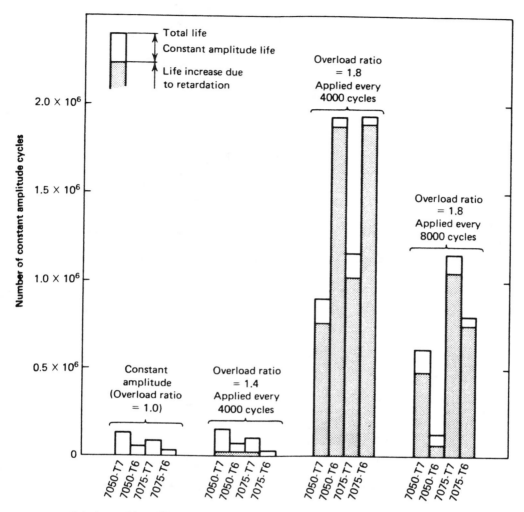

Relative ranking of fatigue life of 7075 and 7050 aluminum alloys under constant amplitude and periodic single overload conditions.

Crack-growth retardation is caused by tension overloading during fatigue testing. The variable-amplitude test is believed to be more sensitive to alloy difference, and it clearly provides more useful information for alloy-development investigations. For example, as illustrated by the data for alloys 7075 and 7050 in the above graph, quite different results are obtained in constant-amplitude tests than in tests with single overloads every 4000 or 8000 cycles. Thus, information on the variation in load level during fatigue cycling is required for correct characterization of the fatigue behavior of aluminum alloys.

Source: J. G. Kaufman and J. S. Santner, "Fracture Properties of Aluminum Alloys," in Application of Fracture Mechanics for Selection of Metallic Structural Materials, James E. Campbell, William W. Gerberich and John H. Underwood, Eds., American Society for Metals, Metals Park OH, 1982, p 197

12-40. Alloy 7075: Effect of Environment and Mode of Loading

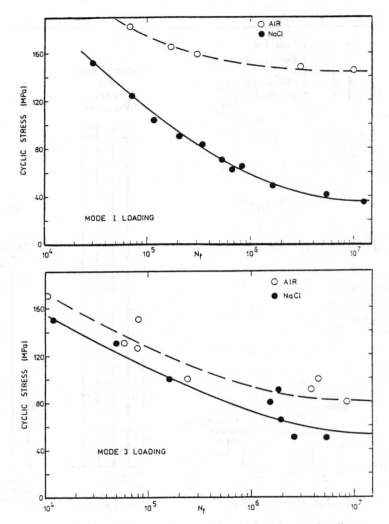

Fatigue behavior of 7075 aluminum alloy in air and aerated sodium chloride solution: (above) under mode 1 loading; (below) under mode 3 loading.

Tests performed on a commercial 7075 alloy in a mode 3 loading condition (torsion) indicated that the reduction in fatigue resistance associated with cathodic charging was considerably less than it was under mode 1 loading (note above charts). Although total immunity to corrosion fatigue was not observed, the slight reduction in fatigue resistance can be associated with conditions that did produce a true mode 3 loading condition both on a micro-scale and on a macro-scale.

To summarize the aluminum alloy results, it appears that corrosion reactions liberate hydrogen, which effectively embrittles the region in the vicinity of a crack tip. The specific details of the embrittlement are not known, but it appears that dislocation transport of the hydrogen is involved. It has been speculated that hydrogen may collect at the semicoherent precipitate-matrix interface, thus explaining the reported fracture plane; however, a great deal more research will have to be performed before a more definitive answer will be available.

Source: D. J. Duquette, "Environmental Effects I: General Fatigue Resistance and Crack Nucleation in Metals and Alloys," in Fatigue and Microstructure, American Society for Metals, Metals Park OH, 1979, p 356

12-41. Alloy 7075-T6: Effects of Corrosion and Pre-Corrosion

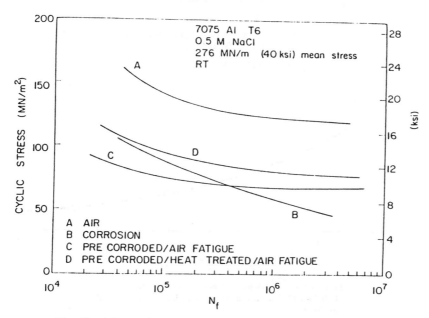

The effects of corrosion and pre-corrosion on the fatigue lives of a 7075-T6 alloy. Note that re-solutionizing and re-aging the alloy after pre-corrosion results in a significant increase in fatigue resistance.

Fatigue resistance of high-strength aluminum alloys is severely affected by corrosive solutions, especially chloride solutions, and this behavior has been attributed either to preferential dissolution at the tips of the growing cracks or to preferential adsorption of damaging ionic species. Experiments on a 7075-T6 commercial alloy and on a high-purity analog of the alloy (Al-5.0Zn-2.5Mg-1.5Cu) indicate that localized hydrogen embrittlement may be responsible for the poor corrosion fatigue resistance of these alloys. For example, the above diagram shows the results of fatigue tests performed on the 7075 alloy under simultaneous exposure to cyclic stresses and a corrosive environment (curve B) compared to tests performed in laboratory air (curve A). If specimens are pre-corroded and tested in laboratory air, there is also a significant reduction in fatigue resistance (curve C). The reduction in life at low N_f is associated with pits which form at nonmetallic inclusions. If the alloy is re-solutionized and aged, equivalent to a low-temperature bake, a significant amount of fatigue resistance is regained, indicating at least partial reversibility of the damaging phenomenon and strongly suggesting a solid-solution effect arising from environmental interaction.

Source: D. J. Duquette, "Fundamentals of Corrosion Fatigue Behavior of Metals and Alloys," in Hydrogen Embrittlement and Stress Corrosion Cracking, R. Gibala and R. F. Hehemann, Eds., American Society for Metals, Metals Park OH, 1984, p 265

12-42. Alloy 7075-T73: Effect of a 3.5% NaCl Environment on Cycles to Failure

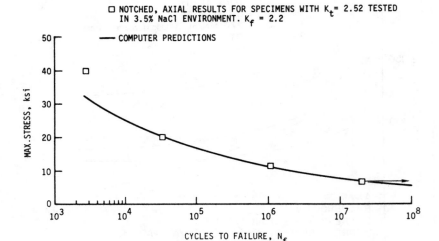

Test results and computer predictions for 0-max stressing of 7075-T73 aluminum alloy in a 3.5 wt % NaCl environment.

A comparison, employing the techniques described, of 0-max stress fatigue results of notched specimens tested in a 3.5 wt % NaCl environment to the predicted behavior is shown in the above S-N data. The agreement between the prediction and test results appears favorable.

Source: M. R. Mitchell, "Fundamentals of Modern Fatigue Analysis for Design," in Fatigue and Microstructure, American Society for Metals, Metals Park OH, 1979, p 435

12-43. Alloy 7075: Effect of Cathodic Polarization on Fatigue Behavior

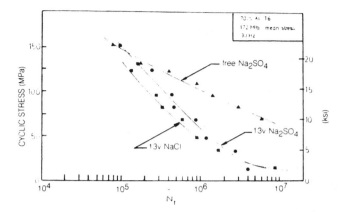

Effect of cathodic polarization on the fatigue behavior of 7075 Al alloy in NaCl and Na_2SO_4.

It had been previously observed that halide ions are particularly damaging to the fatigue behavior of Al alloys; however, if the alloy is cathodically charged during stressing, sulfate ions prove to be equally damaging, particularly at long N_f. At lower N_f the slight decrease observed in Cl^- solutions appears to be associated with damage to the passive film, as shown in the above S-N data. In $SO_4^=$ solutions, a crack must initiate to break the protective film to allow access to the bulk alloy. Cathodic charging of the high-purity analog of the 7075 alloy also shows a reduction in fatigue resistance. In many cases, fatigue crack initiation in the equiaxed-grain high-purity alloy is intergranular, and at more active cathodic potentials there is a tendency toward a higher percentage of transgranular cracking.

Source: D. J. Duquette, "Fundamentals of Corrosion Fatigue Behavior of Metals and Alloys," in Hydrogen Embrittlement and Stress Corrosion Cracking, R. Gibala and R. F. Hehemann, Eds., American Society for Metals, Metals Park OH, 1984, p 266

12-44. Alloy 7075-T6: Effect of Surface Treatments and Notch Designs on Number of Cycles to Failure

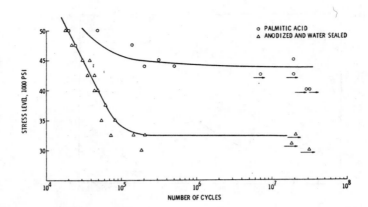

Tension fatigue test of 7075-T6 aluminum alloy sheet, notch factor $K_T = 1$.

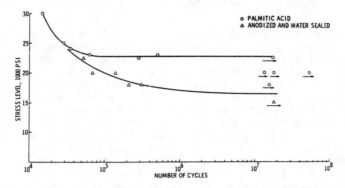

Tension fatigue test of 7075-T6 aluminum alloy sheet, notch factor $K_T = 2.37$.

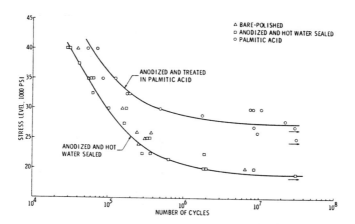

Flexure fatigue test of 7075-T6 aluminum alloy sheet.

The three charts show the effects of notch designs and surface treatments on fatigue properties of aluminum alloy 7075-T6 sheet; the table shows the effects of 17 surface treatments.

Surface treatment	No. of cycles to failure	Surface treatment	No. of cycles to failure
Polished	125,000	Palmitic acid	30,000,000
Anodized and water-sealed	125,000	Stearic acid	8,700,000
Propionic acid	2,800,000	Docosanoic acid	6,000,000
Valeric acid	15,000,000	Sebacic acid	13,700,000
Caproic acid	9,200,000	Octyl alcohol	6,000,000
Octanoic acid	12,300,000	Dodecyl alcohol	7,000,000
Decanoic acid	7,500,000	Dodecylamine	18,500,000
Lauric acid	8,600,000	Hexandeiamine	3,000,000
Myristic acid	11,600,000		

Sheet was anodized: 15% sulfuric acid, 23 °C, 15 amp/sq ft, 40 minutes.
Stress amplitude: 26,000 psi.

Source: Irvin R. Kramer, "Improvement of Metal Fatigue Life by a Chemical Surface Treatment," in Fatigue—An Interdisciplinary Approach, John J. Burke, Norman L. Reed and Volker Weiss, Eds., Syracuse University Press, Syracuse NY, 1964, pp 250, 251

12-45. Alloy 7075-T6: Effect of *R*-Ratio on Fatigue Crack Propagation

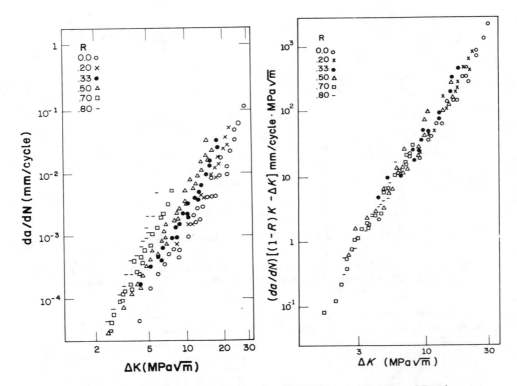

Fatigue crack propagation in aluminum alloy 7075-T6 showing the effect of *R* ratio and the applicability of the Forman, Keraney, and Engle relation. The scatter in the data is much less in the latter.

The above diagrams show that data scatter is much less when the rate da/dN is computed according to the equation due to Foreman et al. $R < 0$. The proposed Foreman equation is:

$$\frac{da}{dN} = \frac{C(\Delta K)^m}{(1-R)K_c - \Delta K}$$

Source: Marc André Meyers and Krishan Kumar Chawla, "Mechanical Metallurgy: Principles and Applications," Prentice-Hall, Inc., Englewood Cliffs NJ, 1984, p 716

12-46. Alloy 7075: Effect of Predeformation on Fatigue Crack Propagation Rates

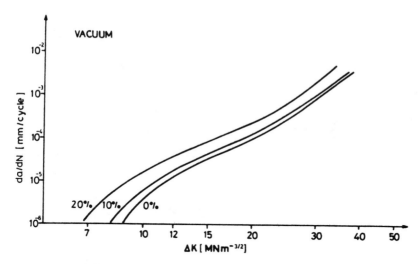

Influence of predeformation by cold rolling on fatigue crack propagation rates for 7075. 1h 100 °C, SEN specimens, vacuum, $R = 0.1, f = 30$ Hz.

Cold deformation also increases the fatigue crack propagation rate as shown in the above graph, which compares an undeformed structure with 10% and 20% cold rolled structures.

Source: G. Lütjering and A. Gysler, "Fatigue and Fracture of Aluminum Alloys," in Aluminum Transformation Technology and Applications (Proceedings of the International Symposium at Puerto Madryn, Chubut, Argentina), C. A. Pampillo, H. Biloni and D. E. Embury, Eds., American Society for Metals, Metals Park OH, 1980, p 207

12-47. Alloys 7075 and 2024-T3: Comparative Fatigue Crack Growth Rates for Two Alloys in Varying Humidity

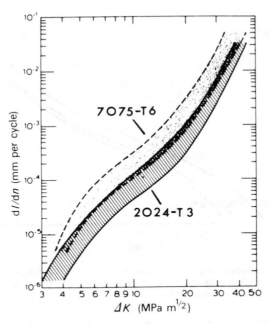

Comparative fatigue crack growth rates for aluminum alloys 2024-T3 and 7075-T6 in air of varying humidity.

Relationships between rate of growth of fatigue cracks and stress intensity for the alloys 2024-T3 and 7075-T6 are shown above. Other 2xxx series alloys show rates of crack propagation similar to that of 2024-T3 over most of the range of test conditions. In general, these alloys have rates of crack growth that are close to one-third those observed in the 7xxx series alloys. It is now common to use precracked specimens to assess comparative resistance of alloys to stress-corrosion cracking, since this type of test avoids uncertainties associated with crack initiation.

Source: I. J. Polmear, Light Alloys, Edward Arnold Ltd, London, England, and American Society for Metals, Metals Park OH, 1981, p 79

12-48. Alloy 7075-T651: Fatigue Life as Related to Harmonic Generation

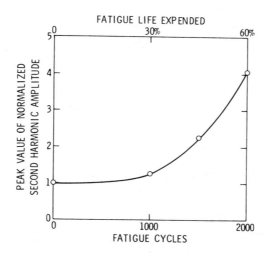

Peak value of normalized second harmonic generation as a function of fatigue life.

Recent experiments on flexural-fatigue specimens (aluminum alloy 7075-T651) clearly show the potential of harmonic generation for fatigue monitoring. The above chart shows the peak value of the harmonic generated as a function of fatigue life. At 60% of the fatigue life expended, the harmonic had increased by about a factor of four.

Source: O. Buck and G. A. Alers, "New Techniques for Detection and Monitoring of Fatigue Damage," in Fatigue and Microstructure, American Society for Metals, Metal Park OH, 1979, p 137

12-49. Alloys 7075-T6 and 7475-T73: Effect of Laser-Shock Treatment on Fatigue Properties

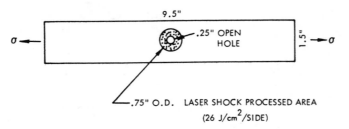

Fatigue Test Specimen Geometry

FATIGUE TEST RESULTS FOR 7075-T6 ALUMINUM
σ_{MAX} = 20 KSI NET R = .1

CONTROL SPECIMENS		LASER-SHOCKED	
SPECIMEN NO.	NO. OF CYCLES TO FAILURE	SPECIMEN NO.	NO. OF CYCLES TO FAILURE
1	51,500	1	473,800
2	72,100	2	520,700
3	385,000		
AVG.	169,600	AVG.	497,250
SCATTER	7.5	SCATTER	1.1

FATIGUE TEST RESULTS FOR 7475-T73 ALUMINUM
σ_{MAX} = 20 KSI NET R = .1

CONTROL SPECIMENS		LASER-SHOCKED	
SPECIMEN NO.	NO. OF CYCLES TO FAILURE	SPECIMEN NO.	NO. OF CYCLES TO FAILURE
1	41,500	1	171,800
2	74,300	2	266,200
3	109,300		
AVG.	75,033	AVG.	218,950
SCATTER	2.63	SCATTER	1.5

The fatigue test specimens were 0.25 inch thick by 1.5 inches wide and approximately 9.5 inches long, as shown in the above sketch. The specimen blanks were laser-shock processed, and then the 0.25-inch-diameter hole was bored through the center of the laser-shock-processed area. The diameter of the laser-shock-processed area is three times the fastener hole diameter. All of the specimens had machined surfaces of less than 125 RMS. All of these open-hole specimens were fatigue tested to failure at a maximum net section stress of 20,000 psi, and an $R = 0.1$ under constant-amplitude load control. Three control specimens for each material were tested to establish the typical fatigue life for the material. Two LSP specimens were tested for each material to establish the degree of improvement due to the laser-shock processing.

The fatigue test results for the 7075-T6 material are summarized in the upper tabulation. The LSP specimens showed three times better fatigue lives on the average and much less scatter than the unprocessed material. The results for the 7475-T73 material are summarized in the lower tabulation; these show the same typically large increases in fatigue life and reduced scatter. It should be noted that the 7075-T6 material shows better fatigue resistance than the 7475-T73 material, whether or not it is laser-shock processed. This is largely due to the differences in dislocation/precipitate interactions that result from the T6 and T73 heat treatments. The dislocations appear to shear through the precipitate particles in the T6 condition. The precipitate particles are apparently so strong in the T73 condition that the dislocations just loop around the particles.

Source: William F. Bates, Jr., "Laser Shock Processing of Aluminum Alloys," in Source Book on Applications of the Laser in Metalworking, Dr. Edward A. Metzbower, Ed., American Society for Metals, Metals Park OH, 1981, pp 256-258

12-50. Alloy 7075-T6: Effect of Laser-Shock Treatment on Hi-Lok Joints

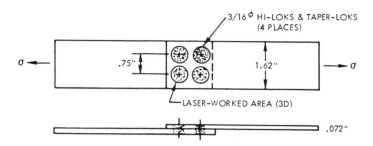

Joint Fatigue Test Specimen Geometry

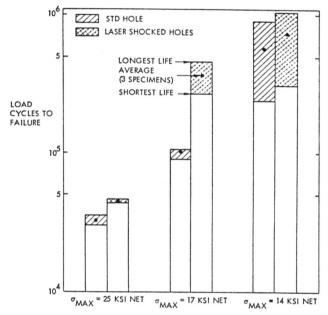

Fatigue Test Results for Laser-Shock-Processed 7075-T6 (Clad) Hi-Lok Joints

The full load transfer joint shown in the above sketch was made from 7075-T6 clad aluminum alloy and fatigue tested. The purpose of this test was to evaluate the fatigue life improvement of laser-shock-processed fastener holes when the holes are loaded by the Hi-Lok fastener in bearing. A secondary purpose was to find out if the cheaper Hi-Lok fastener system in a laser-shock-processed hole would show as good a fatigue life as the much more expensive Taper-Lok fastener system. The above bar chart shows the test results for three different stress levels. At each stress level, three specimens with standard holes and three specimens with laser-shock-processed holes were tested. The specimens tested at the 14-ksi stress level showed severe fretting at the intersection of the hole wall with a badly galled area of the fretted faying surface. All of the fatigue origins occurred at or near the hole wall corners on the faying surface.

Source: William F. Bates, Jr., "Laser Shock Processing of Aluminum Alloys," in Source Book on Applications of the Laser in Metalworking, Dr. Edward A. Metzbower, Ed., American Society for Metals, Metals Park OH, 1981, pp 262-263

12-51. Alloy 7075 (High Purity): Effect of Iron and Silicon on Cycles to Failure

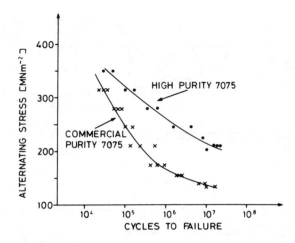

Influence of Fe and Si content on S-N curves ($R = -1$).

The large Fe- and Si-containing inclusions are detrimental to the fatigue life of smooth specimens, because these inclusions serve as easy nucleation sites for cracks. Comparing two alloys, one containing these inclusions (Commercial Purity 7075) and the other one not (High Purity 7075), shows the improvement in fatigue life due to the removal of these inclusions (see the above S-N curves). The alloy termed High Purity 7075 in this figure still contains Cr and therefore the small Cr-containing inclusions. This is important because the removal of these small inclusions would have the opposite effect on fatigue life.

Source: G. Lütjering and A. Gysler, "Fatigue and Fracture of Aluminum Alloys," in Aluminum Transformation Technology and Applications (Proceedings of the International Symposium at Puerto Madryn, Chubut, Argentina), C. A. Pampillo, H. Biloni and D. E. Embury, Eds., American Society for Metals, Metals Park OH, 1980, p 193

12-52. Alloy X-7075: Effect of Grain Size on Cycles to Failure

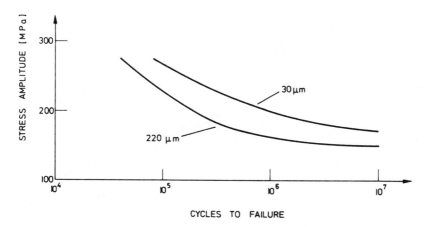

Influence of grain size on S-N curves ($R = -1$, $f = 100$ Hz) for X-7075 with PFZ (20 h at 160 °C, or 320 °F).

The resulting improvement in fatigue life due to the grain size reduction for this crack nucleation mechanism is shown in the above S-N curve. Again, the tensile yield stress was equal for both grain sizes. Also for low-cycle fatigue it was found that reducing the grain size of 7XXX series alloys results in increased fatigue life of smooth specimens in the overaged condition.

Source: G. Lütjering and A. Gysler, "Fatigue and Fracture of Aluminum Alloys," in Aluminum Transformation Technology and Applications (Proceedings of the International Symposium at Puerto Madryn, Chubut, Argentina), C. A. Pampillo, H. Biloni and D. E. Embury, Eds., American Society for Metals, Metals Park OH, 1980, p 192

12-53. Alloy X-7075: Effect of Grain Size on Stress-Life Behavior

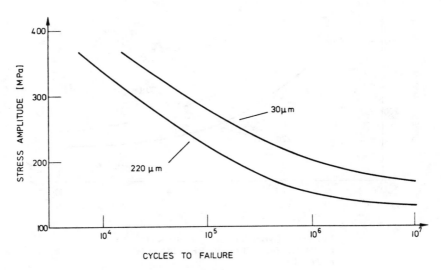

Aluminum alloy X-7075; 24 h at 100 °C (212 °F).

The above chart shows the grain-size effect in a stress-controlled test for a high-purity 7075 alloy (X-7075) aged to contain shearable precipitates. Since the flow stress is determined by the interaction of dislocations with the coherent precipitates, the yield stress is approximately the same for both alloys. Optical examinations of the specimen surfaces show that cracks nucleate much earlier in specimens with the large grain size. Cracks nucleated at intense slip bands for both grain sizes.

Source: Edgar A. Starke, Jr., and Gerd Lütjering, "Cyclic Plastic Deformation and Microstructure," in Fatigue and Microstructure, American Society for Metals, Metals Park OH, 1979, p 225

12-54. Alloy X-7075: Effect of Environment; Air vs Vacuum

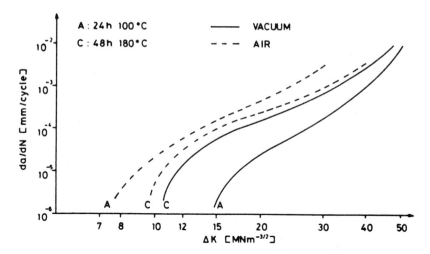

Influence of environment (laboratory air) on fatigue crack propagation rates for underaged (A) and overaged (C) condition. X-7075, CT specimens, $R = 0.1$, $f = 30$ Hz.

A basic correlation between microstructural parameters and fatigue crack propagation rate can only be determined so clearly if the tests are performed with the exclusion of any aggressive environment. To illustrate this point, the above graph shows the comparison between underaged and overaged microstructure also for tests performed in laboratory air. The aggressive environment has a much more pronounced effect on the underaged condition, leading even to an opposite ranking of the alloy conditions. In laboratory air the cracks propagate still along slip bands at low da/dN rates.

Source: G. Lütjering and A. Gysler, "Fatigue and Fracture of Aluminum Alloys," in Aluminum Transformation Technology and Applications (Proceedings of the International Symposium at Puerto Madryn, Chubut, Argentina), C. A. Pampillo, H. Biloni and D. E. Embury, Eds., American Society for Metals, Metals Park OH, 1980, p 204

12-55. Alloy X-7075: Effect of Environment on Two Different Grain Sizes

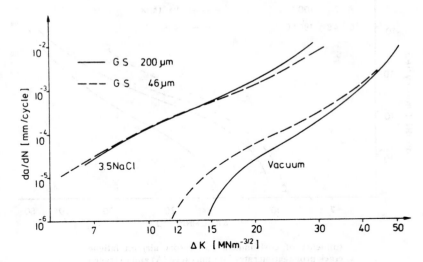

Influence of environment (3.5% NaCl) on fatigue crack propagation rates for two different grain sizes. X-7075, 24 h 100 °C, CT specimens, $R = 0.1$, $f = 30$ Hz.

The same tendency is observed for the grain size dependence of crack propagation if the tests are carried out in a 3.5% NaCl solution (note above curves). The influence of environment is larger for the large grain size. For this highly aggressive environment the cracks propagate at low da/dN rates along grain boundaries in a complete brittle fashion.

Source: G. Lütjering and A. Gysler, "Fatigue and Fracture of Aluminum Alloys," in Aluminum Transformation Technology and Applications (Proceedings of the International Symposium at Puerto Madryn, Chubut, Argentina), C. A. Pampillo, H. Biloni and D. E. Embury, Eds., American Society for Metals, Metals Park OH, 1980, p 204

12-56. Alloy X-7075: Effect of Grain-Boundary Ledges on Cycles to Failure

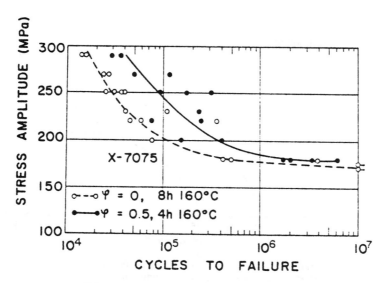

Effect of grain-boundary ledges on the stress-life behavior of an alloy containing nonshearable precipitates and PFZ.

One method that may be employed to reduce the slip length in the PFZ is thermomechanical processing. If enough cold deformation is employed to introduce steps (or "ledges") into the grain boundaries, the effective slip length within the PFZ is drastically reduced (similar to a small grain size) with corresponding improvement in resistance to fatigue-crack nucleation. The above chart shows the results of a stress-controlled test for two high-purity 7075 alloys, one cold-worked 50% to produce grain-boundary steps. The cold work drastically reduced the incidence of grain-boundary cracking and improved the fatigue life at high stress amplitudes. At low stress amplitudes and long fatigue lives, crack nucleation occurred at inclusions for both alloys. This effect is most likely due to stress concentration at inclusions.

Source: Edgar A. Starke, Jr., and Gerd Lütjering, "Cyclic Plastic Deformation and Microstructure," in Fatigue and Microstructure, American Society for Metals, Metals Park OH, 1979, p 230

12-57. Alloys X-7075 and 7075: Effects of Chromium Inclusions on Fatigue Crack Propagation

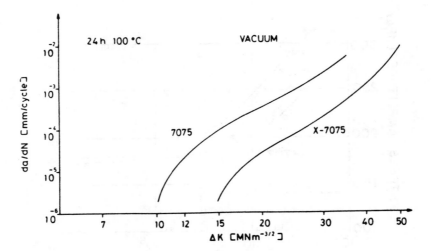

Influence of Cr-containing inclusions on fatigue crack propagation rates by comparing aluminum alloys X-7075 and 7075. 24 h 100 °C, CT specimens, vacuum, $R = 0.1$, $f = 30$ Hz.

As shown above, the small inclusions have a much stronger influence on fatigue crack propagation because they lower the reversibility of slip and they crack within the plastic zone ahead of the crack tip. Furthermore, they normally increase fatigue crack propagation rates also indirectly by their effect on grain size and shape.

Source: G. Lütjering and A. Gysler, "Fatigue and Fracture of Aluminum Alloys," in Aluminum Transformation Technology and Applications (Proceedings of the International Symposium at Puerto Madryn, Chubut, Argentina), C. A. Pampillo, H. Biloni and D. E. Embury, Eds., American Society for Metals, Metals Park OH, 1980, p 207

12-58. Alloy 7475-T6: S-N Diagram for a Superplastic Fine-Grain Alloy

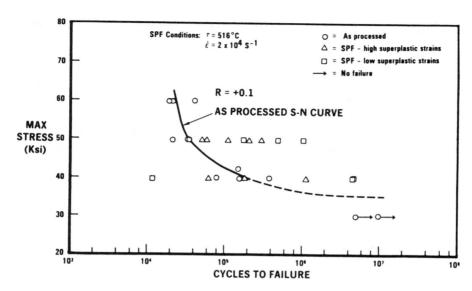

S-N curves for a superplastic aluminum alloy: fine-grain 7475. All testing was done with smooth specimens.

Tests on fine-grain 7475 alloy have shown improved fatigue life as superplastic strain is increased, as shown in the above S-N diagram. An even more dramatic improvement is obtained in damage tolerance.

Source: C. Bampton, F. McQuilkin and G. Stacher, "Superplastic Forming Applications to Bomber Aircraft," in Superplastic Forming, Suphal P. Agrawal, Ed., American Society for Metals, Metals Park OH, 1985, p 77

12-59. Alloy 7475: Effect of Alignment of Grain Boundaries on Cycles to Failure

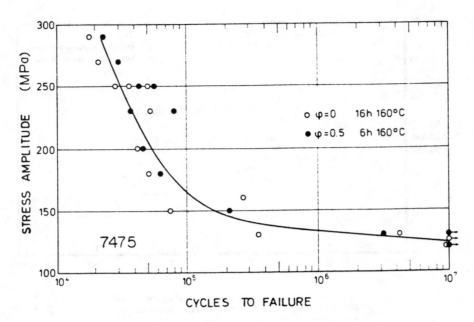

Effect of alignment of grain boundaries—and alignment plus steps in grain boundaries—on the stress-life behavior of a 7475 aluminum alloy containing nonshearable precipitates and PFZ.

If the stress axis is parallel or perpendicular to the long grain dimension, there will be no shear stress parallel to the grain boundary, and preferential deformation within the PFZ will be restricted. Grain-boundary alignment is then as effective in restricting deformation in the PFZ as are steps produced by thermomechanical treatment: this is shown by the stress-life curves in the above graph.

Source: Edgar A. Starke, Jr., and Gerd Lütjering, "Cyclic Plastic Deformation and Microstructure," in Fatigue and Microstructure, American Society for Metals, Metals Park OH, 1979, p 232

12-60. Alloy 7475-T6: Superplastic vs Nonsuperplastic, as Related to Fatigue Crack Growth

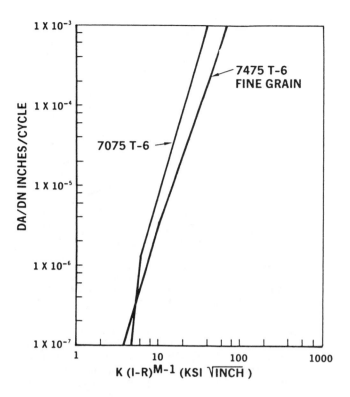

This comparison of conventional, coarse-grain, nonsuperplastic aluminum alloy 7075 with superplastic alloy 7475 shows almost an order-of-magnitude reduction in crack growth for the superplastic material.

Source: C. Bampton, F. McQuilkin and G. Stacher, "Superplastic Forming Applications to Bomber Aircraft," in Superplastic Forming, Suphal P. Agrawal, Ed., American Society for Metals, Metal Park OH, 1985, p 77

12-61. Alloys X-7075 and 7075: Effect of Chromium-Containing Inclusions on Cycles to Failure

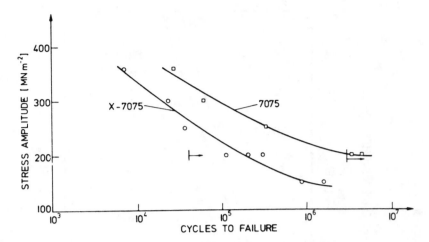

Influence of Cr-containing inclusions on S-N curves ($R = -1$, $f = 100$ Hz) comparing aluminum alloys X-7075 and 7075, 24 h 100°C. (Arrows indicate crack nucleation visible by LM at $\sigma_a = \pm 200$ MNm^{-2}.)

The above S-N curves compare results obtained from testing commercial 7075 alloy with the alloy X-7075 which does not contain Cr. These small inclusions, as in the tensile test, inhibit the formation of intense slip bands, thus retarding crack nucleation as indicated by arrows on the graph. Due to these small Cr-containing inclusions, the grain size of the 7075 alloy was somewhat smaller as compared to that of X-7075, which also may have contributed to the observed improved fatigue behavior.

Source: G. Lütjering and A. Gysler, "Fatigue and Fracture of Aluminum Alloys," in Aluminum Transformation Technology and Applications (Proceedings of the International Symposium at Puerto Madryn, Chubut, Argentina), C. A. Pampillo, H. Biloni and D. E. Embury, Eds., American Society for Metals, Metals Park OH, 1980, p 195

12-62. Aluminum Forging Alloys: Stress Amplitude vs Reversals to Failure

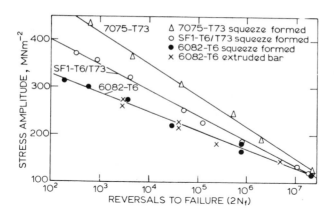

S-N fatigue data for several squeeze-formed forging-type aluminum alloys compared with extruded AA 6082-T6.

The above chart presents results from push-pull, about mean zero, fatigue tests which have been carried out on a servohydraulically controlled machine. The tests have been carried out on samples cut from actual components, not from separately made testpieces. The results from conventionally extruded AA 6082 (H30) are included for reference: in this case, the data are in the longitudinal direction, it not being possible to obtain samples of sufficient size from the transverse direction.

The results show good fatigue properties for squeeze-formed material, which in one case compare favorably with conventionally extruded material. This further substantiates the claim that squeeze formings in general are comparable with forgings with respect to mechanical performance.

Source: G. Williams and K. M. Fisher, "Squeeze Forming of Aluminium-Alloy Components," in Production to Near Net Shape: Source Book, C. J. Van Tyne and B. Avitzur, Eds., American Society for Metals, Metals Park OH, 1983, p 367

12-63. Al-5Mg-0.5Ag: Effect of Condition on Fatigue Characteristics

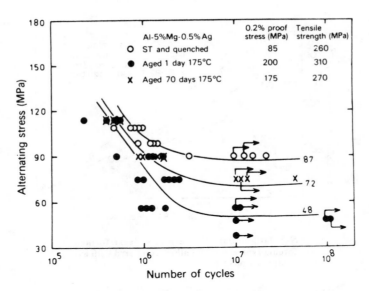

Fatigue (S-N) curves for the alloy Al-5Mg-0.5Ag in different conditions.

The fact that microstructure can have a greater influence upon the fatigue properties of aluminum alloys than the level of tensile properties has been demonstrated for an Al-Mg alloy containing a small addition of silver. It is well known that binary Al-Mg alloys such as Al-5Mg, in which the magnesium is present in solid solution, display a relatively high level of fatigue strength. The same applies for an Al-5Mg-0.5Ag alloy in the as-quenched condition, and the above diagram shows that the endurance limit after 10^8 cycles is ±87 MPa, which approximately equals the 0.2% proof stress. This result is attributed to the interaction of magnesium atoms with dislocations, which minimizes formation of coarse slip bands during fatigue. The silver-containing alloy responds to age hardening at elevated temperatures due to the formation of a finely dispersed precipitate, and the 0.2% proof stress may be raised to 200 MPa after aging for one day at 175 °C (350 °F).

Source: I. J. Polmear, Light Alloys, Edward Arnold Ltd, London, England, and American Society for Metals, Metals Park OH, 1981, p 42

12-64. Al-Zn-Mg and Al-Zn-Mg-Zr: Effect of Grain Size on Strain-Life Behavior

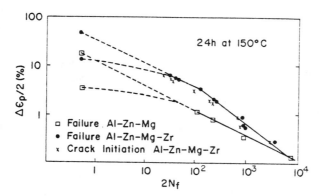

Effect of grain size on the strain-life behavior of an alloy having nonshearable precipitates plus PFZ. The Al-Zn-Mg alloy had large grain size; the Al-Zn-Mg-Zr, small grain size.

The above chart shows Coffin-Manson life plots of two over-aged Al-Zn-Mg alloys. The small-grained Al-Zn-Mg-Zr alloy has a much longer life than does the large-grained Al-Zn-Mg alloy. The improvement in life is attributed to increasing the cycles to crack initiation, as indicated in the chart. A convergence is noted for long lives (low plastic-strain amplitudes) for this strain-controlled test. Since the fine-grained material hardens more than the other at low strains, the stress to enforce the applied strain is greater at long lives, and this affects the life improvement due to the fine grains.

Source: Edgar A. Starke, Jr., and Gerd Lütjering, "Cyclic Plastic Deformation and Microstructure," in Fatigue and Microstructure, American Society for Metals, Metals Park OH, 1979, p 228

12-65. Al-Zn-Mg: Strain-Life Curves of a Large-Grained Alloy

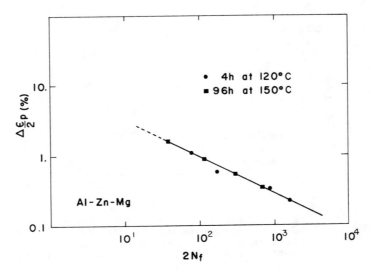

Strain-life curves of large-grained Al-Zn-Mg alloy having shearable precipitates when underaged (4 h at 120 °C, or 250 °F) and nonshearable precipitates plus PFZ when overaged (96 h at 150 °C, or 300 °F).

Since the strain localization occurs in a region free of solute, overaging the matrix precipitates or adding dispersoids does not homogenize the deformation. This is clearly illustrated by comparing the Coffin Manson life curves of underaged and overaged specimens of *large-grained* Al-Zn-Mg alloy (see above chart). The tensile yield strength and strain to fracture are approximately the same for both specimens. The underaged alloy has shearable precipitates, which results in strain localization, the formation of intense slip bands, and early crack nucleation under cyclic loading. Overaging was one method described for homogenizing deformation; however, this method is not effective for large-grained material. Preferential deformation in the PFZ also leads to strain localization and results, for this particular case, in the same fatigue life. Dispersoids distributed throughout the matrix would not inhibit strain localization in the PFZ for the same reason.

Source: Edgar A. Starke, Jr., and Gerd Lütjering, "Cyclic Plastic Deformation and Microstructure," in Fatigue and Microstructure, American Society for Metals, Metals Park OH, 1979, p 227

12-66. Aluminum With a Copper Overlay: Stress Amplitude vs Cycles to Failure

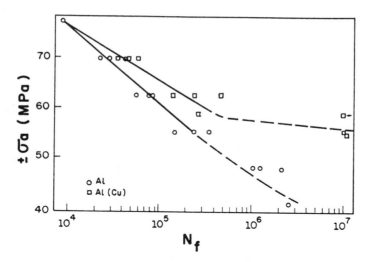

Stress amplitude σ_a versus number of cycles to failure N_f for Al and Al with a Cu layer. Note the pronounced improvement in the latter at large N_f.

The fraction of fatigue life spent in crack nucleation, N_i/N_f, increases with decreasing load amplitude (i.e., at high N's). It would be expected that the treatment suggested above will produce a great effect in large fatigue life regimes (i.e., under conditions where initiation of fatigue crack is more important than its propagation). The above graph shows this phenomenon in the case of pure aluminum and aluminum with a copper surface layer.

Source: Marc André Meyers and Krishan Kumar Chawla, "Mechanical Metallurgy: Principles and Applications," Prentice-Hall, Inc., Englewood Cliffs NJ, 1984, p 707

12-67. P/M Alloys 7090 and 7091 vs Extruded 2024

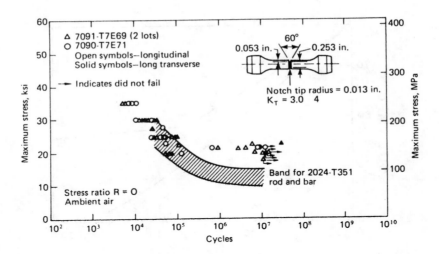

S-N diagram that provides a comparison of notched axial fatigue strength for P/M alloy 7090 and 7091 extrusions vs I/M alloy 2024-T351 rod and bar.

The notched axial fatigue strengths of alloys 7090 and 7091 are 35 to 40% higher than those of alloys 7050, 7075 and 2024 (an I/M alloy often selected for its resistance to fatigue) at one million or more cycles.

Source: Robert H. Graham, "Wrought Aluminum P/M Alloys," in *Powder Metallurgy—Applications, Advantages and Limitations*, Erhard Klar, Ed., American Society for Metals, Metals Park OH, 1983, p 240

12-68. P/M Alloys 7090 and 7091 vs I/M 7050 and 7075 Products

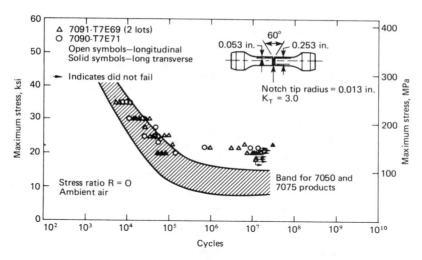

S-N diagram that compares notched axial fatigue strength for P/M alloy 7090 and 7091 extrusions vs I/M 7050 and 7075 products.

The notched axial fatigue strengths of alloys 7090 and 7091 are 35 to 40% higher than those of alloys 7050, 7075 and 2024 (an I/M alloy often selected for its resistance to fatigue) at one million or more cycles, as shown above.

Source: Robert H. Graham, "Wrought Aluminum P/M Alloys," in Powder Metallurgy—Applications, Advantages and Limitations, Erhard Klar, Ed., American Society for Metals, Metals Park OH, 1983, p 240

12-69. P/M Aluminum Alloys: Typical Fatigue Behavior

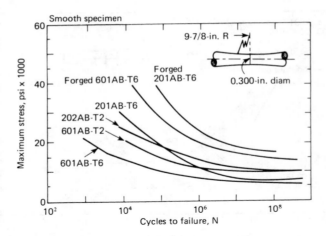

Typical fatigue behavior of alloys 601AB, 201AB and 202AB.

Fatigue is an important design consideration for P/M parts subject to dynamic stresses. The above S-N diagram shows typical fatigue behavior of specimens of alloys 601AB, 201AB and 202AB in the T2 (as-cold-formed after sintering) and/or T6 tempers.

Source: John D. Generous and Wayne C. Montgomery, "Aluminum P/M—Properties and Applications," in Powder Metallurgy— Applications, Advantages and Limitations, Erhard Klar, Ed., American Society for Metals, Metals Park OH, 1983, p 214

12-70. P/M Aluminum Alloys: Comparison With Specimens Made by Ingot Metallurgy

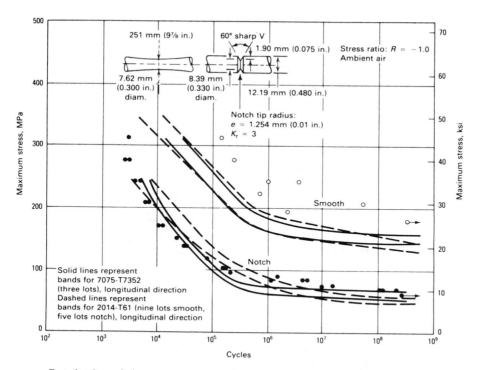

Rotating-beam fatigue strength for die forgings of P/M alloy X7091-T7E76 and ingot metallurgy alloys 7075-T7352 and 2014-T61. For P/M X7091-T7E76: ○, smooth, transverse direction; ●, notched, transverse direction; → denotes test specimen did not fail in number of cycles indicated.

Source: Metals Handbook, 9th Edition, Volume 7, Powder Metallurgy, American Society for Metals, Metals Park OH, 1984, p 469

12-71. P/M Aluminum Alloys: Comparison With Forged 7175 for Cycles to Failure

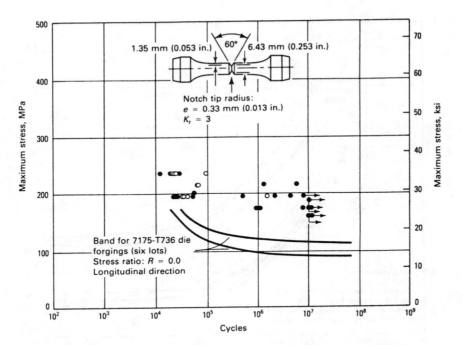

Comparison of axial-stress notch fatigue strength of P/M alloy X7091-T7E69 die forgings and ingot metallurgy alloy 7175-T736 die forgings. ○, longitudinal direction, one lot; ●, short transverse direction, two lots; → denotes test specimen did not fail in number of cycles indicated. Stress ratio: $R = 0.1$.

Source: Metals Handbook, 9th Edition, Volume 7, Powder Metallurgy, American Society for Metals, Metals Park OH, 1984, p 469

12-72. Various Aluminum Alloys: Comparison of Grades for Corrosion-Fatigue Crack Growth Rates; Air vs Salt Water

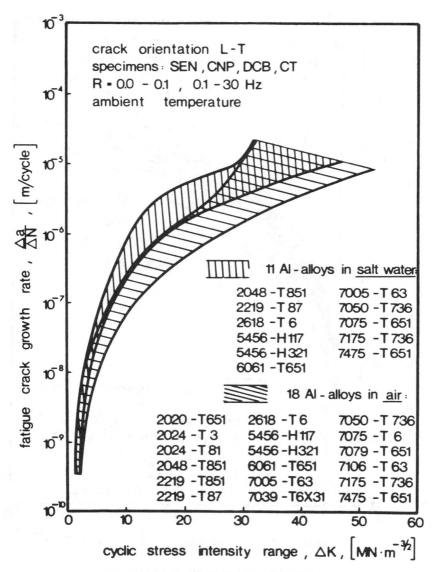

Comparison of scatterbands of corrosion-fatigue crack growth rates and fatigue crack growth rates of many commercial aluminum alloys.

Source: Markus O. Speidel, "Aluminum as a Corrosion Resistant Material," in Aluminum Transformation Technology and Applications (Proceedings of the International Symposium at Puerto Madryn, Chubut, Argentina), C. A. Pampillo, H. Biloni and D. E. Embury, Eds., American Society for Metals, Metals Park OH, 1980, p 615

12-73. Various Aluminum Alloys: Comparison of Grades for Corrosion-Fatigue Crack Growth Rates in Salt Water

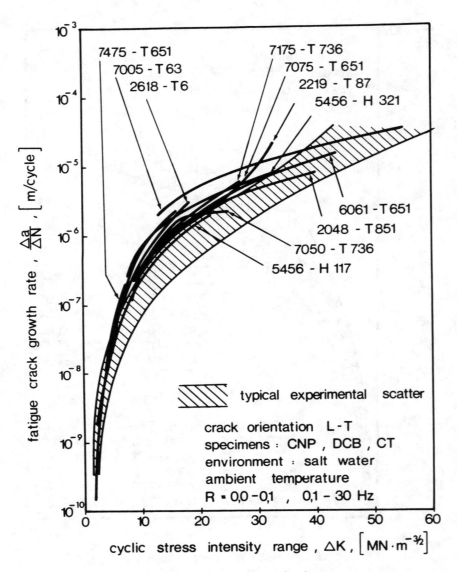

Corrosion-fatigue crack growth rates in salt water for aluminum alloys exceed the scatterband.

As shown in the above graph, curves for growth rate are somewhat higher than the air-test scatterband, but at very low and very high stress-intensity ranges, no significant difference between fatigue and corrosion fatigue crack growth rates is observed.

Source: Markus O. Speidel, "Aluminum as a Corrosion Resistant Material," in Aluminum Transformation Technology and Applications (Proceedings of the International Symposium at Puerto Madryn, Chubut, Argentina), C. A. Pampillo, H. Biloni and D. E. Embury, Eds., American Society for Metals, Metals Park OH, 1980, p 614

12-74. Various Aluminum Alloys: Wrought vs Cast, and Influence of Casting Method on Fatigue Life

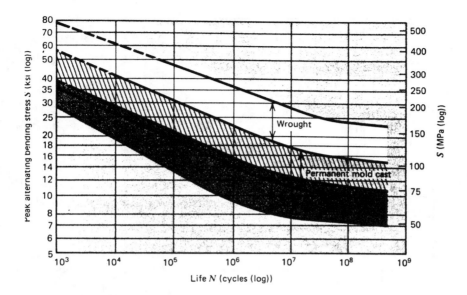

Representative S-N curves for various aluminum alloys are shown in the above graph. Note the absence of a sharply defined "knee" and true endurance limit. This is typical of nonferrous metals. In the absence of an endurance limit, the fatigue strength at 10^8 or 5×10^8 cycles is often used. (To give a "feel" for the time required to accumulate this many cycles, an automobile would typically travel nearly 400,000 miles before any one of its cylinders fired 5×10^8 times.) As is true for most metals and alloys, the wrought versions of aluminum alloys have greater fatigue strength than the cast (see graph). It will also be noted in the graph that there is an overlapping of fatigue strength for the sand and permanent mold casting methods (same alloy).

Source: Robert C. Juvinall, Fundamentals of Machine Component Design, John Wiley & Sons, New York NY, 1983, p 207

12-75. Aluminum Casting Alloy AL-195: Interrelationship of Fatigue Properties With Degree of Porosity

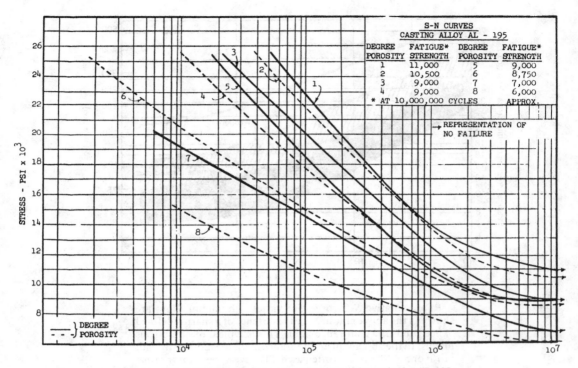

Interrelationship of fatigue properties with degree of porosity for AL-195 casting alloy.

Source: N. E. Promisel, "Evaluation of Non-ferrous Materials," in Materials Evaluation in Relation to Component Behavior (Proceedings of the Third Sagamore Ordnance Materials Research Conference), Syracuse University Research Institute, Syracuse NY, 1956, p 65

12-76. Aluminum Casting Alloy LM25-T6: Squeeze Formed vs Chill Cast; Effect on Reversals to Failure

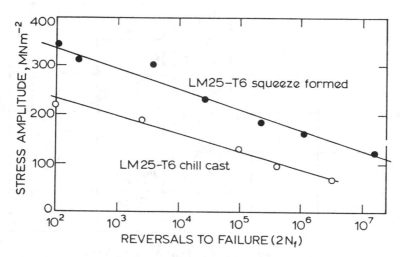

S-N curves for aluminum casting alloy LM25-T6; chill cast versus squeeze formed.

Fatigue tests have been carried out with LM25 samples, which were cut from a bracket component. A servohydraulically controlled fatigue machine was used to execute push-pull tests about mean zero. The results are presented in the above chart, which includes for reference the results of similar tests carried out on conventionally cast LM25. It can be seen that a significant improvement in the fatigue performance has been achieved by squeeze forming this type of alloy.

Source: G. Williams and K. M. Fisher, "Squeeze Forming of Aluminium-Alloy Components," in Production to Near Net Shape: Source Book, C. J. Van Tyne and B. Avitzur, Eds., American Society for Metals, Metals Park OH, 1983, p 367

13-1. Copper: Effect of Air and Water Vapor on Cycles to Failure

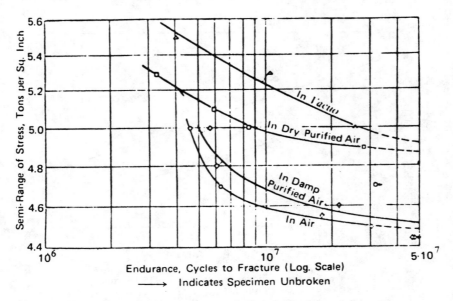

The effect of air and water vapor on the fatigue life of annealed copper.

The effect of atmospheric oxygen on fatigue life of copper has been investigated; oxygen and water vapor reduce fatigue life in copper. Alternate static exposure to air and dynamic exposure to vacuum do not affect fatigue life, and S-N curves diverge as applied stresses are reduced (see graph). Based on these experiments, the investigators concluded that:

1. Fatigue cracks form early, because the majority of life is concerned with crack propagation (environment has little or no effect on nucleation and initial growth).
2. Oxygen and water vapor are the primary damaging constituents in air (water vapor alone is effective).
3. Oxygen must be a gas (preoxidation or intermittent exposure is not effective).

Source: D. J. Duquette, "Environmental Effects I: General Fatigue Resistance and Crack Nucleation in Metals and Alloys," in Fatigue and Microstructure, American Society for Metals, Metals Park OH, 1979, p 336

13-2. Copper: Applied Plastic-Strain Amplitude vs Fatigue Life

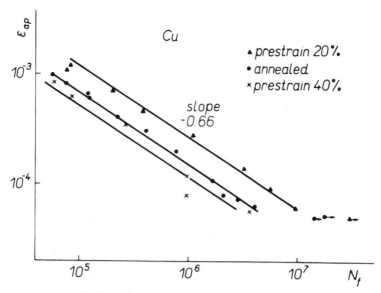

Applied plastic-strain amplitude versus fatigue-life curves for copper at long life.

Helgeland was the first to observe and claim a fatigue limit for copper (actually the plateau stress, although it was not recognized as such at the time). Unfortunately, his results were apparently contradicted by those of Kettunen, who observed failures at stresses down to 17.7 MPa. This difficulty was resolved by Laird, who showed that Lukáš and Klesnil's long-life Coffin-Manson plots showed failures to occur only down to the plastic-strain fatigue limit; at lower strains, no failures were observed in the testing time available (see above chart). However, Lukáš et al. also carried out stress-cycling tests, in which they monitored the plastic strain. Specimens that had been stress-cycled yielded a plot of saturation plastic-strain amplitude versus life, where failures occurred at strains as low as 10^{-5}. The difference between these tests is that in strain cycling, the stress is low in the initial cycles and increases to saturation, whereas in stress cycling, full application of the load in the first cycle causes a large strain in a soft material. This initial large strain creates the PSB cell structures, which would not otherwise form in a constant-strain test. Since Kettunen applied the full load to his specimens, failures were observed at stresses below that of the plateau. Helgeland, on the other hand, although he was stress cycling, imposed a low stress at the start of his tests and increased it gradually to the chosen value.

Source: Campbell Laird, "Mechanisms and Theories of Fatigue," in Fatigue and Microstructure, American Society for Metals, Metals Park OH, 1979, p 195

13-3. Copper Alloy C11000 (ETP Wire): Effect of Temperature on Fatigue Strength

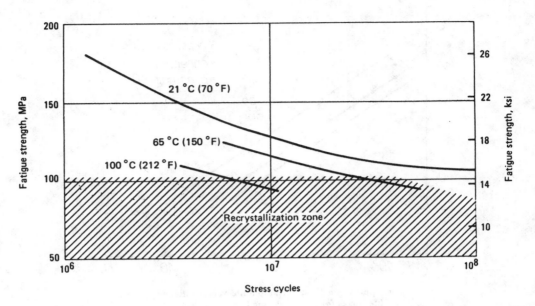

Rotating-beam fatigue strength of electrolytic tough pitch copper, C11000 wire, 2 mm (0.08 in.) diam, H80 temper when tested at various temperatures.

Source: Metals Handbook, 9th Edition, Volume 2, Properties and Selection: Nonferrous Alloys and Pure Metals, American Society for Metals, Metals Park OH, 1979, p 289

13-4. Copper Alloy C26000 (Cartridge Brass): Influence of Grain Size and Cold Work on Cycles to Failure

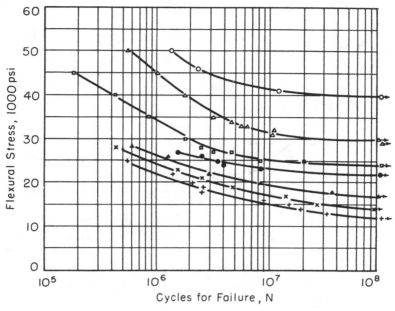

Legend
- ○ Group G (60% Cold Drawn)
- △ Group D (40% Cold Drawn)
- □ Group A (20% Cold Drawn)
- ● Anneal 1 Grain Size 0.012 mm.
- ▲ Anneal 3 Grain Size 0.026 mm.
- × Anneal 4 Grain Size 0.051 mm.
- + Anneal 5 Grain Size 0.131 mm.

Influence of grain size and cold work on fatigue strength of copper alloy C26000 (cartridge brass).

Changes in grain size and in degree of cold work which result in increased tensile strength or hardness usually result in improved fatigue strength. The above S-N curves illustrate this and indicate the influence of grain size and cold work on the fatigue strength of alpha brass. All specimens were prepared from the same heat and, therefore, had the same nominal composition.

Source: George M. Sinclair, "Some Metallurgical Aspects of Fatigue," in Fatigue—An Interdisciplinary Approach, John J. Burke, Norman L. Reed and Volker Weiss, Eds., Syracuse University Press, Syracuse NY, 1964, p 69

13-5. Copper Alloy C83600 (Leaded Red Brass): S-N Curves; Scatter Band

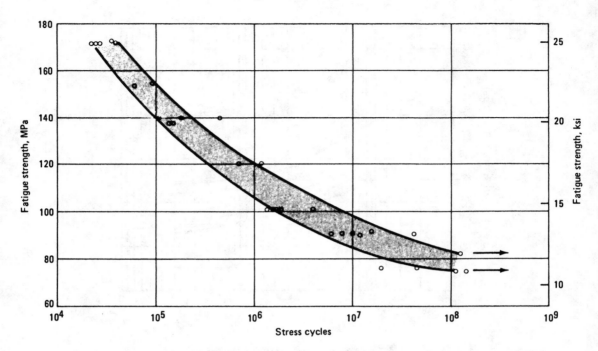

S-N curves (scatter band) for copper alloy C83600 (leaded red brass).

Source: Metals Handbook, 9th Edition, Volume 2, Properties and Selection: Nonferrous Alloys and Pure Metals, American Society for Metals, Metals Park OH, 1979, p 406

13-6. Copper Alloy C86500 (Manganese Bronze): S-N Curves; Scatter Band

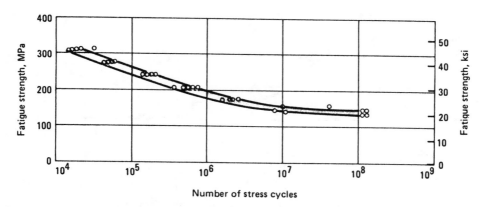

S-N curves (scatter band) for copper alloy C86500 (manganese bronze). All testing was performed at room temperature.

Source: Metals Handbook, 9th Edition, Volume 2, Properties and Selection: Nonferrous Alloys and Pure Metals, American Society for Metals, Metals Park OH, 1979, p 35

13-7. Copper Alloys C87500 and C87800 (Silicon Brasses): S-N Curves; Scatter Band

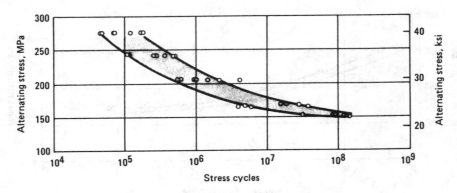

S-N curves (scatter band) for copper alloys C87500 and C87800 (silicon brasses) tested at room temperature.

Source: Metals Handbook, 9th Edition, Volume 2, Properties and Selection: Nonferrous Alloys and Pure Metals, American Society for Metals, Metals Park OH, 1979, p 416

13-8. Copper Alloy C92200 (Navy "M" Bronze): S-N Curves; Scatter Band

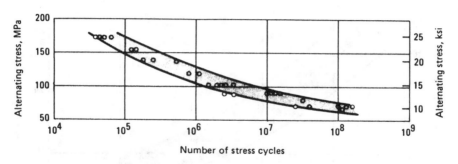

S-N curves (scatter band) for copper alloy C92200 (Navy "M" Bronze, or steam bronze) tested at room temperature.

Source: Metals Handbook, 9th Edition, Volume 2, Properties and Selection: Nonferrous Alloys and Pure Metals, American Society for Metals, Metals Park OH, 1979, p 421

13-9. Copper Alloy C93700 (High-Leaded Tin Bronze): S-N Curves; Scatter Band

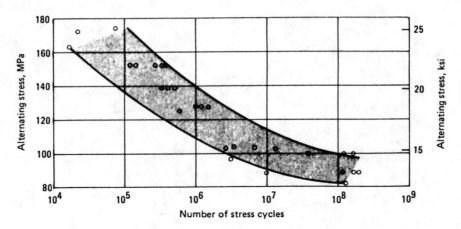

S-N curves (scatter band) for copper alloy C93700 (high-leaded tin bronze) tested at room temperature.

Source: Metals Handbook, 9th Edition, Volume 2, Properties and Selection: Nonferrous Alloys and Pure Metals, American Society for Metals, Metals Park OH, 1979, p 426

13-10. Copper Alloy No. 192: Effect of Salt Spray on Tubes

Results of fatigue tests on copper alloy tubes before and after salt spray exposure.

The tubes made from the copper alloy failed in the range of 10^5 to 10^6 cycles. After exposure for 180 days to salt spray, the fatigue performance level was not lowered (see plot above).

Brazed steel tubes, prior to salt exposure, failed in the same test in the range of 10^5 to 10^7 cycles. After 30-days exposure to salt spray, the resistance to fatigue was 10^5 to 10^6 cycles. After 90-days exposure, the steel tubes showed no fatigue strength in this particular test.

Source: Donald K. Miner, "An Effective Solution to the Problem of Hydraulic Brake Line Corrosion," in Source Book on Copper and Copper Alloys, American Society for Metals, Metals Park OH, 1979, p 356

13-11. Copper Alloy 955: Goodman-Type Diagram

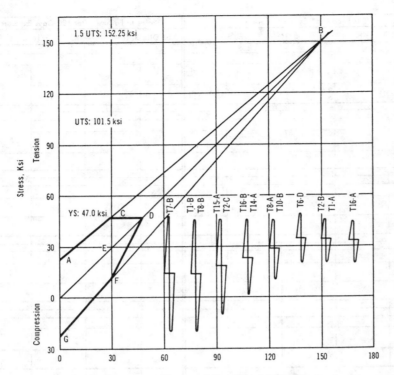

Goodman-type diagram (after Creech) for annealed copper alloy 955.

Two variable-speed, flat-plate testing machines of the fixed-deflection type were used for the test work. These machines have a speed range of 750-2000 cpm with a maximum deflection of 1 in. The yield, ultimate tensile strength and 1.5 times the ultimate are plotted in the graph above. The fatigue limit at zero mean stress was determined and was found to be 22.0 ksi.

Source: J. M. Cieslewicz, "A Modified Goodman Diagram to Predict the Fatigue Limits of Copper Alloy 955," in Source Book on Copper and Copper Alloys, American Society for Metals, Metals Park OH, 1979, p 40

14-1. Magnesium Casting Alloy QE22A-T6: Effects of Notches and Testing Temperature

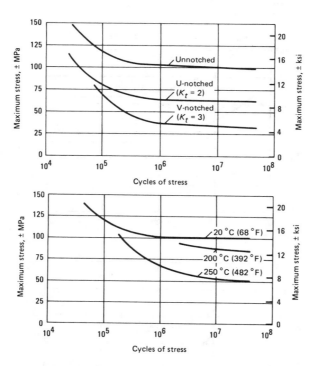

S-N curves for magnesium alloy sand castings, QE22A-T6, showing effects of type of notches (upper graph) and testing temperature (lower graph). Rotating-beam (Wohler) tests. Machine speed was 2960 Hz.

Source: Metals Handbook, 9th Edition, Volume 2, Properties and Selection: Nonferrous Alloys and Pure Metals, American Society for Metals, Metals Park OH, 1979, p 589

14-2. Magnesium Casting Alloy QH21A-T6: S-N Curves; Effects of Notches and Testing Temperature

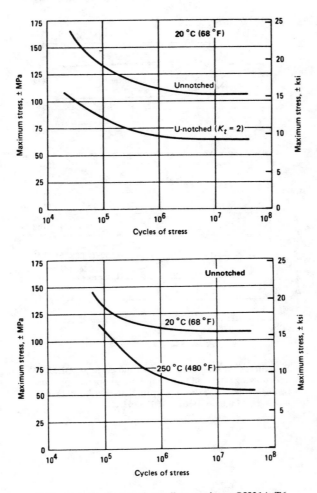

S-N curves for magnesium alloy castings, QH21A-T6, showing effects of notches (upper graph) and testing temperature (lower graph). Rotating-beam (Wohler) tests; machine speed 2960 Hz.

Source: Metals Handbook, 9th Edition, Volume 2, Properties and Selection: Nonferrous Alloys and Pure Metals, American Society for Metals, Metals Park OH, 1979, p 590

14-3. Mg-Al-Zn Casting Alloys: Effects of Surface Conditions on Fatigue Properties

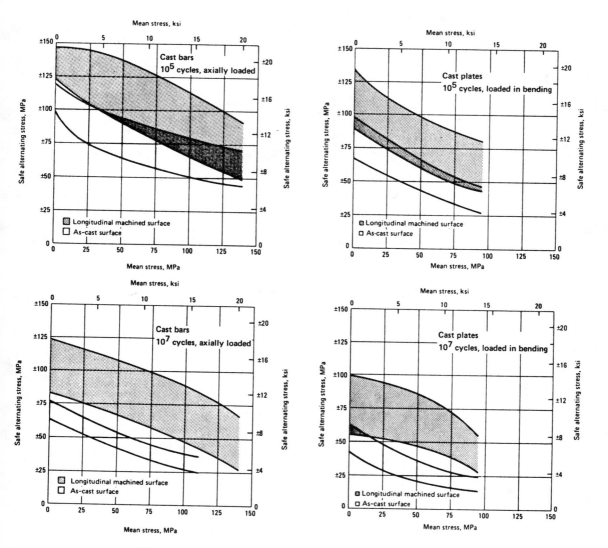

Effect of type of surface on fatigue properties of cast Mg-Al-Zn alloys.

Machining improves fatigue properties of castings, as shown in the above curves. Small radii, notches or fretting corrosion are more likely to reduce fatigue life than are minor variations in composition or heat treatment.

Source: Metals Handbook, 9th Edition, Volume 2, Properties and Selection: Nonferrous Alloys and Pure Metals, American Society for Metals, Metals Park OH, 1979, p 532

15-1. Molybdenum: Fatigue Limit Ratio vs Temperature

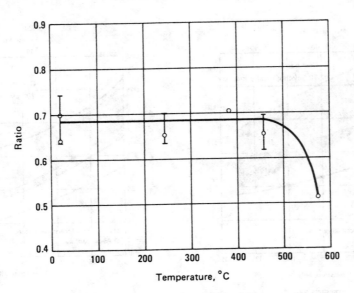

Ratio of the fatigue limit of molybdenum at various temperatures to its tensile strength at the same temperature.

Source: Metals Handbook, 9th Edition, Volume 2, Properties and Selection: Nonferrous Alloys and Pure Metals, American Society for Metals, Metals Park OH, 1979, p 774

16-1. Tin-Lead Soldering Alloy: S-N Data for Soldered Joints

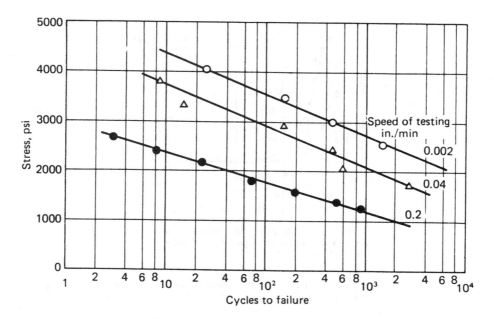

The fatigue strength of soldered joints is a complex and difficult subject to examine. Because solder alloys are strain-rate sensitive and have large elongation capabilities, the performance of fatigue tests under constant stress causes progressive and rapid relaxation of the joint, and conversely, tests under constant strain do not reflect a practical application situation. The influence of the rate of stress cycling in terms of rate of straining on the fatigue life of copper soldered joints with 60%Sn-40%Pb alloy is presented in the above graph.

Source: Metals Handbook, 9th Edition, Volume 6, Welding, Brazing, and Soldering, American Society for Metals, Metals Park OH, 1983, p 1095

16-2. Babbitt: Variation of Bearing Life With Babbitt Thickness

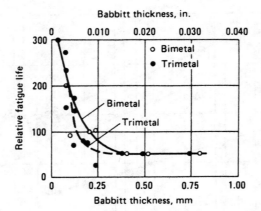

Variation of bearing life with babbitt thickness for lead or tin babbitt bearings. Bearing load was 14 MPa (2000 psi) for all tests.

One of the most useful concepts in bearing-material design came with the recognition that the effective load capacities and fatigue strengths of lead and tin alloys were sharply increased when these alloys were used as thin layers intimately bonded to strong bearing backs of bronze or steel. Use is made of this principle (see graph above), in both two-layer and three-layer constructions, in which the surface layer is composed of a lead or tin alloy, usually no more than 0.13 mm (0.005 in.) thick.

Source: Metals Handbook, 9th Edition, Volume 3, Properties and Selection: Stainless Steels, Tool Materials and Special-Purpose Metals, American Society for Metals, Metals Park OH, 1980, p 806

16-3. SAE12 Bearing Alloy: Effect of Temperature on Fatigue Life

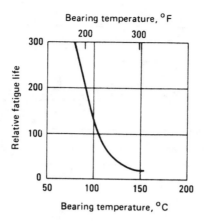

Variation of bearing life with temperature for SAE12 bimetal bearings.

The alloy lining was 0.05 to 0.13 mm (0.002 to 0.005 in.) thick, on steel backing. Bearing load: 14 MPa (2000 psi). As indicated, operating temperature markedly influences fatigue life.

Source: Metals Handbook, 9th Edition, Volume 3, Properties and Selection: Stainless Steels, Tool Materials and Special-Purpose Metals, American Society for Metals, Metals Park OH, 1980, p 813

17-1. Unalloyed Titanium, Grade 3: S-N Curves for Annealed vs Cold Rolled

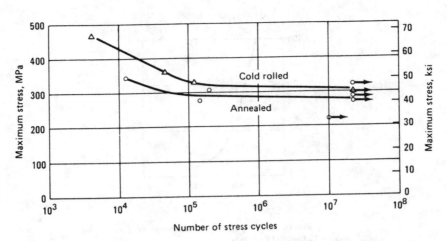

S-N curve for unalloyed grade 3 titanium. Data were obtained by rotating-beam testing at room temperature.

Source: Metals Handbook, 9th Edition, Volume 3, Properties and Selection: Stainless Steels, Tool Materials and Special-Purpose Metals, American Society for Metals, Metals Park OH, 1980, p 376

17-2. Unalloyed Titanium, Grade 4: S-N Curves for Three Testing Temperatures

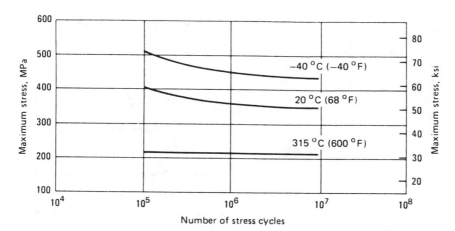

S-N curves for unalloyed titanium, grade 4, at subzero, room, and elevated temperatures. Data were obtained by rotating-beam testing of unnotched, polished specimens machined from annealed bar stock.

Source: Metals Handbook, 9th Edition, Volume 3, Properties and Selection: Stainless Steels, Tool Materials and Special-Purpose Metals, American Society for Metals, Metals Park OH, 1980. p 378

17-3. Ti-24V and Ti-32V: Stress Amplitude vs Cycles to Failure

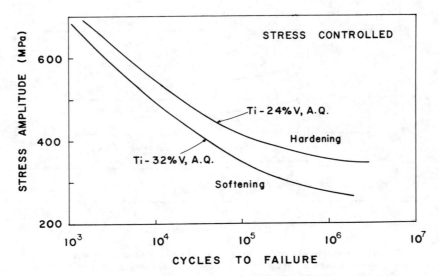

Stress-life curves of two Ti-V alloys that undergo cyclic hardening (Ti-24%V) and cyclic softening (Ti-32%V).

Cyclic-response curves indicate that the Ti-24%V alloy undergoes extensive cyclic hardening, whereas Ti-32%V undergoes cyclic softening, as indicated above. Hardening is caused by incomplete reversibility of twinning.

Source: Edgar A. Starke, Jr., and Gerd Lütjering, "Cyclic Plastic Deformation and Microstructure," in Fatigue and Microstructure, American Society for Metals, Metals Park OH, 1979, p 236

17-4. Ti-5Al-2.5Sn: Effects of Notches and Types of Surface Finish

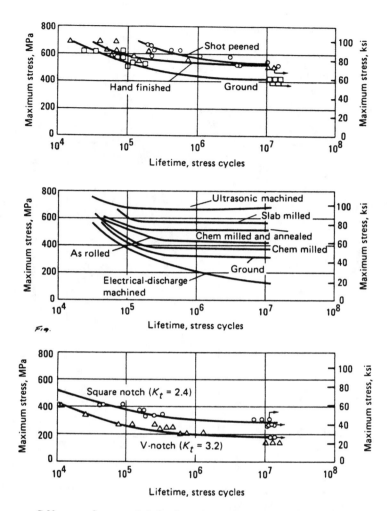

S-N curves for annealed titanium alloy Ti-5Al-2.5Sn (rotating-beam tests). Top and center graphs show fatigue strength for different types of surface finish. Bottom graph shows fatigue strength as affected by type of notch.

Source: Metals Handbook, 9th Edition, Volume 3, Properties and Selection: Stainless Steels, Tool Materials and Special-Purpose Metals, American Society for Metals, Metals Park OH, 1980, p 382

17-5. Ti-5Al-2.5Sn and Ti-6Al-4V: Fatigue Crack Growth Rates

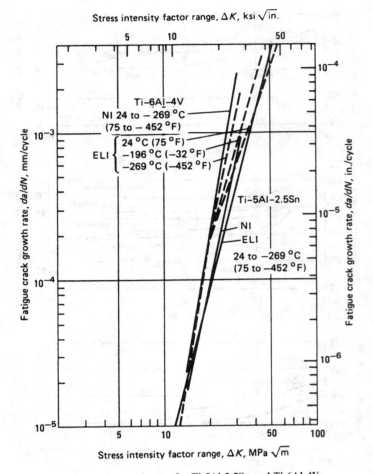

Fatigue crack growth rates for Ti-5Al-2.5Sn and Ti-6Al-4V.

Data on fatigue crack growth rates for Ti-5Al-2.5Sn and Ti-6Al-4V alloys are plotted above. These data indicate that the exposure temperature has no effect on the fatigue crack growth rates for Ti-5Al-2.5Sn and Ti-6Al-4V(NI). However, over part of the ΔK range, the fatigue crack growth rates for Ti-6Al-4V(ELI) are higher at cryogenic temperatures than at room temperature at the same ΔK values.

Source: Metals Handbook, 9th Edition, Volume 3, Properties and Selection: Stainless Steels, Tool Materials and Special-Purpose Metals, American Society for Metals, Metals Park OH, 1980, p 765

17-6. Ti-6Al-6V-2Sn: Effects of Machining and Grinding

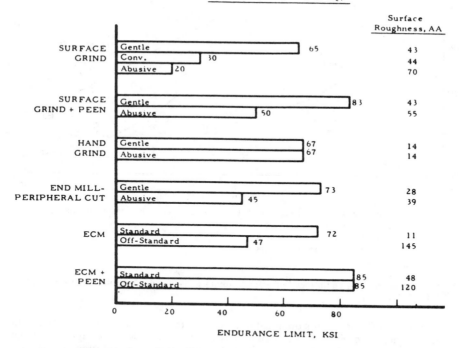

Bar chart presentation showing effects of various machining and grinding operations on fatigue characteristics of titanium alloy Ti-6Al-6V-2Sn.

Source: Norman Zlatin and Michael Field, "Procedures and Precautions in Machining Titanium Alloys," in Titanium and Titanium Alloys: Source Book, Matthew J. Donachie, Jr., Ed., American Society for Metals, Metals Park OH, 1982, p 355

17-7. Ti-6Al-6V-2Sn (HIP): S-N Curves for Titanium Alloy Powder Consolidated by HIP

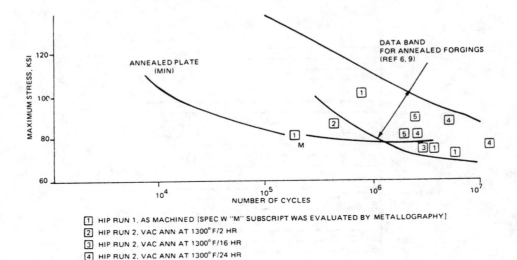

1. HIP RUN 1, AS MACHINED [SPEC W "M" SUBSCRIPT WAS EVALUATED BY METALLOGRAPHY]
2. HIP RUN 2, VAC ANN AT 1300°F/2 HR
3. HIP RUN 2, VAC ANN AT 1300°F/16 HR
4. HIP RUN 2, VAC ANN AT 1300°F/24 HR
5. HIP RUN 4, VAC ANN AT 1300°F/24 HR

S-N curves showing endurance limits for titanium alloy powder consolidated by HIP at 900 °C (1650 °F).

Note that most data points obtained in this phase fell within the representative data band for annealed forgings. In the specimen designated with an "M" subscript, low fatigue endurance was apparently associated with failure initiation at an inclusion. This shows that a clean powder is required for parts that are fatigue-critical and must operate with the equivalent of fully forged properties.

Source: R. H. Witt and W. T. Highberger, "Experience With Net-Shape Processes for Titanium Alloys," in Production to Near Net Shape: Source Book, C. J. Van Tyne and B. Avitzur, Eds., American Society for Metals, Metals Park OH, 1982, p 277

17-8. Ti-6Al-6V-2Sn (HIP): S-N Curves for Annealed Plate vs HIP

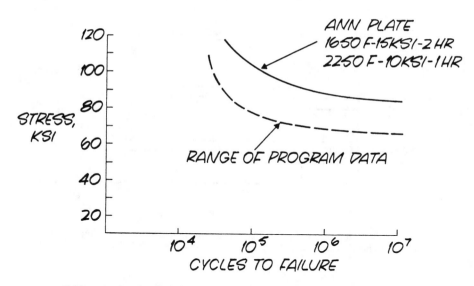

S-N curves showing that HIP Ti-6Al-6V-2Sn is equivalent to annealed plate of the same composition.

Source: W. Theodore Highberger, "Manufacture of Titanium Components by Hot Isostatic Pressing," in Production to Near Net Shape: Source Book, C. J. Van Tyne and B. Avitzur, Eds., American Society for Metals, Metals Park OH, 1983, p 304

17-9. Ti-6Al-2Sn-4Zr-2Mo: Bar Chart Presentation on Effects of Machining and Grinding

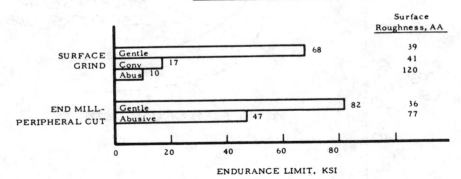

Bar chart presentation showing the effects of specific machining and grinding operations on fatigue characteristics of titanium alloy Ti-6Al-2Sn-4Zr-2Mo.

Source: Norman Zlatin and Michael Field, "Procedures and Precautions in Machining Titanium Alloys," in Titanium and Titanium Alloys: Source Book, Matthew J. Donachie, Jr., Ed., American Society for Metals, Metals Park OH, 1982, p 355

17-10. Ti-6Al-2Sn-4Zr-2Mo: Constant-Life Fatigue Diagram

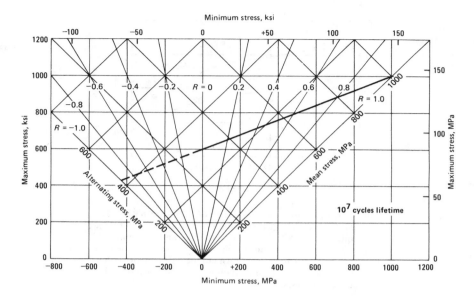

Constant-life fatigue diagram for duplex annealed Ti-6Al-2Sn-4Zr-2Mo sheet, 1 mm (0.04 in.) thick.

Source: Metals Handbook, 9th Edition, Volume 3, Properties and Selection: Stainless Steels, Tool Materials and Special-Purpose Metals, American Society for Metals, Metals Park OH, 1980, p 385

17-11. Ti-6Al-2Sn-4Zr-6Mo: Low-Cycle Axial Fatigue Curves

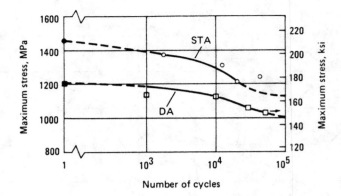

Low-cycle axial fatigue curves for Ti-6Al-2Sn-4Zr-6Mo. STA (solution treated and aged) condition: 1 h at 870 °C (1600 °F), water quench, age 8 h at 595 °C (1100 °F) and air cool. DA (duplex annealed) condition: 15 min at 870 °C, air cool, then 8 h at 540 °C (1000 °F) and air cool. All fatigue tests conducted at a stress ratio of $R = 0.1$. Open symbols indicate fatigue tests; solid symbols, tension tests.

Source: Metals Handbook, 9th Edition, Volume 3, Properties and Selection: Stainless Steels, Tool Materials and Special-Purpose Metals, American Society for Metals, Metals Park OH, 1980, p 395

17-12. Ti-8Mo-2Fe-3Al: S-N Curves; Solution Treated and Aged Condition

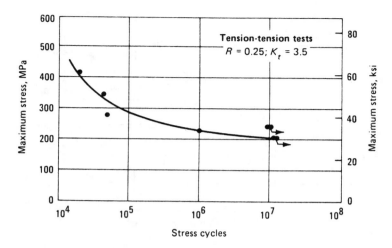

S-N curve for Ti-8Mo-2Fe-3Al titanium alloy in the solution treated and aged condition. Data are for 1.5-mm (0.060-in.) thick sheet solution treated 10 min at 790 °C (1450 °F), air cooled, and aged 8 h at 480 °C (900 °F).

Source: Metals Handbook, 9th Edition, Volume 3, Properties and Selection: Stainless Steels, Tool Materials and Special-Purpose Metals, American Society for Metals, Metals Park OH, 1980, p 403

17-13. Ti-10V-2Fe-3Al: S-N Curves; Notched vs Unnotched Specimens in Axial Fatigue

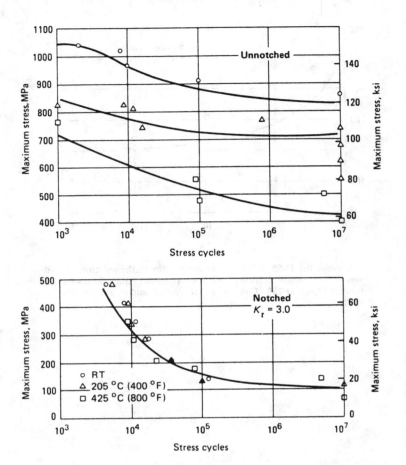

Axial fatigue of Ti-10V-2Fe-3Al bar stock in the STOA (solution treated and overaged) condition. Specimens were taken from round bars 75 mm (3 in.) in diameter that had been solution treated 1 h at 760 °C (1400 °F), furnace cooled, overaged 8 h at 565 °C (1050 °F) and air cooled. Tests were conducted at a stress ratio of $R = 0.1$ and a frequency of 20 Hz. Top: results of unnotched bars tested at room temperature. Bottom: fatigue characteristics of notched specimens tested at elevated temperature.

Source: Metals Handbook, 9th Edition, Volume 3, Properties and Selection: Stainless Steels, Tool Materials and Special-Purpose Metals, American Society for Metals, Metals Park OH, 1980, p 399

17-14. Ti-10V-2Fe-3Al and Ti-6Al-4V: Comparison of Fatigue Crack Growth Rates

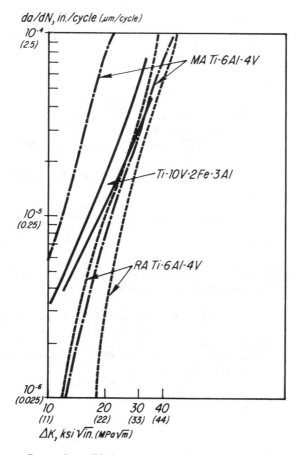

Comparison of fatigue crack growth rates. Data for Ti-10V-2Fe-3Al, $R = 0.05$, $F = 1$ to 30 Hz; for MA Ti-6Al-4V, $R = 0.08$, $F = 1$ to 25 Hz; for RA Ti-6Al-4V, $R = 0.08$, $F = 6$ Hz.

Fatigue crack growth rates in air have been found to lie in the scatter band for mill annealed (MA) Ti-6Al-4V, as shown above. At high ΔK values, Ti-10V-2Fe-3Al approaches the performance of Ti-6Al-4V in the recrystallized annealed (RA) condition.

Source: Wayne A. Reinsch and Harry W. Rosenberg, "Three Recent Developments in Titanium Alloys," in Titanium and Titanium Alloys: Source Book, Matthew Donachie, Jr., Ed., American Society for Metals, Metals Park OH, 1982, p 375

17-15. Ti-10V-2Fe-3Al: S-N Curve; Notched Bar Fatigue Life for a Series of Forgings Compared With Ti-6Al-4V Plate

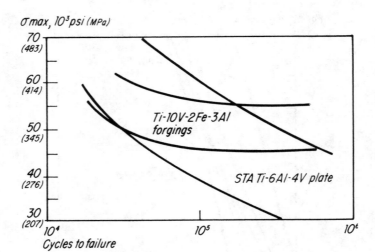

Comparison of notched fatigue lives for Ti-10V-2Fe-3Al forgings and Ti-6Al-4V plate. Data for Ti-10V-2Fe-3Al, $R = 0.05$, $F = K_T = 2.9$; for STA Ti-6Al-4V plate, $R = 0.1$, $K_T = 3$.

Fatigue characteristics of Ti-10V-2Fe-3Al are equal to or superior to those of Ti-6Al-4V. Notched fatigue results are shown above. Data from a series of die forgings have shown that the mean value for fracture toughness is 49.1 ksi $\sqrt{\text{in.}}$ (54 MPa $\sqrt{\text{m}}$), with a standard deviation of 2.3 ksi $\sqrt{\text{in.}}$ (2.5 MPa $\sqrt{\text{m}}$). K_{Iscc} in 3.5% NaCl is typically about 90% of the K_{Ic}.

Source: Wayne A. Reinsch and Harry W. Rosenberg, "Three Recent Developments in Titanium Alloys," in Titanium and Titanium Alloys: Source Book, Matthew Donachie, Jr., Ed., American Society for Metals, Metals Park OH, 1982, p 375

17-16. Ti-13V-11Cr-3Al: Constant-Life Fatigue Diagrams

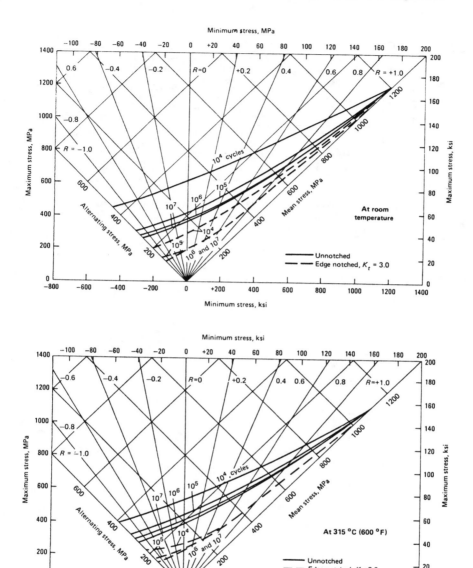

Constant-life fatigue diagrams for Ti-13V-11Cr-3Al, STA (solution treated and aged) condition, longitudinal orientation. Data are for axial fatigue of edge-polished sheet specimens of material solution treated and aged to room-temperature tensile strength of 1203 MPa (174.5 ksi). Corresponding yield strength was 1080 MPa (156.7 ksi); at 315 °C (600 °F), the tensile strength was 1078 MPa (156.3 ksi) and the yield strength was 876 MPa (127.0 ksi). Tests were conducted at a speed of 60 Hz.

Source: Metals Handbook, 9th Edition, Volume 3, Properties and Selection: Stainless Steels, Tool Materials and Special-Purpose Metals, American Society for Metals, Metals Park OH, 1980, p 401

17-17. Ti-6Al-4V: Effect of Condition and Notches on Fatigue Characteristics

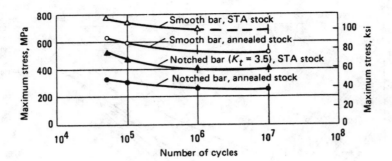

S-N curves for titanium alloy Ti-6Al-4V (rotating beam) showing effects of STA (solution treated and aged) versus annealed conditions, and effect of notches.

Source: Metals Handbook, 9th Edition, Volume 3, Properties and Selection: Stainless Steels, Tool Materials and Special-Purpose Metals, American Society for Metals, Metals Park OH, 1980, p 389

17-18. Ti-6Al-4V: Effect of Direction on Endurance

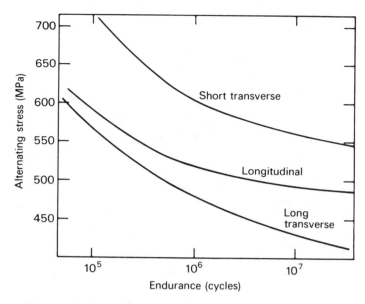

Rotating-cantilever fatigue (*S-N*) curves for three testing directions in 57 mm thick, forged and annealed Ti-6Al-4V bar.

These curves show that fatigue properties are lowest in the long transverse direction. This result has been attributed to the fact that Poisson's ratios are also sensitive to crystal orientation, these ratios being higher in the longitudinal and short transverse directions because stressing occurs parallel to the basal planes. Higher ratios imply greater constraint, which means that the levels of strain will be reduced and the fatigue strength enhanced in these two directions. The differences observed in fatigue strengths in the longitudinal and short transverse directions have been attributed to relative changes in grain shapes that also occur during processing.

Source: I. J. Polmear, Light Alloys, Edward Arnold Ltd, London, England, and American Society for Metals, Metals Park OH, 1981, p 193

17-19. Ti-6Al-4V: Effect of Isothermally Rolled vs Extruded Material on Cycles to Failure

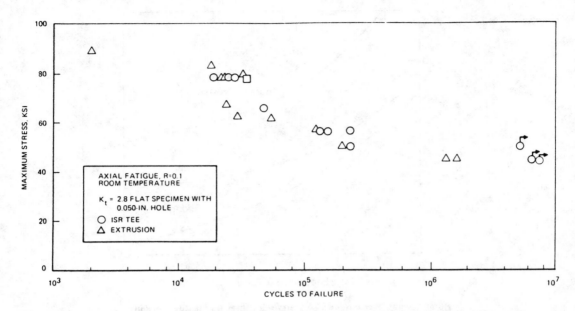

S-N fatigue data for isothermally rolled tees versus extruded material. The notched fatigue behavior of the ISR tees is as good as or slightly better than that of the extrusion.

Source: W. T. (Ted) Highberger, Govind R. Chanani and Gregory V. Scarich, "Advanced Titanium Metallic Materials and Processes for Application to Naval Aircraft Structures," in Production to Near Net Shape: Source Book, C. J. Van Tyne and B. Avitzur, Eds., American Society for Metals, Metals Park OH, 1983, p 124

17-20. Ti-6Al-4V: Comparison of Wrought vs Isostatically Pressed Material for Cycles to Failure

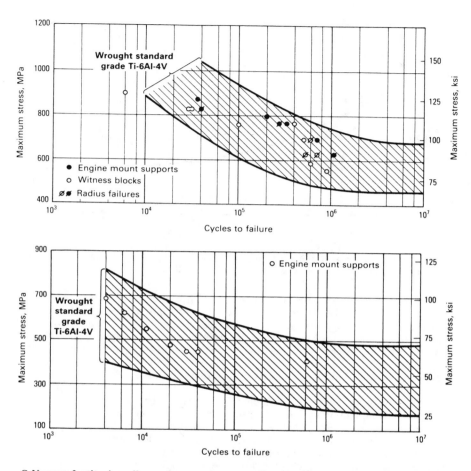

S-N curves for titanium alloy engine mount supports. Top: Data are for the standard wrought grade; $R = 0.1$, $K_r = 1.0$, load controlled smooth specimens. Bottom: Data are for isostatically pressed alloy powder, notched specimens; $R = 0.1$, $K_r = 3$.

Source: Metals Handbook, 9th Edition, Volume 7, Powder Metallurgy, American Society for Metals, Metals Park OH, 1984, p 654

17-21. Ti-6Al-4V: Effect of Fretting and Temperature on Cycles to Failure

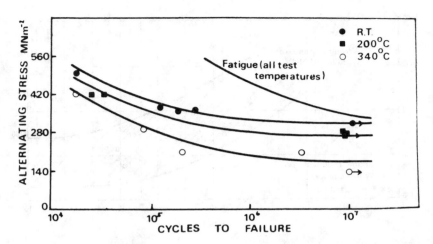

Effect of simultaneous fretting and fatigue of shot-peened Ti-6Al-4V at room temperature, 200, and 340 °C at a mean tensile stress of 140 MNm^{-2}.

Room temperature fretting was found to have little effect on the fatigue strength at 10^7 cycles. Fretting at 200 °C lowered the fatigue strength by approximately 15%; furthermore, the fretting fatigue life in the overstress region (70 MNm^{-2} above the nonfretted run out stress) was lowered by two orders of magnitude compared with results in the absence of fretting. At 340 °C, fretting similarly reduced specimen life at overstress levels; however, more importantly, the fretting fatigue strength at 10^7 cycles was reduced to approximately 40% of that found under room temperature conditions.

The gross result of fretting normally is fatigue failure brought about by surface damage in conjunction with normal or transient high stresses in a component. It should be said at the outset that visual assessment of fretting mildness or severity is inconclusive by itself, in that the presence of more or less fretting debris on a microscopic examination is not necessarily relevant to the loss of surface integrity. It may in fact be misleading and should not be relied upon for assessing the severity of fatigue life degradation. It is the stress state acting in concert with stress raisers (e.g., pits, tears, cracks) which determines the actual fatigue propensity.

Source: Practical Observations of Fretting Fatigue Cracks, p 180

17-22. Ti-6Al-4V (Beta Rolled): Effect of Finishing Operations on Cycles to Failure

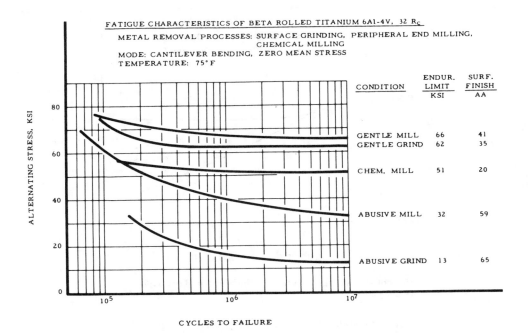

S-N curves for beta-rolled titanium alloy Ti-6Al-4V. Curves show the effects of the various finishing operations on fatigue.

Source: Norman Zlatin and Michael Field, "Procedures and Precautions in Machining Titanium Alloys," in Titanium and Titanium Alloys: Source Book, Matthew J. Donachie, Jr., Ed., American Society for Metals, Metals Park OH, 1982, p 354

17-23. Ti-6Al-4V: Effect of Yield Strength on Stress-Life Behavior

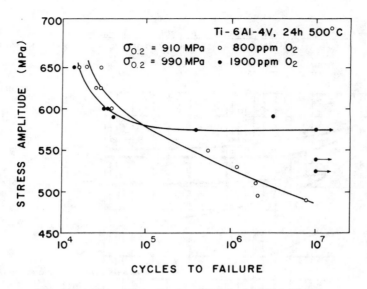

Effect of yield strength on the stress-life behavior of two Ti-6Al-4V alloys.

In order to establish microstructural effects on fatigue behavior, comparisons should be made on materials having the same yield stress, especially for stress-controlled tests. This is illustrated above where it is shown that two titanium samples having different yield strengths have different stress-life behavior when tested at 500 °C (930 °F).

Source: Edgar A. Starke, Jr., and Gerd Lütjering, "Cyclic Plastic Deformation and Microstructure," in Fatigue and Microstructure, American Society for Metals, Metals Park OH, 1979, p 237

17-24. Ti-6Al-4V: Effect of Stress Relief on Cycles to Failure

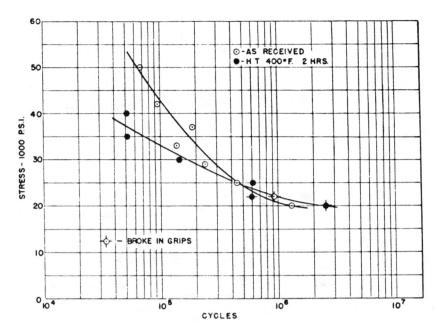

Flexural fatigue tests of titanium sheet (123,000 T.S.).

Flexural tests of the sheet specimen were made at 1725 CPM. Results are indicated by the open points in the above S-N diagram. The endurance limit was not reached at stresses as low as 20,000 psi. If an estimated limit of 19,000 psi is chosen, the endurance ratio would be only 0.155, a value considerably lower than for any other known metal or alloy. Most investigators have obtained normal values around 0.3 in similar tests.

Several of the sheet fatigue specimens developed fatigue cracks away from the milled specimen edges. The cracks did not appear to be associated with any visible surface imperfection. For these reasons, it was assumed that the sample was abnormal, rather than the test procedure. Very careful oil-powder and fluorescent powder tests, supplemented by metallographic examination, failed to reveal any surface cracks, even when the sheet was flexed to open any incipient hairline defects.

It was considered possible, though not probable, that residual stresses from cold rolling were acting in a deleterious manner. If so, a moderate temperature stress relief might help. Brief experiments soon disclosed that temperatures at least as high as 400 °F (205 °C) did not lower the hardness; in fact, the hardness may have increased very slightly. Knowing this, a set of sheet fatigue specimens was stress relieved for two hours at 400 °F (205 °C). The solid points in the graph above represent the results obtained with these specimens. The endurance limit was not altered significantly. A definite shift to the left in the upper portion of the curve was evident, although the direction of shift was opposite to that, had the heat treatment released undesirable stresses.

Source: Titanium Symposium, Office of Naval Research, p 97

17-25. Ti-6Al-4V: Interrelationship of Machining Practice and Cutting Fluids on Cycles to Failure

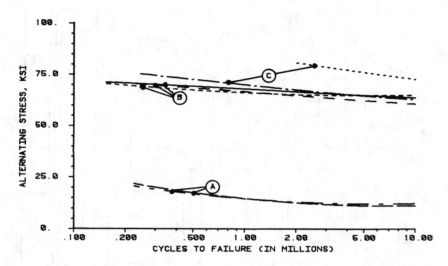

Alternating stress vs cycles to failure in high cycle fatigue of machined titanium surfaces using neutral, chlorinated, and sulfurized soluble cutting fluids. A = abusive grinding; B = low stress grinding; C = end milling.

Influence of Chlorinated and Sulfurized Cutting Fluids On High Cycle Fatigue Properties of Ti-6Al-4V Machined Surfaces at 75° F

Sol. Cutting Fluids	10^7 Cycle Fatigue Strength (ksi)		
	End Milling	Low Stress Grinding	Abusive Grinding
Neutral	75	62.5	12.5
Chlorinated	65	57.5	12.5
Sulfurized	--	67.5	12.5

Source: V. A. Tipnis and J. D. Christopher, "Machinability Testing for Industry," in Machinability Testing and Utilization of Machining Data, American Society for Metals, Metals Park OH, 1979, p 26

17-26. Ti-6Al-4V: Relative Effects of Machining and Grinding Operations on Endurance Limit

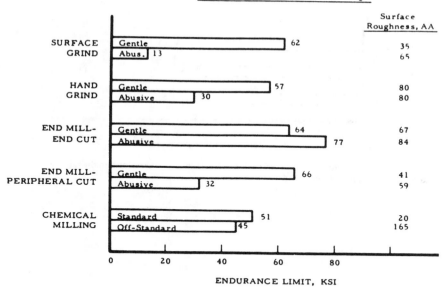

Bar chart presentation showing relative effects of various machining and grinding operations on fatigue characteristics of titanium alloy Ti-6Al-4V.

Source: Norman Zlatin and Michael Field, "Procedures and Precautions in Machining Titanium Alloys," in Titanium and Titanium Alloys: Source Book, Matthew J. Donachie, Jr., Ed., American Society for Metals, Metals Park OH, 1982, p 354

17-27. Ti-6Al-4V: Effects of Various Metal Removal Operations on Endurance Limit

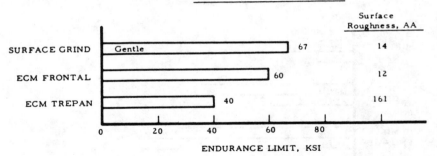

Bar chart presentation showing effects of various metal removal operations on the fatigue characteristics of titanium alloy Ti-6Al-4V.

Source: Norman Zlatin and Michael Field, "Procedures and Precautions in Machining Titanium Alloys," in Titanium and Titanium Alloys: Source Book, Matthew J. Donachie, Jr., Ed., American Society for Metals, Metals Park OH, 1982, p 355

17-28. Ti-6Al-4V: Effect of Texture on Fatigue Strength

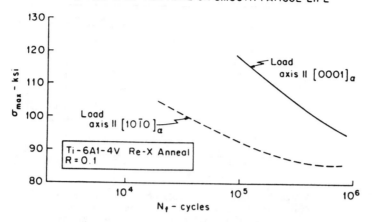

S-N curves showing the effect of texture on the fatigue strength of Ti-6Al-4V. Fatigue strength is greater when the stress axis coincides with the direction of a high density of basal poles.

Source: J. C. Williams and E. A. Starke, Jr., "The Role of Thermomechanical Processing in Tailoring the Properties of Aluminum and Titanium Alloys," in Deformation, Processing, and Structure, George Krauss, Ed., American Society for Metals, Metals Park OH, 1984, p 334

17-29. Ti-6Al-4V: Effect of Complex Texture on Cycles to Failure

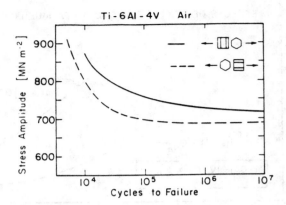

S-N curves showing the effect of more complex texture on fatigue strength of Ti-6Al-4V. These data show that a mixed texture lowers the high-stress end of the S-N curve preferentially.

Source: J. C. Williams and E. A. Starke, Jr., "The Role of Thermomechanical Processing in Tailoring the Properties of Aluminum and Titanium Alloys," in Deformation, Processing, and Structure, George Krauss, Ed., American Society for Metals, Metals Park OH, 1984, p 335

17-30. Ti-6Al-4V: Effect of Texture and Environment on Cycles to Failure

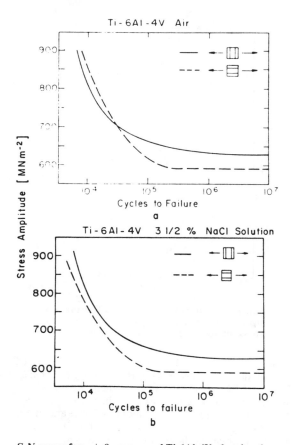

S-N curves for α + β − processed Ti-6Al-4V, showing the effects of texture and environment on fatigue strength. (a) Tested in air. (b) Tested in 3½% NaCl. These data show that testing in an aqueous 3½% NaCl solution reduces fatigue strength when the stress axis is along [0001].

Source: J. C. Williams and E. A. Starke, Jr., "The Role of Thermomechanical Processing in Tailoring the Properties of Aluminum and Titanium Alloys," in Deformation, Processing, and Structure, George Krauss, Ed., American Society for Metals, Metals Park OH, 1984, p 335

17-31. Ti-6Al-4V: Fatigue Crack Growth Rates

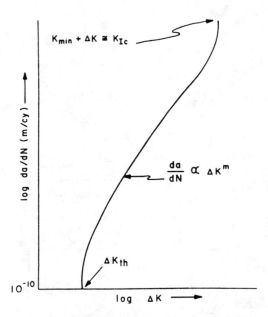

Schematic plot showing characteristic shape of fatigue crack growth rate (da/dN) versus cyclic-stress-intensity (ΔK) curves.

It can be seen that at higher growth rates there is a linear portion of the curve. This linear portion was represented as ΔK^m by Paris and Erdogan and is now frequently referred to as the Paris law regime of fatigue crack growth. Most structural materials show variations in near-threshold FCP rate and in ΔK_{th} but fewer show significant variations in FCP rate in the Paris law regime. In contrast, Ti alloys show significant variations in FCP rate over the entire range. At the highest crack growth rates shown above, the FCP rate curve bends upward. This is controlled by fracture toughness. However, since crack growth rates are uncontrollably rapid in this latter regime, it is of little interest and will not be discussed further here. Moreover, since the majority of the lifetime of a crack component is spent in the low-FCP-rate regime, factors which control FCP at rates less than $\sim 10^{-6}$ m/cycle are probably most important. These factors include microstructure and texture.

Source: J. C. Williams and E. A. Starke, Jr., "The Role of Thermomechanical Processing in Tailoring the Properties of Aluminum and Titanium Alloys," in Deformation, Processing, and Structure, George Krauss, Ed., American Society for Metals, Metals Park OH, 1984, p 338

17-32. Ti-6Al-4V: Fatigue Crack Growth Rates for ISR Tee, and Extrusions

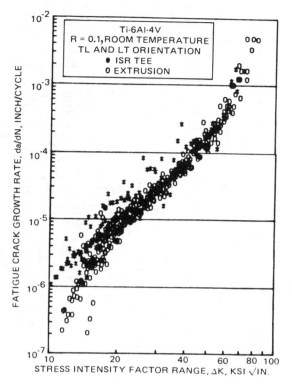

Fatigue crack growth results for ISR tee, and extrusions.

Data for both TL and LT orientations are shown above, along with data for the extrusion. In comparing individual results, no differences were seen between TL and LT. In the chart, it can be seen that particularly at lower stress intensities the fatigue crack rate for the (ISR) isothermally rolled tee is faster than that for the extrusions. This is probably due to the extrusions being beta formed while the ISR tees are alpha-beta formed.

Source: W. T. (Ted) Highberger, Govind R. Chanani and Gregory V. Scarich, "Advanced Titanium Metallic Materials and Processes for Application to Naval Aircraft Structures," in Production to Near Net Shape: Source Book, C. J. Van Tyne and B. Avitzur, Eds., American Society for Metals, Metals Park OH, 1983, p 124

17-33. Ti-6Al-4V: Fatigue Crack Growth Rates

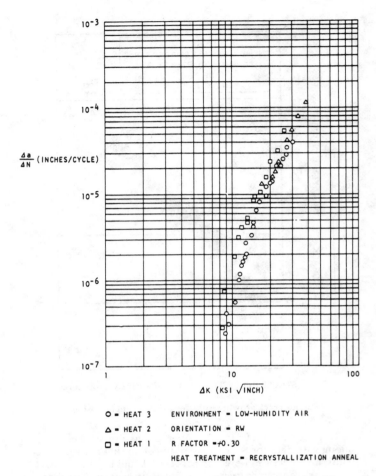

○ = HEAT 3 ENVIRONMENT = LOW-HUMIDITY AIR
△ = HEAT 2 ORIENTATION = RW
□ = HEAT 1 R FACTOR = +0.30
HEAT TREATMENT = RECRYSTALLIZATION ANNEAL

Fatigue crack growth rates for three different heats of Ti-6Al-4V titanium alloy.

The fatigue crack growth rate in the RW orientation for this alloy, when recrystallization-annealed, behaved similarly with decreasing oxygen and aluminum. The crack growth rate is shown as a function of ΔK tested at an R factor of +0.30.

Source: M. J. Harrigan, M. P. Kaplan and A. W. Sommer, "Effect of Chemistry and Heat Treatment on the Fracture Properties of Ti-6Al-4V Alloy," in Titanium and Titanium Alloys: Source Book, Matthew J. Donachie, Jr., Ed., American Society for Metals, Metals Park OH, 1982, p 65

17-34. Ti-6Al-4V: Effect of Final Cooling on Fatigue Crack Growth Rates

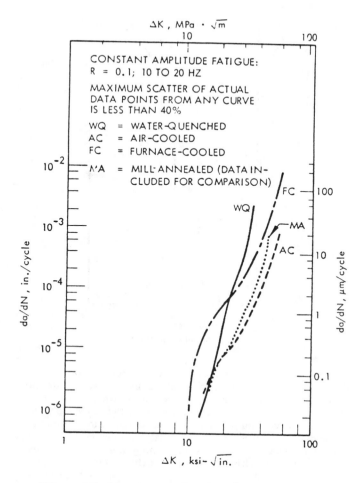

Effects of final cooling rate on fatigue crack growth rate in duplex-annealed Ti-6Al-4V, 1-in. plate, 1775 °F (968.3 °C), 1/2 h, air cooled; and 1450 °F (787.4 °C), 1 h, cooled as noted.

From the data presented above, it can be seen that air cooling, per se, produced little or no change in the cyclic crack growth compared to the mill-annealed base material. The slightly decreased crack growth rates above a ΔK of 20 ksi $\sqrt{\text{in.}}$ (22 MPa · $\sqrt{\text{m}}$) are, more probably than not, the result of the higher fracture toughness of the air-cooled material. However, both water quenching and furnace cooling resulted in fatigue crack growth rates noticeably different from those measured for the mill-annealed base material. As shown, furnace cooling had a consistently detrimental effect on the crack growth rate while water quenching produced greatly increased crack growth rates above a stress-intensity range of 18 ksi $\sqrt{\text{in.}}$ (20 MPa · $\sqrt{\text{m}}$). The accelerated growth rate above 18 ksi $\sqrt{\text{in.}}$ (20 MPa · $\sqrt{\text{m}}$) may be attributed to the proximity of the maximum stress intensity to the critical value. The critical stress-intensity value for water quenching was an exceptionally low 38 ksi $\sqrt{\text{in.}}$ (42 MPa · $\sqrt{\text{m}}$).

Source: R. E. Lewis, J. G. Bjeletich, T. M. Morton and F. A. Crossley, "Effect of Cooling Rate on Fracture Behavior of Mill-Annealed Ti-6Al-4V," in Titanium and Titanium Alloys: Source Book, Matthew J. Donachie, Jr., Ed., American Society for Metals, Metals Park OH, 1982, p 90

17-35. Ti-6Al-4V: Effect of Dwell Time on Fatigue Crack Growth Rates

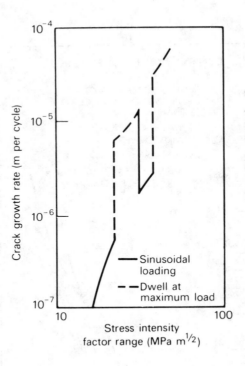

A phenomenon which may be unique to certain titanium alloys is the effect of dwell periods at maximum load on rates of growth of fatigue cracks. This effect is shown schematically here, and increases in the rate of crack growth of as much as 50 times may occur compared with results obtained in tests on the same material subjected only to sinusoidal stress cycles. Dwell effects are maximized in alloys containing substantial amounts of the α-phase which have a preferred texture such that stressing is normal to the basal planes, whereas they appear to be insignificant if stressing occurs parallel to the basal planes of the α-phase, or if the microstructure is homogeneous and fine grained. Particular attention has been paid to α/β alloys, e.g., Ti-6Al-4V, in which dwell effects have also been found to decrease with increasing amounts of the β-phase in the microstructure. In all cases, dwell effects disappear when stressing occurs at temperatures above 75 °C (165 °F), and they are generally considered to arise from the preferential diffusion of hydrogen, during the dwell period, to regions of localized hydrostatic tension ahead of an advancing crack. Such an accumulation of hydrogen would tend to embrittle this region, and it has even been suggested that brittle plates of TiH_2 may be formed.

Source: I. J. Polmear, Light Alloys, Edward Arnold Ltd, London, England, and American Society for Metals, Metals Park OH, 1981, p 200

17-36. Ti-6Al-4V: Fatigue Crack Growth Data

- Annealed 2 hours at 705°C, air-cooled after forging
- α/β transus 1005°C
- Axial loading: smooth specimens, $K_t = 1.0$

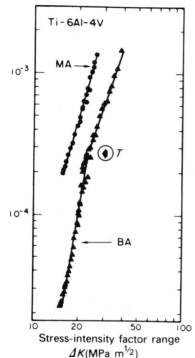

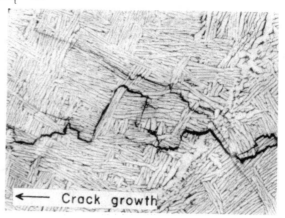

Left: Fatigue crack growth rates for Ti-6Al-4V rolled plates in the β-annealed (BA) and mill annealed (MA) conditions. BA = 0.5 h 1038°C, air-cool to room temperature. Tests conducted at 5 Hz using compact tension specimens. Ratio of minimum to maximum load = 0.1. Above: Branching of fatigue cracks within the Widmanstätten packets of the α-laths.

Work on Ti-6Al-4V rolled plate has indicated that the superior fatigue performance with the β-annealed condition is associated with relatively slower rates of crack propagation (above graph). This effect, in turn, is attributed to the slower progress of cracks through the Widmanstätten microstructure, particularly at stress intensities below a critical value at which desirable crack branching occurs within packets of the α-laths.

Source: I. J. Polmear, Light Alloys, Edward Arnold Ltd, London, England, and American Society for Metals, Metals Park OH, 1981, p 179

17-37. Ti-6Al-4V P/M: Comparison of HIP'd Material With Alpha-Beta Forgings for Cycles to Failure

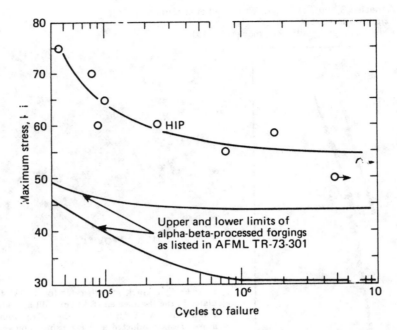

Notched fatigue strength of HIP'd P/M Ti-6Al-4V compared with fatigue strength of alpha-beta processed forgings. $K_t = 3$; Hz = 30; $R = 0.1$.

Source: J. H. Moll, V. C. Petersen and E. J. Dulis, "Powder Metallurgy Parts for Aerospace Applications," in Powder Metallurgy—Applications, Advantages and Limitations, Erhard Klar, Ed., American Society for Metals, Metals Park OH, 1983, p 286

17-38. Ti-6Al-4V P/M: Comparisons of HIP'd Material With Annealed Plate for Cycles to Failure

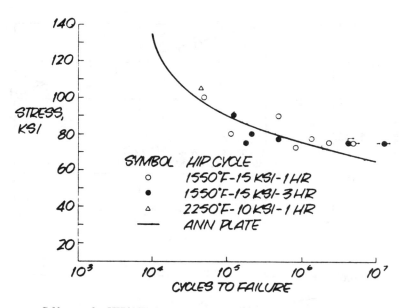

S-N curve for HIP'd Ti-6Al-4V and annealed plate. According to the above data, fatigue results for Ti-6Al-4V are within the required range for plate properties from MIL-T-9046.

Source: W. Theodore Highberger, "Manufacture of Titanium Components by Hot Isostatic Pressing," in Production to Near Net Shape: Source Book, C. J. Van Tyne and B. Avitzur, Eds., American Society for Metals, Metals Park OH, 1983, p 304

17-39. Ti-6Al-4V P/M: Effect of Powder Mesh Size on Fatigue Properties

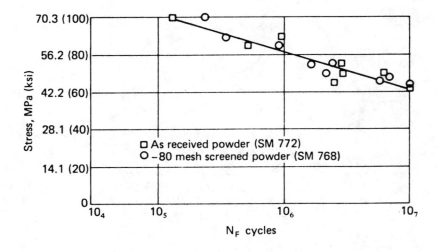

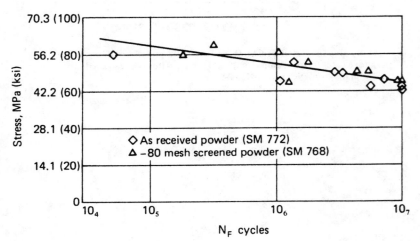

Top: Room-temperature properties. Bottom: Properties at 700 °F (370°C).

High-cycle fatigue (HCF) data were developed on Ti-6Al-4V (Std) by Williams International in a program to apply near-net-shape HIP technology to a compressor rotor part for the F-107 cruise missile engine. In this study, two size fractions of powder were used: −35 mesh (as-received) and −80 mesh. There was no difference in HCF test results between the two sizes. Room-temperature and 700 °F (370 °C) S/N curves are shown above.

Source: J. H. Moll, V. C. Petersen and E. J. Dulis, "Powder Metallurgy Parts for Aerospace Applications," in Powder Metallurgy—Applications, Advantages and Limitations, Erhard Klar, Ed., American Society for Metals, Metals Park OH, 1983, p 286

17-40. Ti-6Al-4V P/M: Comparison of Blended Elemental, Prealloyed and Wrought Material for Effect on Cycles to Failure

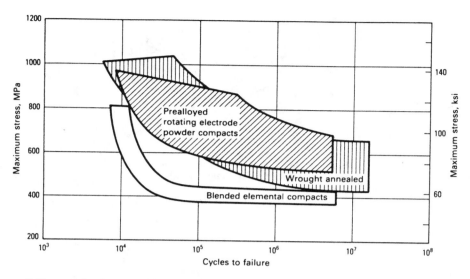

S-N curves showing comparison of smooth axial fatigue behavior of Ti-6Al-4V blended elemental and prealloyed P/M compacts with wrought annealed material. Tested at room temperature, $R = 0$.

The fatigue behavior of titanium P/M compacts is compared to wrought products in the graph above. The blended elemental material is inferior to prealloyed compacts and I/M materials. This is caused by residual chlorides and consequent porosity; also, chemcial heterogeneity may lead to areas of similarly aligned alpha plates. Blended elemental compacts, however, compete well with many titanium alloy castings in fatigue strength. Prealloyed powder compacts exhibit fatigue behavior equivalent to that of I/M materials. This situation is achieved by careful control of cleanliness (powder handling) and microstructure. Cleanliness depends on the environment in which the powder is produced, conditions of subsequent handling, and microstruture developed by compaction. Cleanliness dictates the amount of contamination contained in the final product; microstructure determines the ability of the compact to accommodate foreign particles and resist crack initiation.

Source: Metals Handbook, 9th Edition, Volume 7, Powder Metallurgy, American Society for Metals, Metals Park OH, 1984, p 753

17-41. Ti-6Al-4V: P/M Compacts vs I/M Specimens: Cycles to Failure

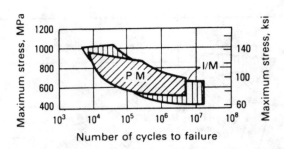

Fatigue chart presentation showing a comparison of fatigue behavior of Ti-6Al-4V compacts with ingot metallurgy material.

17-42. Ti-6Al-4V: Comparison of Specimens Processed by Various Fabrication Processes for Cycles to Failure

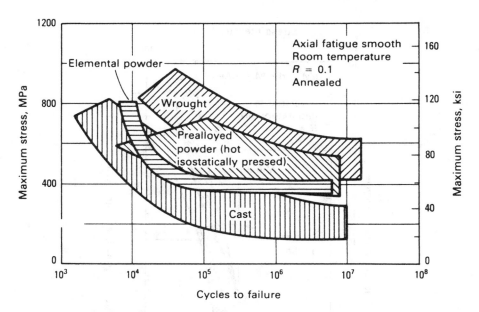

S-N curves (bands) for titanium alloy Ti-6Al-4V processed by various fabrication processes. The inconsistent fatigue life of the hot isostatically pressed product is usually casued by inclusions in the compact.

Source: Metals Handbook, 9th Edition, Volume 7, Powder Metallurgy, American Society for Metals, Metals Park OH, 1984, p 439

17-43. Ti-6Al-4V: Comparison of Fatigue Crack Growth Rate, P/M vs I/M

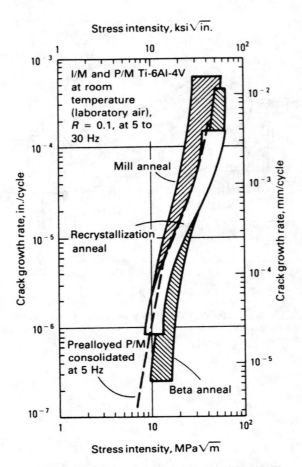

Comparison of fatigue crack growth rate of Ti-6Al-4V P/M compact with I/M material heat treated to various conditions. The fatigue crack growth rate of blended elemental and prealloyed compacts is equivalent to I/M material with the same microstructure.

Source: Metals Handbook, 9th Edition, Volume 7, Powder Metallurgy, American Society for Metals, Metals Park OH, 1984, p 752

17-44. Ti-6Al-4V: Base Metal vs SSEB-Welded Material for Cycles to Failure

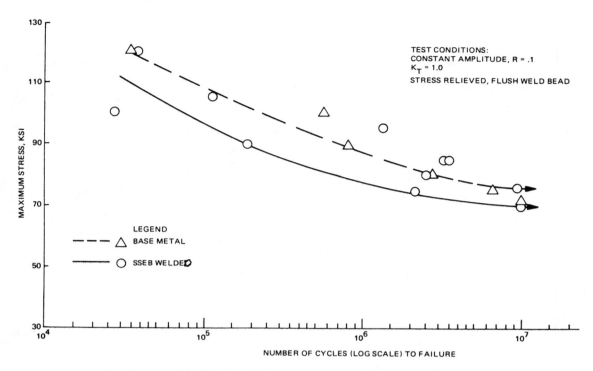

S-N curve for titanium alloy plate—base metal versus SSEB-welded. Results show that the constant-amplitude fatigue life of SSEB weldments in 0.440-in.-thick plate equals that of the base metal.

Source: R. H. Witt, J. G. Maciora and H. P. Ellison, "Sliding-Seal Electron-Beam Welding of Titanium," in Source Book on Electron Beam and Laser Welding, Melvin M. Schwartz, Ed., American Society for Metals, Metals Park OH, 1981, p 87

17-45. Ti-6Al-4V: Base Metal vs SSEB-Welded Material for Cycles to Failure

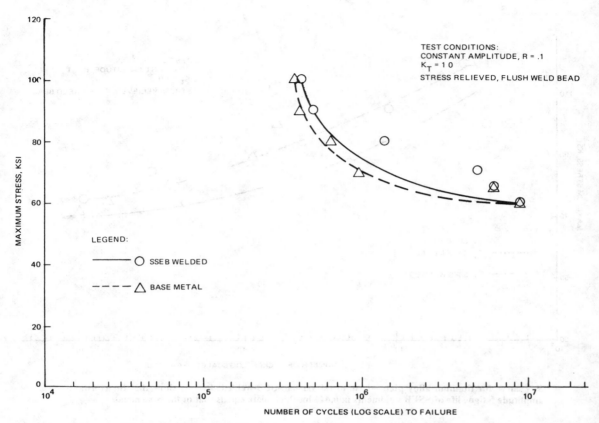

S-N curves for titanium alloy plate—base metal versus SSEB-welded. Results show that the constant-amplitude fatigue life of SSEB weldments in 0.940-in.-thick plate equals that of the base metal.

Source: R. H. Witt, J. G. Maciora and H. P. Ellison, "Sliding-Seal Electron-Beam Welding of Titanium," in Source Book on Electron Beam and Laser Welding, Melvin M. Schwartz, Ed., American Society for Metals, Metals Park OH, 1981, p 87

17-46. Ti-6Al-4V EB Weldments: Base Metal Compared With Flawless Weldments

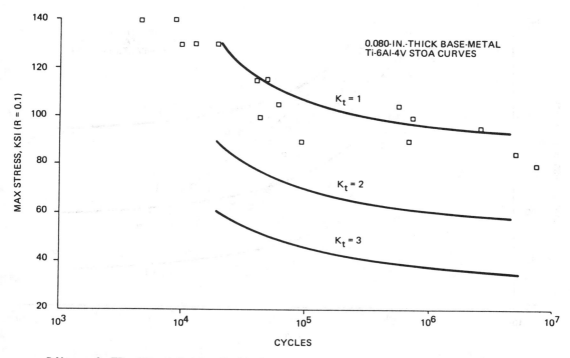

S-N curves for EB weldments that were flawless (lower two curves). Upper curve shows scatter band for base metal (0.080-in.-thick Ti-6Al-4V STOA.).

Source: R. Witt, A. Flescher and O. Paul, "Weldability and Quality of Titanium Alloy Weldments," in Titanium and Titanium Alloys: Source Book, Matthew J. Donachie, Jr., Ed., American Society for Metals, Metals Park OH, 1982, p 313

17-47. Ti-6Al-4V EB Weldments: Effects of Porosity on Cycles to Failure

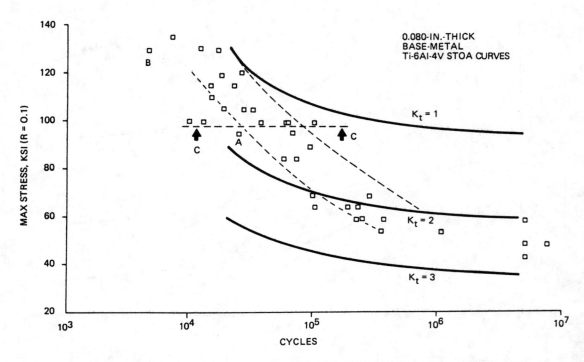

S-N curves for EB-welded Ti-6Al-4V titanium alloy showing effects of porosity.

Above are shown experimental data obtained for porosity-containing EB welds which are superimposed on a set of curves for the base material (0.080-in.-thick Ti-6Al-4V STOA sheet) at various K_t factors. For the points within the boundaries of the band, radiography indicated scattered porosity (0.003 to 0.005 in. in diameter). For points below the lower boundary of the band, radiography indicated either linear or heavily scattered porosity.

Source: R. Witt, A. Flescher and O. Paul, "Weldability and Quality of Titanium Alloy Weldments," in Titanium and Titanium Alloys: Source Book Matthew J. Donachie, Jr., Ed., American Society for Metals, Metals Park OH, 1982, p 312

17-48. Ti-6Al-4V Gas Metal-Arc Weldments: Effects of Porosity on Cycles to Failure

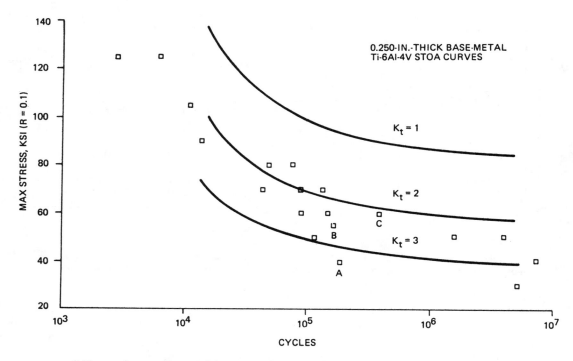

S-N curves for porosity-containing gas metal-arc welds. In the above graph the experimental fatigue data for porosity-containing GMAW weldments are superimposed on *S-N* graphs for Ti-6Al-4V STOA material (0.25 in. thick) for various K_t factors.

Source: R. Witt, A. Flescher and O. Paul, "Weldability and Quality of Titanium Alloy Weldments," in *Titanium and Titanium Alloys: Source Book*, Matthew J. Donachie, Jr., Ed., American Society for Metals, Metals Park OH, 1982, p 313

17-49. Ti-6Al-4V: Unwelded vs Electron Beam Welded Material for Cycles to Failure

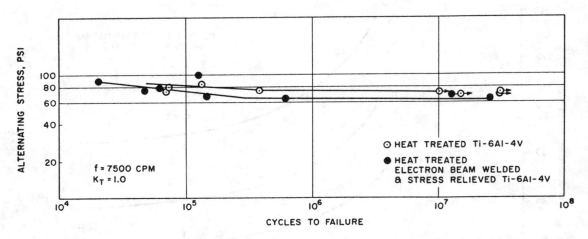

Room temperature rotating-beam fatigue life of unwelded and electron-beam-welded Ti-6Al-4V titanium alloy in fully heat-treated condition. Decrease in fatigue strength of the weldment relative to the parent metal did not exceed 12%.

Source: S. M. Silverstein, V. Strautman and W. R. Freeman, "Application of Electron Beam Welding to Rotating Gas Turbine Components," in Source Book on Electron Beam and Laser Welding, Melvin M. Schwartz, Ed., American Society for Metals, Metals Park OH, 1981, p 169

17-50. Ti-6Al-4V: S-N Diagram for Laser-Welded Sheet

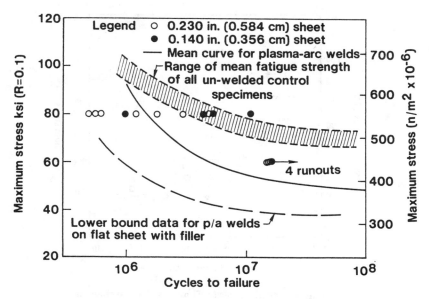

S-N diagram for laser-welded titanium alloy sheet.

The fatigue properties of welds as shown above indicate that under proper welding conditions, laser welds can be made in Ti-6Al-4V which exhibit base metal fatigue characteristics. The best laser weld failures initiated at sites in the base metal, whereas other weld failures originated at undetected small pores. Where failures initiated in the base metal, it was concluded that no porosity or weld defects of sufficient size to preferentially initiate fatigue fractures was present.

Source: E. M. Breinan, C. M. Banas and M. A. Greenfield, "Laser Welding—The Present State-of-the-Art," in Source Book on Electron Beam and Laser Welding, Melvin M. Schwartz, Ed., American Society for Metals, Metals Park OH, 1981, p 289

17-51. Ti-6Al-4V (Cast): S-N Diagram for Notched Specimens

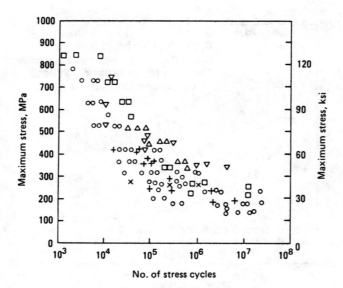

Notch fatigue strength of as-cast Ti-6Al-4V. Each symbol represents fatigue data from a different source. Stress ratio, R, typically was +0.1; stress concentration factor, K_t, was mostly 3.0, but a few tests were run at $K_t = 1.0$.

Source: Metals Handbook, 9th Edition, Volume 3, Properties and Selection: Stainless Steels, Tool Materials and Special-Purpose Metals, American Society for Metals, Metals Park OH, 1980, p 411

18-1. Zirconium 702: Effects of Notches and Testing Temperature on Cycles to Failure

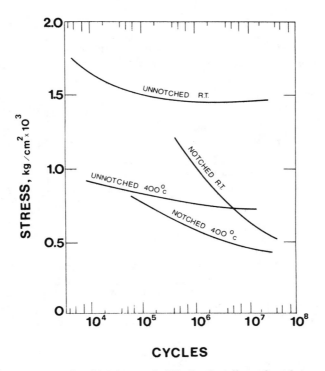

S-N curves for zirconium grade 702, showing effects of notches and elevated temperature (400 °C, or 750 °F) on fatigue characteristics.

As indicated above, zirconium and its alloys exhibit a fatigue limit behavior similar to most ferrous alloys.

Source: Donald R. Knittel, "Zirconium," in Corrosion and Corrosion Protection Handbook, Philip A. Schweitzer, Ed., Marcel Dekker, Inc., New York NY, 1983, p 198

19-1. Steel Castings (General): Effect of Design and Welding Practice on Fatigue Characteristics

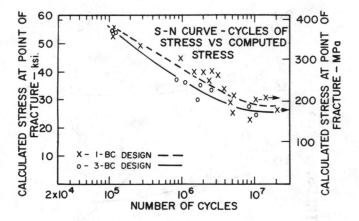

S-N curve for cast box designs.

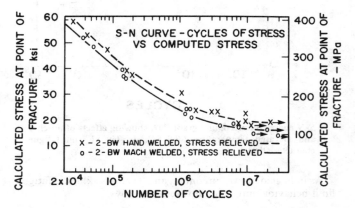

S-N curves of box weldments, comparing hand weldments with machine weldments. All weldments were stress relieved at 1100°F (593°C).

The *S-N* curves shown above indicate that: (1) the welding practice is of no great importance; and (2) the cast steel box design is superior to a weldment design.

Source: Steel Castings Handbook, 5th Edition, Peter F. Weiser, Ed., Steel Founders' Society of America, Rocky River OH, 1980, p 7-6

19-2. Steel Castings (General): Effects of Discontinuities on Fatigue Characteristics

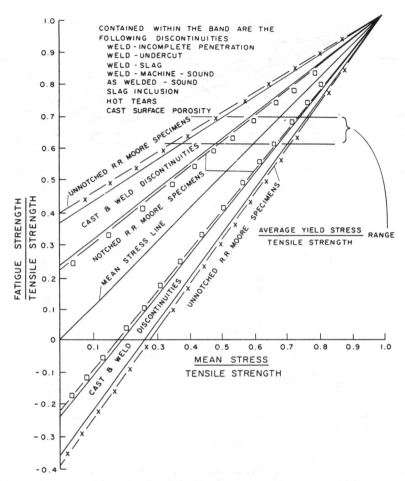

Goodman diagram for bending fatigue for normalized and tempered 8630 cast steel. (Machined notch of R. R. Moore specimen: 60° included angle, 0.0015-in. (0.0381-mm) root radius.)

Surface condition has a significant effect on fatigue life and fatigue limit. A highly polished smooth test specimen can exhibit twice the fatigue strength of a rough machined sample. A good design approach is to use the notched fatigue limit as a design value. For cast steels a 0.0015-in. (0.0381-mm) root radius circumferential notch in a rotating beam fatigue specimen reduces the fatigue limit by about 0.7 of the unnotched value. This is sufficient to account for variations in surface finish and minor surface discontinuities. The above diagram shows that even severe surface discontinuities, not normally permitted by workmanship standards, do not reduce the fatigue limit by much more than the 0.7 value.

The above emphasis on surface discontinuities is due to the fact that subsurface discontinuities which do not have a crack-like sharpness and which do not significantly reduce the load-bearing area of a component generally have little effect on fatigue performance.

Source: Steel Castings Handbook, 5th Edition, Peter F. Weiser, Ed., Steel Founders' Society of America, Rocky River OH, 1980, p 7-6

20-1. Closed-Die Steel Forgings: Effect of Surface Condition on Fatigue Limit

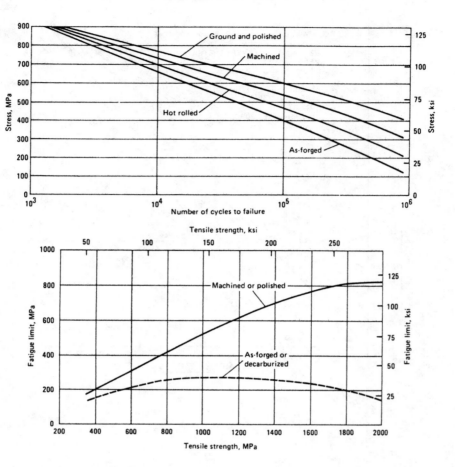

Application of small-scale laboratory fatigue testing to the analysis of components or assemblies introduces additional variables. One is the effect of surface condition. The curves in the top curve above demonstrate that the fatigue strength of steel specimens varies markedly, depending on whether the surface is polished, machined, hot rolled, or as-forged. The steel tested was an unidentified wrought low-alloy steel heat treated to 269 to 285 HB, equivalent to a tensile strength of 876 MPa (127 ksi) and a yield strength of 696 MPa (101 ksi). Sample preparation required that the specimens be machined and polished after heat treatment and that rolling or forging precede heat treatment. For a fatigue life of one million cycles, the fatigue limit was 393 MPa (57 ksi) for the ground specimens, 317 MPa (46 ksi) for the machined specimens, 207 MPa (30 ksi) for the as-rolled specimens, and only 152 MPa (22 ksi) for the as-forged specimens.

The curves in the bottom graph apply to steels with tensile strength ranging from 345 to 2070 MPa (50 to 300 ksi) and are approximations from several independent investigations. Sample preparation for "as-forged or decarburized" specimens at the 965 MPa (140 ksi) tensile-strength level include 4140-type steels rough machined from bar stock, heated to approximately 900 °C (1650 °F) in a gas-fired muffle for 20 to 30 min, very lightly swaged from an original 7.47-mm (0.294-in.) diameter to a final diameter of 7.16 mm (0.282 in.), and air cooled. Heat treatment consisted of austenitizing in a salt bath at approximately 830 °C (1525 °F) for 45 min, oil quenching, tempering in air for 1 h at approximately 620 °C (1150 °F), and water quenching. Forging and heat treating produced a surface decarburized to a depth of about 0.06 mm (0.0025 in.). These specimens exhibited a fatigue strength, at 10^6 cycles, of about 310 MPa (45 ksi), compared with 470 MPa (68 ksi) for samples that were not forged but were machined or polished and free of decarburization. Decarburization lowers the strength levels obtained by heat treatment.

Source: Metals Handbook, 9th Edition, Volume 1, Properties and Selection: Irons and Steels, American Society for Metals, Metals Park OH, 1978, p 355

21-1. P/M: Relation of Density to Fatigue Limit and Fatigue Ratio

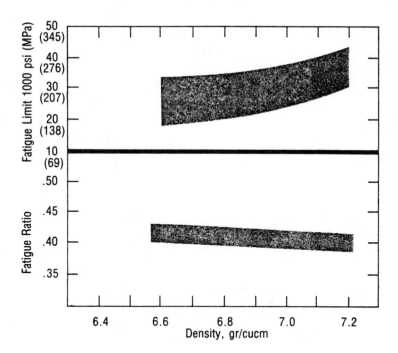

The relationship of fatigue strength to density is shown above. Fatigue strength is best at high densities. For similar P/M and wrought parts, the ultimate tensile strength to fatigue strength ratios are the same. However, fatigue strengths of P/M parts generally are more stable and uniform than for wrought parts. Parts containing nickel show improved fatigue resistance compared to iron-carbon steels, and high-density nickel steel parts can be case hardened to improve wear and fatigue properties.

Source: Kurt H. Miska, "Powder Metal Parts," in Source Book on Powder Metallurgy, Samuel Bradbury, Ed., American Society for Metals, Metals Park OH, 1979, p 3

21-2. P/M: Relation of Fatigue Limit to Tensile Strength for Sintered Steels

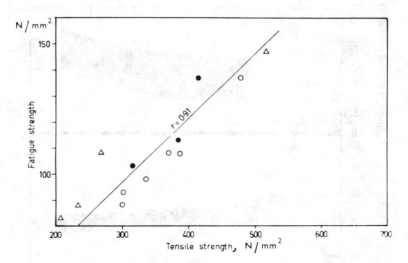

Fatigue limit of different sintered steels as a function of tensile strength. Triangles are values for materials without phosphorus; open circles correspond to PNC materials, closed circles to PASC materials.

Source: Per Lindskog, "The Effect of Phosphorus Additions on the Tensile, Fatigue, and Impact Strength of Sintered Steels Based on the Sponge Iron Powder and High-Purity Atomized Iron Powder," in Source Book on Powder Metallurgy, Samuel Bradbury, Ed., American Society for Metals, Metals Park OH, 1979, p 46

21-3. P/M (Nickel Steels): As-Sintered vs Quenched and Tempered for Cycles to Failure

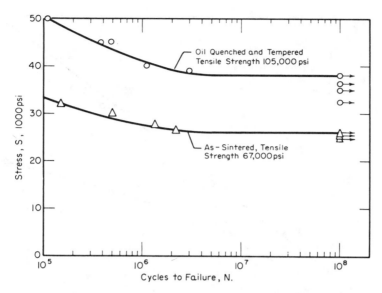

S-N diagrams representing fatigue behavior of 7.0 g/cm^3 density, 4 Ni-0.48 C steels, and effect of quenching and tempering on tensile and fatigue strength.

One of the characteristics of the fatigue behavior of wrought steels is that the *S-N* curve usually shows a distinct fatigue limit. This is most marked in wrought plain carbon steels and usually occurs between 10^5 and 10^7 cycles. A typical *S-N* curve for an as-sintered nickel steel is shown above. As-sintered nickel steels possess distinct fatigue limits occurring between 10^6 and 10^8 cycles.

Source: A. F. Kravic and D. L. Pasquine, "Fatigue Properties of Sintered Nickel Steels," in Source Book on Powder Metallurgy, Samuel Bradbury, Ed., American Society for Metals, Metals Park OH, 1979, p 28

21-4. P/M (Nickel Steels): Relation Between Fatigue Limit and Tensile Strength for Sintered Steels

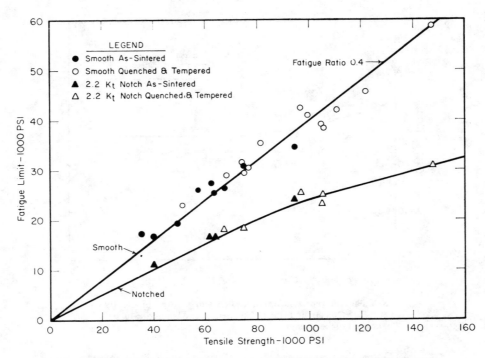

Relation between fatigue limit and tensile strength (fatigue ratio) of sintered nickel steels.

A plot of the fatigue ratio (above) indicates an average smooth value of 0.4 up to 150,000 psi tensile strength. Thus the average fatigue ratio for sintered nickel steel is 0.4 which is apparently independent of density level, alloy content, and state of heat treatment and therefore can be used to predict the fatigue behavior of other sintered nickel steels.

Source: A. F. Kravic and D. L. Pasquine, "Fatigue Properties of Sintered Nickel Steels," in Source Book on Powder Metallurgy, Samuel Bradbury, Ed., American Society for Metals, Metals Park OH, 1979, p 30

21-5. P/M (Nickel Steels): Effect of Notches on Cycles to Failure for the As-Sintered Condition

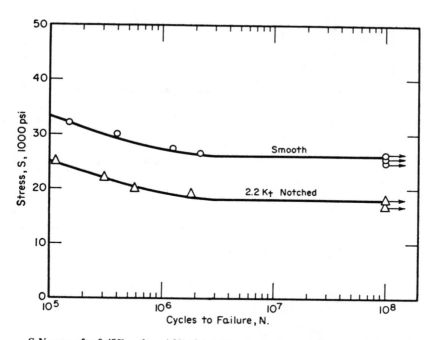

S-N curves for 0.48% carbon-4.0% nickel alloy steel in the as-sintered condition (7.0 g/cm³ density). The two curves demonstrate the effect of a notch on fatigue characteristics.

Source: A. F. Kravic and D. L. Pasquine, "Fatigue Properties of Sintered Nickel Steels," in Source Book on Powder Metallurgy, Samuel Bradbury, Ed., American Society for Metals, Metals Park OH, 1979, p 33

21-6. P/M (Nickel Steels): Effect of Notches on Cycles to Failure for the Quenched and Tempered Condition

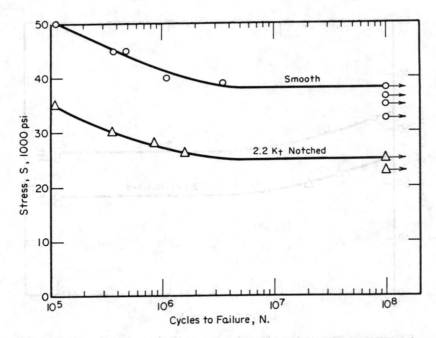

S-N curves for 0.48% carbon-4.0% nickel alloy steel in the quenched and tempered condition (7.0 g/cm³ density). The two curves demonstrate the effect of a notch on fatigue characteristics.

Source: A. F. Kravic and D. L. Pasquine, "Fatigue Properties of Sintered Nickel Steels," in Source Book on Powder Metallurgy, Samuel Bradbury, Ed., American Society for Metals, Metals Park OH, 1979, p 34

21-7. P/M (Low-Carbon, 1-5%Cu): Effects of Notches and Nitriding on Cycles to Failure

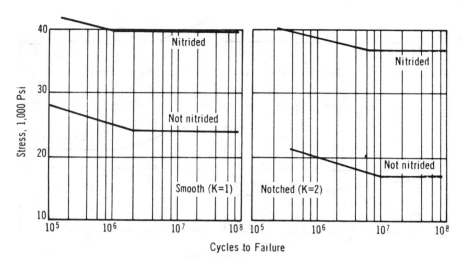

S-N curves for sintered powders (low-carbon; 1 to 5% copper, 7.1 g/cm³ density). As shown above, notches greatly lower fatigue strength, particularly of those that were not nitrided.

Source: "Nitriding Improves Fatigue Resistance of P/M Parts," in Source Book on Nitriding, American Society for Metals, Metals Park OH, 1977, p 292

21-8. P/M (Sintered Iron, Low-Carbon, No Copper): Effect of Density and Nitriding on Cycles to Failure

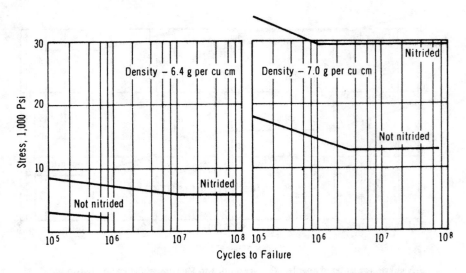

S-N curves for P/M parts. As shown above, the greater the density, the higher the fatigue strength of sintered iron powder (low carbon; no copper; notched; $K = 2$). Nitriding in a salt bath is especially beneficial, it will be noted. Bath temperature was 565 °C (1050 °F); nitriding time was two hours.

Source: "Nitriding Improves Fatigue Resistance of P/M Parts," in Source Book on Nitriding, American Society for Metals, Metals Park OH, 1977, p 292

21-9. P/M: Effect of Nitriding on Ductile Iron and Sintered Iron (3%Cu) for Cycles to Failure

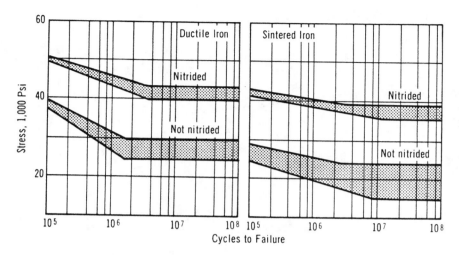

Left: Effect of salt bath nitriding on fatigue strength of ductile iron. Right: Effect of salt bath nitriding on P/M parts. Specimens were made from a 3% copper sintered iron ranging from 6.2 to 7.0 g/cm³ in density. All specimens were unnotched and were heated in a nitriding salt at 565 °C (1050 °F) for two hours.

Source: "Nitriding Improves Fatigue Resistance of P/M Parts," in Source Book on Nitriding, American Society for Metals, Metals Park OH, 1977, p 291

22-1. Brass/Mild Steel Composite: Comparison of Brass-Clad Mild Steel With Brass and Mild Steel for Cycles to Failure

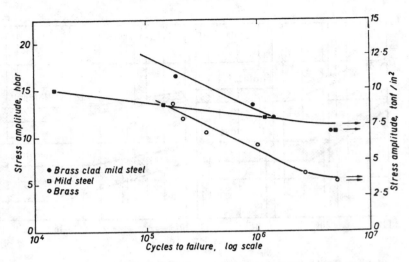

S-N curves for composite of mild steel clad (by the explosion technique) with brass.

Initially, the composite has greater fatigue strength than either brass or mild steel alone, but above about 10^6 cycles the values for the composite drop to about that of mild steel but still remain substantially higher than for brass alone.

Source: S. K. Banerjee and B. Crossland, "Mechanical Properties of Explosively-Cladded Plates," in Source Book on Innovative Welding Processes, Melvin M. Schwartz, Ed., American Society for Metals, Metals Park OH, 1981, p 148

22-2. Stainless Steel/Mild Steel Composite: Comparison of Stainless-Clad Mild Steel With Stainless Steel and Mild Steel for Cycles to Failure

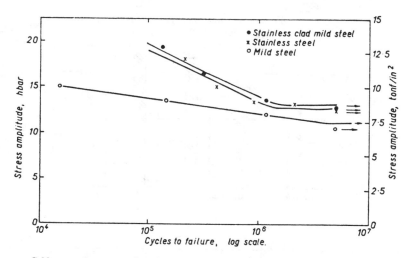

S-N curves for composite of mild steel clad (by the explosion technique) with austenitic stainless steel. Here it is seen that fatigue characteristics of the composite are nearly the same as for stainless steel, and substantially higher than the fatigue strength of the unclad mild steel.

Source: S. K. Banerjee and B. Crossland, "Mechanical Properties of Explosively-Cladded Plates," in Source Book on Innovative Welding Processes, Melvin M. Schwartz, Ed., American Society for Metals, Metals Park OH, 1981, p 148

23-1. Carbon and Alloy Steels (Seven Grades): Effects of Nitrocarburizing on Fatigue Strength

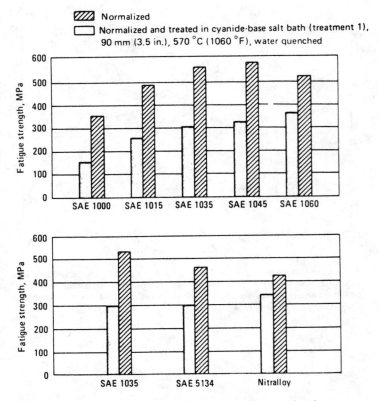

Bar charts showing increases in fatigue limit that may be obtained by nitrocarburizing (gas or liquid processes).

The amount of improvement in fatigue strength of nitrocarburized materials, as determined with unnotched Wöhler test specimens, depends on the hardness and depth of the diffusion zone. The potential for improvement in fatigue strength lessens with increasing carbon and alloy content.

Source: Metals Handbook, 9th Edition, Volume 4, Heat Treating, American Society for Metals, Metals Park OH, 1981, p 269

23-2. Carbon and Alloy Steels (Seven Grades): Effects of Tufftriding on Fatigue Characteristics

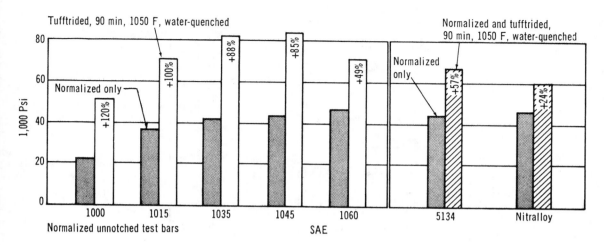

Basic fatigue characteristics are directly related to carbon content, as indicated in the above bar charts for carbon and alloy steels (unnotched test bars). Tufftriding these steels shows results which prove that fatigue strength increases inversely with carbon content; that is, the lower the carbon, the greater percentage increase in fatigue strength by Tufftriding.

Source: Edward Taylor, "Tufftride: Only Skin Deep?," in Source Book on Nitriding, American Society for Metals, Metals Park OH, 1977, p 280

23-3. Carbon and Alloy Steels (Six Grades): Effects of Nitriding on Fatigue Strength

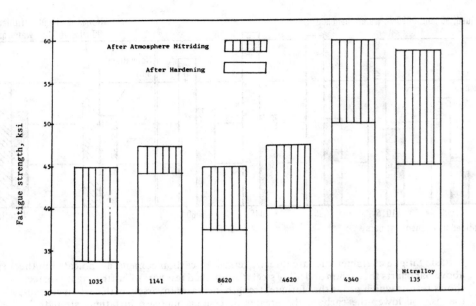

Atmosphere nitriding studies showing the interrelationships of steel composition and nitriding by the gas process, and the effect on fatigue strength from nitriding.

Source: J. A. Riopelle, "Short Cycle Atmosphere Nitriding," in Source Book on Nitriding, American Society for Metals, Metals Park OH, 1977, p 286

23-4. Carbon-Manganese Steel: Effects of Nickel Coating on Fatigue Strength

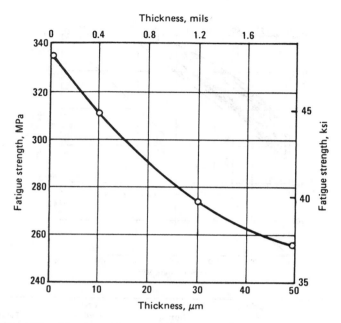

Effect of coating thickness on the fatigue strength of a carbon-manganese steel.

The reduction in fatigue strength produced by electroless nickel deposits is affected by the thickness of the coating. Thicker deposits have the greatest effect on fatigue strength. This is illustrated in the above graph, which shows the reduction in strength of a carbon-manganese steel (Werkstoff St52) produced by different thicknesses of a 5% boron-nickel.

Source: Metals Handbook, 9th Edition, Volume 5, Surface Cleaning, Finishing, and Coating, American Society for Metals, Metals Park OH, 1982, p 232

24-1. Coil Springs, Music Wire (Six Sizes): Data Presented by Means of a Goodman Diagram

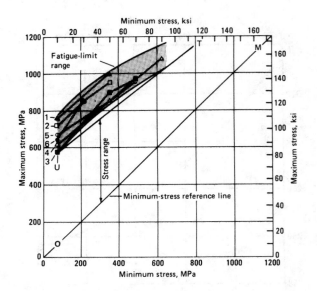

Spring No.	Wire diam		Spring OD		Spring index	Free length		Total turns	Active turns	Total tested
	mm	in.	mm	in.		mm	in.			
1	0.81	0.032	9.52	0.375	10.7	22.10	0.87	6.0	4.2	16
2	0.81	0.032	6.35	0.250	6.8	26.97	1.062	7.0	5.2	28
3	1.22	0.048	15.88	0.625	12.0	44.45	1.75	7.0	5.2	38
4	2.59	0.102	22.22	0.875	7.6	60.20	2.37	7.0	5.2	43
5	3.07	0.121	22.22	0.875	6.2	57.15	2.25	7.5	5.7	35
6	4.50	0.177	22.22	0.875	4.9	57.15	2.25	7.5	5.7	25

Data are average fatigue limits from S-N curves for 185 unpeened springs of various wire diameters run to 10 million cycles of stress. All stresses were corrected for curvature using the Wahl correction factor. The springs were automatically coiled, with one turn squared on each end, then baked at 260 °C (500 °F) for 1 h, after which the ends were ground perpendicular to the spring axis. The test load was applied statically to each spring and a check made for set three times before fatigue testing. The springs were all tested in groups of six on the same fatigue testing machine at ten cycles per second. After testing, the unbroken springs were again checked for set and recorded. Number 4 springs, tested at 1070 MPa (155 ksi) max stress, had undergone about 2¼% set after 10 million stress cycles, but the stresses were not recalculated to take this into account. None of the other springs showed appreciable set. The tensile strengths of the wires were according to ASTM A228.

By means of the Goodman diagram many fatigue-limit test results can be shown on the same diagram as indicated above. In this diagram, line OM represents the minimum stress for the cycle; the plotted points represent fatigue limits for the respective minimum stresses used. The vertical distances between these points and the minimum stress reference line represent the stress ranges. Some scatter may be expected, at least partly attributed to normal changes of tensile strength with wire diameter. Line UT is usually drawn to intersect line OM at the average ultimate shear strength of the various sizes of wire.

Source: Metals Handbook, 9th Edition, Volume 1, Properties and Selection: Irons and Steels, American Society for Metals, Metals Park OH, 1978, p 293

24-2. Coil Springs: *S-N* Data for Oil-Tempered and Music Wire Grades

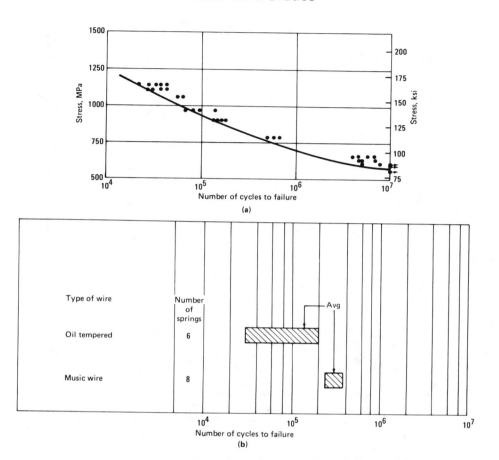

(a) Springs were made of minimum quality music wire 0.59 mm (0.022 in.) in diameter. Spring diameter was 5.21 mm (0.205 in.); D/d was 8.32. Minimum stress was zero. Stresses corrected by Wahl factor. (b) Life of springs used in a hydraulic transmission. They were made of oil-tempered wire (ASTM A229) and music wire (A228). Wire diameter was 4.75 mm (0.187 in.), outside diameter of spring was 44.45 mm (1.750 in.), with 15 active coils in each spring. The springs were fatigue tested in a fixture at a stress of 605 MPa (88 ksi), corrected by the Wahl factor.

The upper graph is a typical *S-N* diagram showing results of compression testing coil springs, where the minimum stress is zero while maximum stress is shown by points on the chart (see spring and testing details given in caption). The lower graph shows an alternate method of presenting fatigue data for steel springs.

Source: Metals Handbook, 9th Edition, Volume 1, Properties and Selection: Irons and Steels, American Society for Metals, Metals Park OH, 1978, p 292

24-3. Coil Springs: Effects of Shot Peening on Cycles to Failure

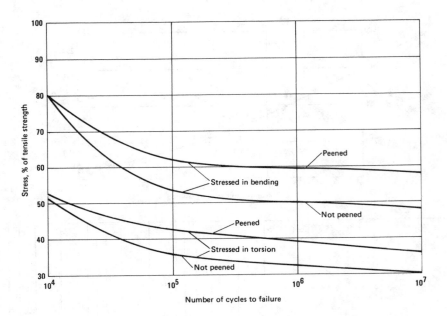

Shot peening is often used to improve fatigue strength of springs by pre-stressing the surface in compression as indicated in the chart above. Shot peening can be applied to wire 1.6 mm (1/16 in.) or more in diameter, and slightly smaller wire using special techniques. The kind of shot used is important; better results are obtained with carefully graded shot having only a few broken, angular particles. Shot size may be optimum at roughly 20% of the wire diameter. However, for larger wire, it has been found that excessive roughening during peening with coarse shot lessens the benefits of peening, apparently by causing minute fissures. Also, peening too deeply leaves little material in residual tension in the core; this negates the beneficial effect of peening, which requires internal tensile stress to balance the surface compression.

Shot peening is effective in largely overcoming the stress-raising effects of shallow pits and seams. Proper peening intensity is an important factor, but more important is the need for both the inside and outside surfaces of the spring to be thoroughly covered. An Almen test strip necessarily receives the same exposure as the outside of the spring, but to reach the inside, the shot must pass between the coils and is thereby much restricted. Thus, for springs with closely spaced coils, a coverage of 400% on the outside may be required to achieve 90% coverage on the inside.

Cold wound steel springs normally are stress relieved after peening to restore the yield point. A temperature of 230 °C (450 °F) is common because higher temperatures degrade or eliminate the improvement in fatigue strength.

The extent of improvement in fatigue strength to be gained by shot peening, according to one prominent manufacturer of cold wound springs, is shown in the above diagram.

Source: Metals Handbook, 9th Edition, Volume 1, Properties and Selection: Irons and Steels, American Society for Metals, Metals Park OH, 1978, p 297

24-4. Coil Springs, 8650 and 8660 Steels: Relation of Design Stresses and Probability of Failure

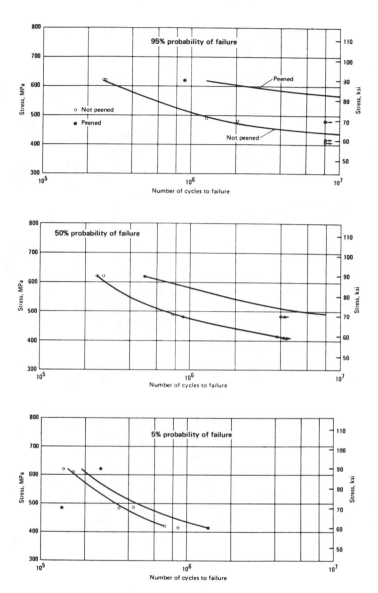

Design stresses. Springs were made from 15.9 to 27.0-mm (⅝ to 1-1/16-in.) diam 8650 and 8660 hot rolled steel and heat treated to between 429 and 444 HB. Springs were shot peened to an average arc height of 0.008 in. on the type C almen strip at 90% visual coverage.

The desirability of conservative design in cyclical service is illustrated in the three charts above, in which the minimum stress used was low. Such data on springs hot wound from bars with as-rolled surfaces are limited, and interpretation is therefore difficult. The value of peening, however, is made quite apparent. Pertinent test data are given above.

Source: Metals Handbook, 9th Edition, Volume 1, Properties and Selection: Irons and Steels, American Society for Metals, Metals Park OH, 1978, p 304

24-5. Coil Springs, HSLA Steels: Effects of Corrosion on Cycles to Failure

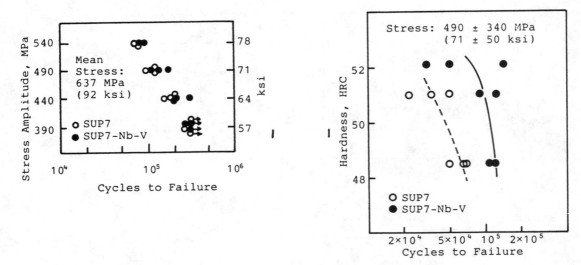

Fatigue life of coil springs: (left) not corroded and (right) corroded.

Compositions of HSLA Springs Tested

	C	Si	Mn	P	S	Cu	Cr	Al	Nb	V
SUP7	0.58	2.09	0.83	0.014	0.008	0.09	0.14	0.025	0	0
SUP7-Nb-V	0.56	1.94	0.79	0.014	0.008	0.09	0.09	0.021	0.15	0.18

Fatigue tests on coil springs at a hardness of 50 HRC were performed to examine the feasibility of SUP7-Nb-V to the actual suspension coil springs. When the coil springs were free of corrosion, the result was as shown above (left), in which SUP7-Nb-V has comparable fatigue life to that of SUP7 in any stress amplitude. When the coil springs were corroded, on the other hand, the result was rather different. The corroding condition was as follows: an exposure in a chamber filled with saltwater mist for 1.08×10^4 s (3 h) and a keeping in the atmosphere for 7.56×10^4 s (21 h). After 10 and 20 cycles of the corroding, the coil springs were loaded with the surface stress of 490 ± 340 MPa (71 ± 50 ksi). The fatigue life of the coil springs subjected to 20 cycles of the corroding are shown above (right). This time, different from the case in the graph at left, there appears a remarkable difference between SUP7 and SUP7-Nb-V. Measurement of the surface corrosion depth of the two steels showed no difference.

Source: Toshiro Yamamoto, Ryohei Kobayashi, Toshio Ozone and Mamoru Kurimoto, "Precipitation Strengthened Spring Steel for Automotive Suspensions," in HSLA Steels—Technology & Applications, American Society for Metals, Metals Park OH, 1984, p 1022

24-6. Leaf Springs, 5160 Steel: Maximum Applied Stress vs Cycles to Failure

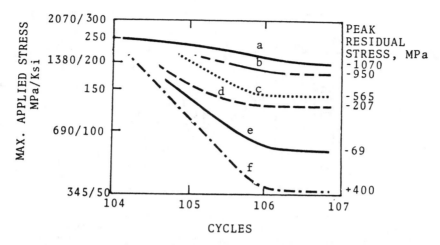

Residual stress and unidirectional bending fatigue data for strain-peened SAE 5160 steel. Applied strain during peening curve a, -0.60%; curve b, -0.30%; curve c, 0% (conventional peening); curve d, preset only; curve e, $+0.30\%$; and curve f, $+0.60\%$.

Leaf spring specimens of SAE 5160 steel quenched and tempered to 48 HRC were shot peened under various conditions of applied strain to introduce a wide range of residual stress; then the S-N curves (see above) were obtained from the same samples by testing in unidirectional bending. In this illustration, the endurance limit corresponding to the specimen strain-peened to produce a residual stress of -565 MPa (-82 ksi) will be used to develop a stress-free ASR diagram for 5160 spring steel (48 HRC). This stress-free ASR diagram will be used to predict the endurance limit for the other specimens containing peak stress values of -1070 (-155), -950 (-138), -207 (-30), -69 (-10), and $+400$ MPa ($+58$ ksi). The predicted endurance limit will be compared with the values determined experimentally.

Source: V. K. Sharma and D. H. Breen, "Some Aspects of Incorporating Residual Stresses in Gear Design," in Residual Stress for Designers and Metallurgists, Larry J. Vande Walle, Ed., American Society for Metals, Metals Park OH, 1981, p 82

24-7. Front Suspension Torsion Bar Springs, 5160H Steel: Distribution of Fatigue Results for Simulated Service Testing

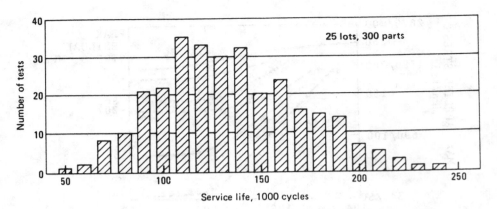

Here are results from simulated service fatigue tests of front suspension torsion bar springs of 5160H steel. Size of hexagonal bar section was 32 mm (1.25 in.). Mean service life, 134,000 cycles; standard deviation, 37,000 cycles; coefficient of variations, 0.28. It must always be considered that results from actual or simulated service testing are likely to vary considerably from results of laboratory testing as shown above.

Source: Metals Handbook, 9th Edition, Volume 1, Properties and Selection: Irons and Steels, American Society for Metals, Metals Park OH, 1978, p 677

24-8. Gears, Carburized Low-Carbon Steel: Relation of Life Factor to Required Life

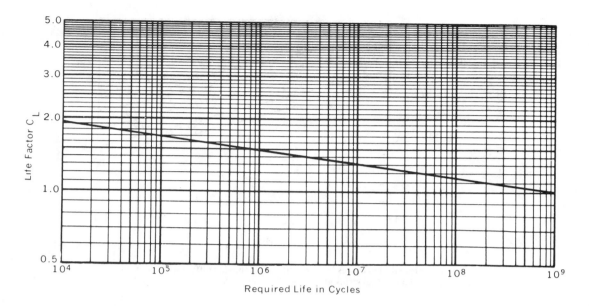

The life factor depends on the required life in cycles. For a single mesh the number of revolutions and the number of cycles are equal. For a gear which has more than one mating member, the life must be equal to the required number of revolutions multiplied by the number of mating gears.

Source: "Bending and Contact Stresses in Hypoid Gear Teeth," in Source Book on Gear Design, Technology and Performance, Maurice A. H. Howes, Ed., American Society for Metals, Metals Park OH, 1980, p 127

24-9. Gears, Carburized Low-Carbon Steel: Bending Stress vs Cycles to Failure

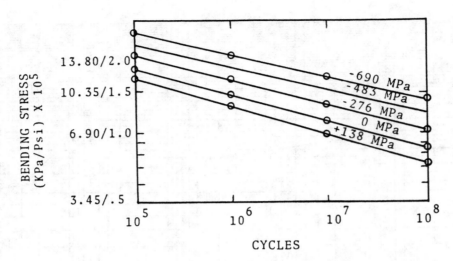

Bending fatigue design curves for carburized gears having different amounts of circumferential residual stress at the root-fillet surface.

The bending fatigue design curves for case-carburized gears with the circumferential root-fillet residual stress varying from +138 MPa (20 ksi) to −690 MPa (100 ksi) are shown above. From these curves the residual stress factors at various life cycles were calculated as the ratio of the allowable bending stress for gears with −483 MPa (70 ksi) residual compression to the allowable stress for gears with +138 (20), 0, −276 (40), and −690 MPa (100 ksi) residual stresses.

Source: V. K. Sharma and D. H. Breen, "Some Aspects of Incorporating Residual Stresses in Gear Design," in Residual Stress for Designers and Metallurgists, Larry J. Vande Walle, Ed., American Society for Metals, Metals Park OH, 1981, p 86

24-10. Gears, Carburized Low-Carbon Steel: Effect of Shot Peening on Cycles to Failure

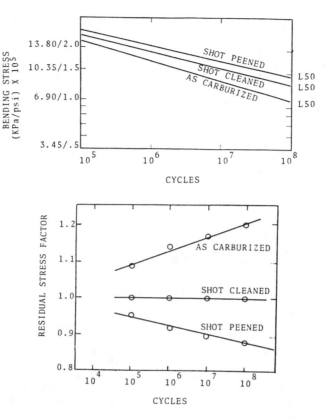

Top: Allowable (L50) bending stress design curves for as-carburized, shot cleaned, and shot peened carburized gears. Bottom: Residual stress factor computed from the upper chart. $K\sigma$ for carburized shot cleaned gears equals 1; that is, the allowable S-N curve for carburized shot cleaned gears is used for design purposes.

$K\sigma$ Based on Dynamometer Tests

The dynamometer test data obtained on testing sets of gears with different magnitudes of residual stresses can be used to develop the S-N curves necessary to calculate the $K\sigma$ factor. The L50 design curves for as-carburized, shot cleaned, and shot peened gears are shown in the upper chart. The data for as-carburized and shot cleaned gears were obtained on testing six-pitch test pinions on a Four Square Dynamometer. The S-N curve for shot peened gears is derived from the results published by Alman and Black. The lower chart shows the residual stress factors calculated from the S-N curves in the upper chart. It is assumed that the S-N curve for shot cleaned gears is used for design purposes; that is, $K\sigma$ for shot cleaned gears equals one. According to these results the effective bending stress for shot cleaned gears at 10^8 cycles is approximately 20% higher than as-carburized gears and approximately 15% lower than shot peened gears. The value of $K\sigma$ deviates from unity with increasing cycles, indicating a more significant effect of residual stress at higher life cycles. At low cycles, the residual stress factor seems to approach one, which means the residual stress has almost no influence on the fatigue properties of a material at high loads. This is obviously because of the stress relaxation caused by the cyclic plastic deformation accompanying low-cycle fatigue.

Source: V. K. Sharma and D. H. Breen, "Some Aspects of Incorporating Residual Stresses in Gear Design," in Residual Stress for Designers and Metallurgists, Larry J. Vande Walle, Ed., American Society for Metals, Metals Park OH, 1981, pp 77, 78

24-11. Gears, Carburized Low-Carbon Steel: Probability-Stress-Life Design Curves

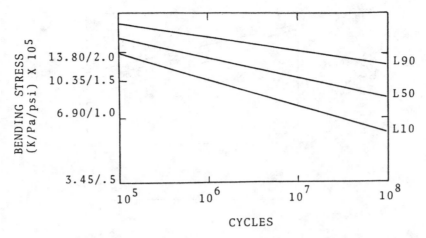

Probability-stress-life design curves for shot cleaned carburized steel gears having a root-fillet surface finish of 5μ in.; i.e., $K\mu = 1$.

Bending Stresses in Gears

In designing gears for a new application, a designer usually begins with a preliminary selection of the tooth widths and other design parameters based on past practices and empirical approaches recommended by AGMA. The applied root-fillet bending stress is then calculated to predict the gear life from the stress-life design curves such as those shown above. The procedure is reiterated to optimize the design so that the calculated life is just equal to the required life with an appropriate level of safety. The allowable stress-life diagram characteristic for each material, heat treatment, and surface treatment is usually obtained on testing acceptable commercial quality gears on a dynamometer.

Source: V. K. Sharma and D. H. Breen, "Some Aspects of Incorporating Residual Stresses in Gear Design," in Residual Stress for Designers and Metallurgists, Larry J. Vande Walle, Ed., American Society for Metals, Metals Park OH, 1981, p 74

24-12. Gears, 8620H Carburized: Bending or Contact Stress vs Cycles to Fracture or Pitting

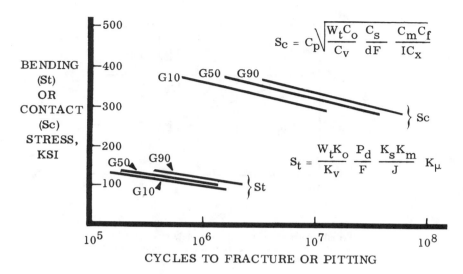

S-N curves showing the wide difference in cycles to failure between bending and contact stress.

Source: D. H. Breen, "Fundamental Aspects of Gear Strength Requirements," in Source Book on Gear Design, Technology and Performance, Maurice A. H. Howes, Ed., American Society for Metals, Metals Park OH, 1980, p 66

24-13. Gears, 8620H Carburized: A Weibull Analysis of Bending Fatigue Data

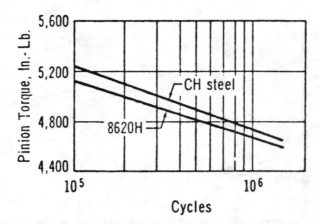

Weibull analysis of bending fatigue data from gear tests indicates that gears made from either the experimental CH steel or 8620H have equivalent durability.

Metallurgical data gathered on these gears established the adequacy of the experimental steel (a CHAT steel—computer harmonizing by application tailoring). Although the experimental steel had a significantly lower case hardenability, it quenched out to a 100% martensite plus austenitic structure at the root-fillet surface. Obviously, it had adequate, though not excessive, case hardenability, thus representing an efficient use of alloy hardenability in CHAT steels.

Source: G. H. Walter, "Computer Oriented Gear Steel Design Procedure," in Source Book on Gear Design, Technology and Performance, Maurice A. H. Howes, Ed., American Society for Metals, Metals Park OH, 1980, p 85

24-14. Gears, 8620H Carburized: *T-N* Curves for Six-Pinion, Four-Square Tests

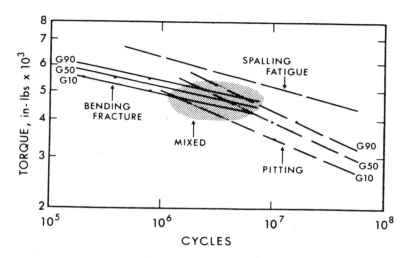

T-N curves for carburized six-pitch pinion, four-square gear tests.

The above fatigue data show torque versus cycles to breakage, pitting and spalling for a six-pitch pinion test. Note that there is a mixed area where failure can occur from any one or a mixture of the three modes.

Source: D. H. Breen, "Fundamental Aspects of Gear Strength Requirements," in Source Book on Gear Design, Technology and Performance, Maurice A. H. Howes, Ed., American Society for Metals, Metals Park OH, 1980, p 66

24-15. Hypoid Gears, 8620H Carburized: Minimum Confidence Level; Stress vs Cycles to Rupture

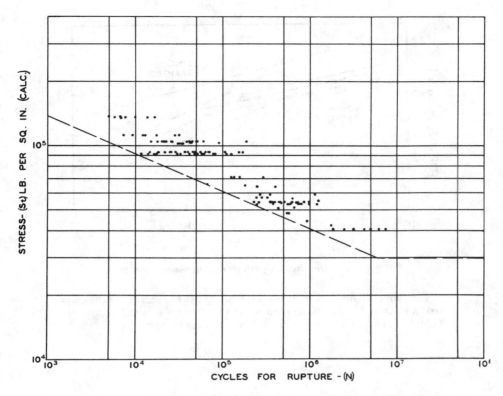

Fatigue life data for hypoid gears. Sloping line indicates minimum confidence level.

Source: "Gleason Method for Estimating the Fatigue Life of Bevel Gears and Hypoid Gears," in Source Book on Gear Design, Technology and Performance, Maurice A. H. Howes, Ed., American Society for Metals, Metals Park OH, 1980, p 386

24-16. Hypoid, Zerol and Spiral Bevel Gears, 8620H Carburized: S-N Scatter Band and Minimum Confidence Level

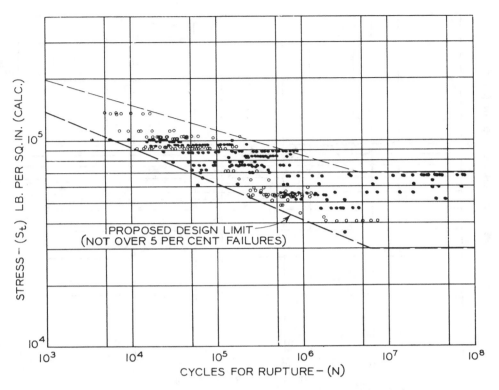

Fatigue data—composite for results obtained by testing various gear designs.

Source: "Gleason Method for Estimating the Fatigue Life of Bevel Gears and Hypoid Gears," in Source Book on Gear Design, Technology and Performance, Maurice A. H. Howes, Ed., American Society for Metals, Metals Park OH, 1980, p 386

24-17. Spiral Bevel and Zerol Bevel Gears, 8620H Carburized: S-N Scatter Band and Minimum Confidence Level

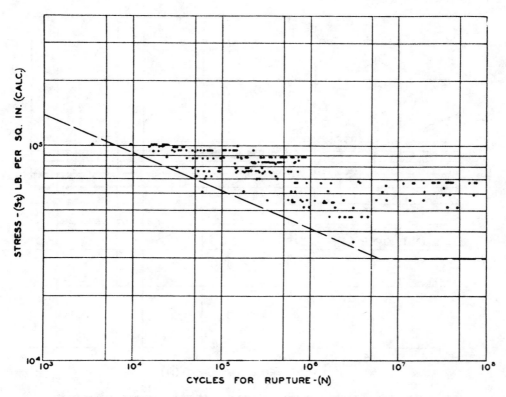

Graph of fatigue life for spiral bevel and Zerol bevel gears. Sloping line indicates the minimum confidence level.

Source: "Gleason Method for Estimating the Fatigue Life of Bevel Gears and Hypoid Gears," in Source Book on Gear Design, Technology and Performance, Maurice A. H. Howes, Ed., American Society for Metals, Metals Park OH, 1980, p 385

24-18. Gears, 8620H Case Hardened: Relation of Life Factor to Cycles to Rupture

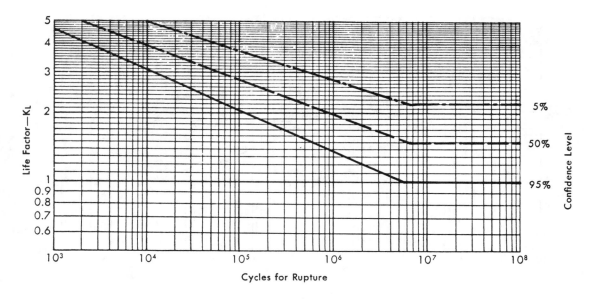

Both strength and durability are fatigue phenomena and therefore display a relationship between stress and life. The life factor for strength may be obtained from the above data.

Source: "Bending and Contact Stresses in Hypoid Gear Teeth," in Source Book on Gear Design, Technology and Performance, Maurice A. H. Howes, Ed., American Society for Metals, Metals Park OH, 1980, p 127

24-19. Bevel Gears, Low-Carbon Steel Case Hardened: Relation of Life Factor to Cycles to Rupture for Various Confidence Levels

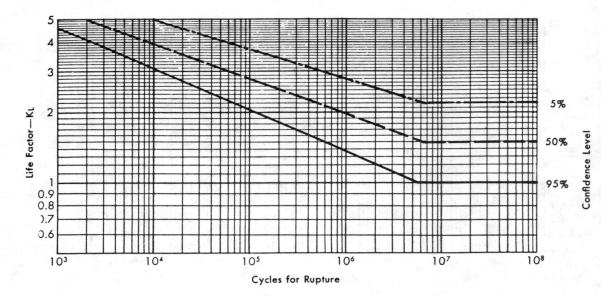

The life factor is obtained from the graph above. This depends upon the required life in cycles. For a single mesh the number of revolutions and the number of cycles are the same. For a gear which has more than one mating member the life must be equal to the required number of revolutions multiplied by the number of mating gears. When the required life is less than 6,000,000 cycles on the pinion, the life factors will be different on gear and mating pinion.

In cases where the load varies, the designer may wish to determine the equivalent life at maximum torque. One suggested method is as follows:

$$L_{CP} = 60 L_H \left[k_1 n_{P1} + k_2 n_{P2} \left(\frac{T_2}{T_1} \right)^{5.68} + k_3 n_{P3} \left(\frac{T_3}{T_1} \right)^{5.68} + \ldots + k_n n_{Pn} \left(\frac{T_n}{T_1} \right)^{5.68} \right]$$

where L_{CP} = required equivalent life in pinion cycles at maximum torque.
L_H = required total life in hours.
$k_1, k_2, k_3 \ldots k_n$ = proportion of time at torque loads $T_1, T_2, T_3 \ldots T_n$ respectively.
$n_{P1}, n_{P2}, n_{P3} \ldots n_{Pn}$ = pinion rpm corresponding to torque loads $T_1, T_2, T_3 \ldots T_n$ respectively.
$T_1, T_2, T_3 \ldots T_n$ = torque loads where T_1 is maximum torque and T_n is minimum torque which will produce a stress above the endurance limit.

The required equivalent life in gear cycles at maximum torque may be obtained by multiplying the life in pinion cycles by the gear ratio:

$$L_{CG} = L_{CP} \frac{N_P}{N_G}$$

Source: "Bending Stresses in Bevel Gear Teeth," in Source Book on Gear Design, Technology and Performance, Maurice A. H. Howes, Ed., American Society for Metals, Metals Park OH, 1980, p 149

24-20. Gears, AMS 6265: S-N Data for Cut vs Forged

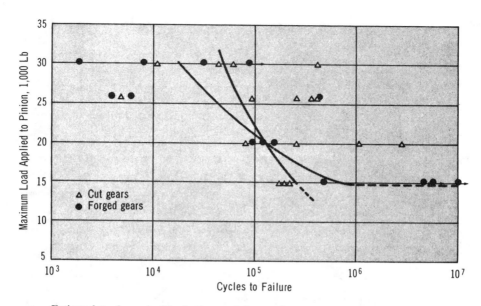

Fatigue data shown in this chart proved that teeth on precision forged AMS 6265 helicopter pinions have a higher fatigue limit than cut teeth. Loads shown are applied actuator loads. Tooth loads are approximately 33% greater.

Source: "How Gearmaking Methods Compare," in Source Book on Powder Metallurgy, Samuel Bradbury, Ed., American Society for Metals, Metals Park OH, 1979, p 347

24-21. Spur Gears, 8620H: S-N Data for Cut vs Forged

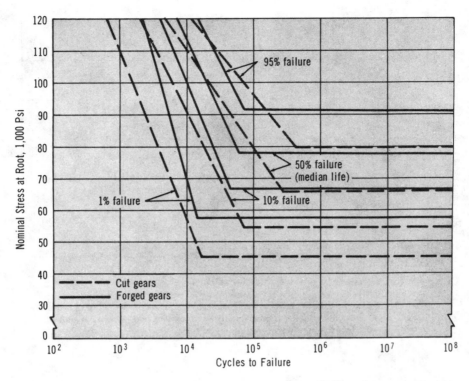

S-N curves for spur gears forged and cut from 8620H steel.

As shown above, results of beam fatigue tests indicate that precision flow-forged gear teeth are about 20% higher in fatigue strength than cut teeth.

Source: "How Gearmaking Methods Compare," in Source Book on Powder Metallurgy, Samuel Bradbury, Ed., American Society for Metals, Metals Park OH, 1979, p 346

24-22. Gears and Pinions: P/M 4600V vs 4615; Weibull Distributions

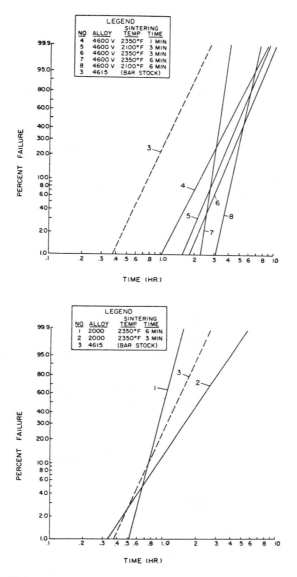

Top: Weibull distribution charts for fatigue testing of actual gears and pinions made from 4600V alloy with various sintering times and temperatures as shown, compared with cut pinions (4615 bar stock). Bottom: Similar to graph at top except for alloy 2000 and 4615 bar stock.

Source: P. C. Eloff and L. E. Wilcox, "Fatigue Behavior of Hot Formed Powder Differential Pinions," in Source Book on Powder Metallurgy, Samuel Bradbury, Ed., American Society for Metals, Metals Park OH, 1980, p 308

24-23. Gears and Pinions: P/M Grades 4600V and 2000 vs 4615; Percent Failure vs Time

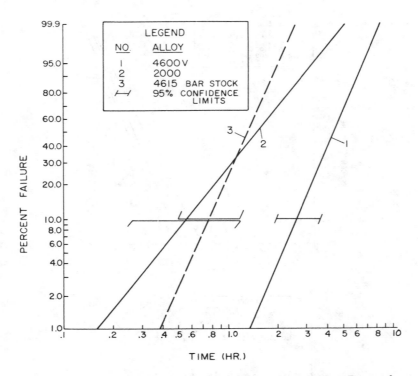

Fatigue data for actual gears tested in specially designed machines. Presented here are Weibull distributions for the three types of alloys tested.

Since the data from the two powder alloys fell into two groups, it was decided to fit one Weibull curve to all of the data points from each alloy. This was done to obtain more data points for each curve. The results are shown in the graph above, which also graphically indicates the 95% confidence limits on the B_{10} lives. It is plain that the 4600V pinions have superior fatigue life at the stress level of 92,400 psi, and the slope of the Weibull curve indicates uniform deoxidation of preforms and therefore less scatter (steeper Weibull slope) in the fatigue data. In the case of the 4600V alloy, the sintering temperature should have little effect on deoxidation, since the major alloying constituents, nickel and molybdenum, are readily reducible by CO at temperatures even below 1150 °C (2100 °F).

Source: P. C. Eloff and L. E. Wilcox, "Fatigue Behavior of Hot Formed Powder Differential Pinions," in Source Book on Powder Metallurgy, Samuel Bradbury, Ed., American Society for Metals, Metals Park OH, 1980, p 313

24-24. Gear Steel AMS 6265: Parent Metal vs Electron Beam Welded

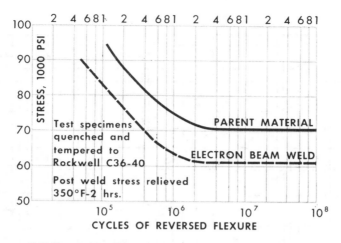

S-N diagram for AMS 6265—parent metal versus electron beam welded.

The welded specimens failed in the weld zone at 86% joint efficiency.

In the weld evaluations made, excellent mechanical properties were found. Other gear materials tested resulted in comparable weld joint efficiencies. In general, it was demonstrated that electron beam mechanical properties were comparable or better than welds made with other fusion welding processes such as gas tungsten arc and metallic arc welding.

Source: N. F. Bratkovich, W. L. McIntire and Robert E. Purdy, "Electron Beam Welding—Applications and Design Considerations for Aircraft Turbine Engine Gears," in Source Book on Electron Beam and Laser Welding, Melvin M. Schwartz, Ed., American Society for Metals, Metals Park OH, 1981, p 199

24-25. Gears, 42 CrMo4 (German Specification): S-N Curves for Various Profiles

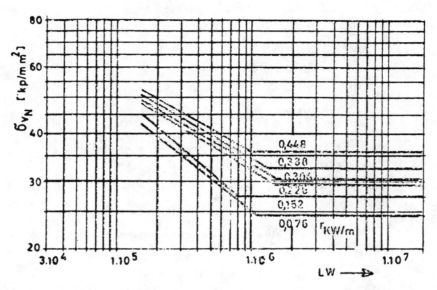

S-N curves for various tooth profiles (50% survival probability in the short life range).

The Woehler curves shown above are based (in the sloping section) on a survival probability of 50% at the number of cycles indicated. The horizontal sections of the curves are based on the highest load that can be carried for a minimum life of five million cycles.

Source: H. Winter and M. Hirt, "The Measurement of Actual Strains at Gear Teeth, Influence of Fillet Radius on Stresses and Tooth Strength," in Source Book on Gear Design, Technology and Performance, Maurice A. H. Howes, Ed., American Society for Metals, Metals Park OH, 1980, p 102

24-26. Gears, 42 CrMo4 (German Specification): Endurance Test Results in the Weibull Distribution Diagram

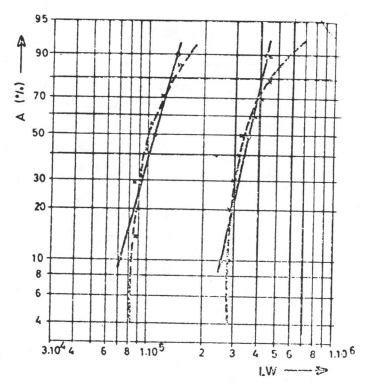

Endurance test results in the Weibull distribution diagram.

The parameters γ, λ are adjusted to the test points. An example is shown in the graph above, which represents in such a probability grid the test results for gears of one tooth form. Scale of ordinates is the failure probability $A = 1 - W = i/(n+1)$ for test i out of n test results sorted to the number of load cycles at which fracture occurred. The test points are approximated with a straight line. From this curve we are able to read, respectively, life values L_{10}, L_{50}, and L_{90} for 10, 50, and 90 percent failure probability, or 90, 50 and 10 percent survival probability.

A more adequate approximation by the theoretical distribution is achieved by a three-parameter Weibull distribution. This formulation produces a minimum endurance L_0, which is reached by all test pieces. Also, the above chart shows the compensating curve which results from the formulation of the three-parameter Weibull distribution.

Source: H. Winter and M. Hirt, "The Measurement of Actual Strains at Gear Teeth, Influence of Fillet Radius on Stresses and Tooth Strength," in Source Book on Gear Design, Technology and Performance, Maurice A. H. Howes, Ed., American Society for Metals, Metals Park OH, 1980, p 102

24-27. Bolts, 1040 and 4037 Steels: Maximum Bending Stress vs Number of Stress Cycles

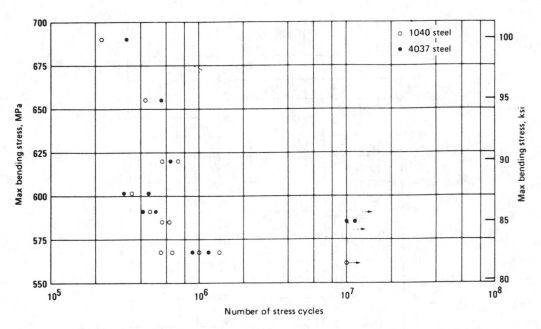

The bolts (⅜ by 2 in., 16 threads to the inch) had a hardness of 35 HRC. Tensile properties of the 1040 steel at three-thread exposure were: yield strength, 1060 MPa (154 ksi); tensile strength (axial), 1200 MPa (175 ksi); tensile strength (wedge), 1190 MPa (173 ksi). For the 4037 steel: yield strength, 1110 MPa (161 ksi); tensile strength (axial), 1250 MPa (182 ksi); tensile strength (wedge), 1250 MPa (182 ksi).

In general, if bolts made of two different steels have equivalent hardnesses throughout identical sections, their fatigue strengths will be similar (see above S-N data), as long as other factors such as mean stress, stress range, and surface condition are the same. If the results of fatigue tests on standard test specimens were interpreted literally, high-carbon steels would be selected for bolts. Actually, steels of high carbon content (more than 0.55% carbon) are unsuitable because they are notch sensitive.

The principal design feature of a bolt is the threaded section, which establishes a notch pattern inherent in the part because of its design. The form of the threads, plus any mechanical or metallurgical condition that also creates a surface notch, is much more important than steel composition in determining the fatigue resistance of a particular lot of bolts.

Source: ASM Committee on Carbon and Alloy Steels, "Threaded Steel Fasteners," in Quality Control Source Book, A. K. Hingwe, Ed., American Society for Metals, Metals Park OH, 1982, p 206

24-28. Bolts: S-N Data for Roll Threading Before and After Heat Treatment

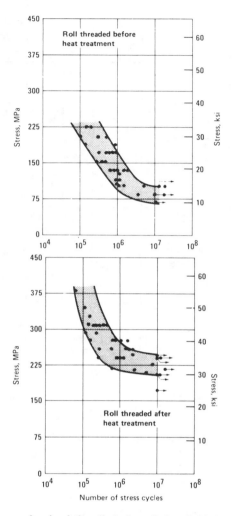

S-N curves showing fatigue limits for roll-threaded bolts. Upper graph represents four different lots of bolts that were roll threaded, then heat treated to average hardness of 22.7, 26.6, 27.6, and 32.6 HRC. Lower graph represents five different lots that were heat treated to average hardnesses of 23.3, 27.4, 29.6, 31.7, and 33.0 HRC, then roll threaded. Bolts having higher hardnesses in each category had higher fatigue strengths.

Other factors being equal, a bolt with threads properly rolled after heat treatment—that is, free from mechanical imperfections—has a higher fatigue limit than one with cut threads. This is true for any strength category. The cold work of rolling increases the strength at the weakest section (the thread root) and imparts residual compressive stresses, similar to those imparted by shot peening.

Source: ASM Committee on Carbon and Alloy Steels, "Threaded Steel Fasteners," in Quality Control Source Book, A. K. Hingwe, Ed., American Society for Metals, Metals Park OH, 1982, p 202

24-29. Power Shafts, AMS 6382 and AMS 6260: Electron Beam Welded vs Silver Brazed Joints

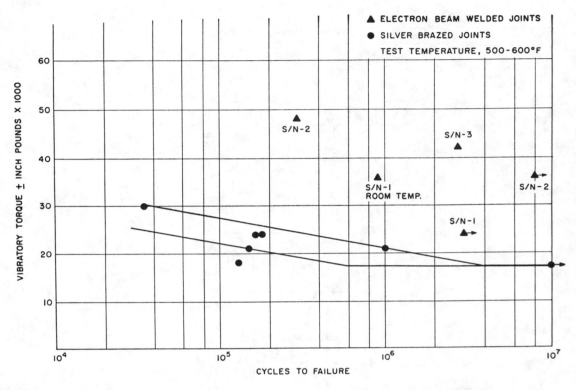

S-N curves for electron beam welded versus silver brazed power shafts made from AMS 6382 and 6260 alloy steels.

In the welded shafts, failures occurred apart from the weld, while in the brazed units all failures occurred in the brazed joints.

Source: S. M. Silverstein, V. Strautman and V. R. Freeman, "Application of Electron Beam Welding to Rotating Gas Turbine Components," in Source Book on Electron Beam and Laser Welding, Melvin M. Schwartz, Ed., American Society for Metals, Metals Park OH, 1981, p 187

24-30. Axle Shafts, 1046, 1541 and 50B54 Steels: *S-N* Data for Induction Hardening vs Through Hardening

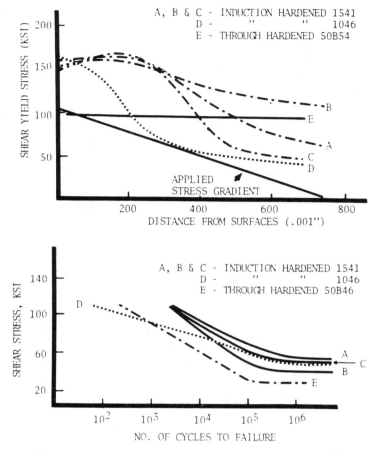

Top: Axle-shaft strength gradients in terms of shear yield strength. Bottom: Fatigue performances of axle shafts as a function of strength gradient.

Induction hardened 0.40% carbon steel axle shafts were developed to replace through hardened alloy steel shafts for both product- and cost-improvement purposes. This was accomplished after a rather comprehensive bench-test program, which examined variables such as surface hardness, core hardness, gradient strength, distortion, composition, and surface-condition effects. The more promising approach was then subjected to chassis, proving-ground, and in-service testing. Some interesting reflections can result from examining some of the fatigue data that were generated. That the through hardened concept was vulnerable can be surmised by considering the stresses developed in a full-floating splined shaft loaded in torsion. The stresses are a maximum at the surface and drop linearly to zero at the center. At the spline, the stresses drop more rapidly at the onset due to the stress concentration caused by the spline. The upper graph shows the stress gradient in the body area when the shaft is loaded to 110,000-psi shear stress. Also plotted on this graph are the shear yield-strength gradients (converted from hardness) of the production alloy shaft, along with several experimental induction hardened shafts. One would expect the through hardened shaft to have a surface-origin failure and to be lower in strength, since its surface is the lowest hardness. Also, the high strength of the center of the shaft is essentially wasted, since it is lowly stressed. Gradient strengthening by induction hardening provides a means of providing a better strength match for the stress gradient. The lower graph gives the fatigue curves established for shafts having the strength gradients shown in the upper graph.

Source: D. H. Breen and E. M. Wene, "Fatigue in Machines and Structures—Ground Vehicles," in Fatigue and Microstructure, American Society for Metals, Metals Park OH, 1979, p 88

24-31. Steel Rollers, 8620H Carburized: Effects of Carburizing Temperature and Quenching Practice on Surface Fatigue

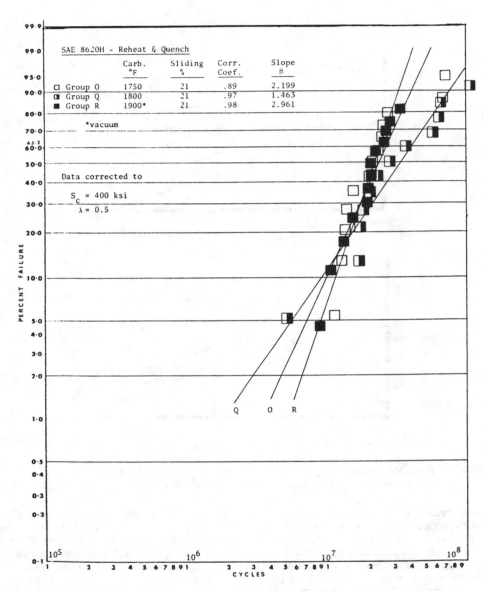

S-N data, Weibull probability plot: Effect of three carburizing temperatures on surface fatigue for carburized 8620H steel. All were slow cooled and reheated for quenching. This technique improved fatigue characteristics compared with direct-quenched rollers.

Source: S. L. Rice, "Pitting Resistance of Some High Temperature Carburized Cases," in Source Book on Gear Design, Technology and Performance, Maurice A. H. Howes, Ed., American Society for Metals, Metals Park OH, 1980, p 234

24-32. Steel Rollers, 8620H Carburized: Effects of Carburizing Temperature and Quenching Practice on Surface Fatigue

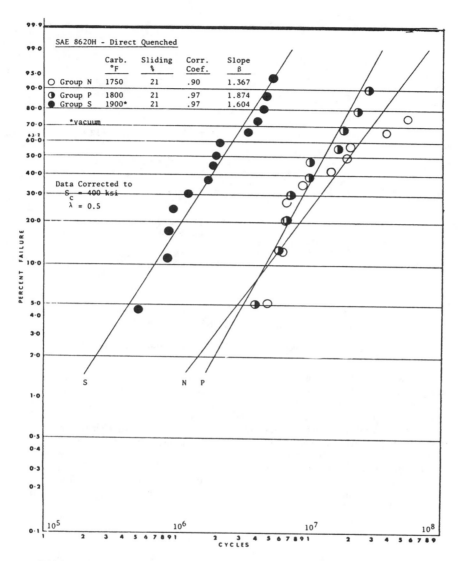

S-N data, Weibull probability factor: Effect of the three carburizing temperatures on surface fatigue for carburized 8620H steel. All were direct quenched from the temperatures shown above.

Source: S. L. Rice, "Pitting Resistance of Some High Temperature Carburized Cases," in Source Book on Gear Design, Technology and Performance, Maurice A. H. Howes, Ed., American Society for Metals, Metals Park OH, 1980, p 233

24-33. Linkage Arm, Cast Low-Carbon Steel: Starting Crack Size vs Cycles to Failure

Fatigue life of a linkage arm as a function of starting crack size.

The variation of the fatigue life, N_f, with the starting flaw size a_i, is shown in the diagram above. The fatigue life increases dramatically at very small a_i values. The far curve shows that in the long life regime the final crack size has only a small effect on N_f. This is because fatigue crack growth rates are very low at low ΔK values and hence the greatest fraction of fatigue life is spent at the smaller crack sizes. Since the controlling parameter is ΔK, low life for small crack sizes is possible at high cyclic stresses. The second set of curves shows that doubling the cyclic stress range reduces the fatigue life by about an order of magnitude. Also, if the starting ΔK value is high, the final crack size has a larger effect on the cyclic life. The above diagram shows the importance of adjusting both the cyclic stress and starting flaw size to optimize the fatigue life.

Source: Steel Castings Handbook, 5th Edition, Peter F. Weiser, Ed., Steel Founders' Society of America, Rocky River OH, 1980, p 4-17